The Systemic Effects of Advanced Cancer

Swarnali Acharyya
Editor

The Systemic Effects of Advanced Cancer

A Textbook on Cancer-Associated Cachexia

 Springer

Editor
Swarnali Acharyya
Columbia University Irving Medical Center
New York, NY, USA

ISBN 978-3-031-09789-8   ISBN 978-3-031-09518-4   (eBook)
https://doi.org/10.1007/978-3-031-09518-4

This Springer imprint is published by the registered company Springer Nature Switzerland AG
The registered company address is: Gewerbestrasse 11, 6330 Cham, Switzerland

Contents

Part IV   Preventing and Targeting Cachexia in Cancer

A Systemic View of Metastatic Disease: Inter-Organ Crosstalk and Therapeutic Implications

Swarnali Acharyya

Abstract

As cancer progresses, the communication and crosstalk between cancer cells and the tumor microenvironment become a key determinant of successful cancer cell growth and survival. These interactions arise through either local contact-mediated signaling or the systemic release of soluble factors, extracellular vesicles and particles (EVP) by tumor cells. Systemic effects that are initiated early during cancer progression, when cancer cell growth is still confined to the primary tumor, establish the pre-metastatic niche (PMN). PMN formation lays the groundwork for cancer-cell colonization of distant organs by altering the extracellular matrix, immune-cell composition, and vascular permeability of distant organs. Tumors also independently reprogram host physiology and metabolism, which often results in a debilitating skeletal-muscle-wasting syndrome known as cachexia. Indeed, over 80% of advanced cancer patients develop cachexia and consequently suffer from reduced tolerance to anti-neoplastic therapies and premature death. This introductory chapter summarizes the various systemic effects of cancer that are detrimental to the host. The design of treatments that can prevent and/or reverse these systemic changes is ultimately expected to improve the quality of life and survival of cancer patients.

Preface

As primary tumors evolve, cancer cells are able to escape from the tumor and enter the systemic circulation. However, the vast majority of these cancer cells die, and less than 0.01% are able to colonize distant organs and generate overt metastatic lesions [1, 2]. The success of metastatic colonization and growth depends on cell-intrinsic programs in cancer cells as well as on the crosstalk between cancer cells and the distant microenvironment [3]. These tumor-microenvironment communications are important during multiple stages of the invasion-metastasis cascade and are often mediated by either the contact-mediated signaling or the release of soluble factors and extracellular vesicles and particles (EVPs) from the tumor [4]. But when does this interaction begin? Studies show that the partnership between the tumor and the microenvironment evolves early, and is already underway during the primary tumor stage. In fact, signaling molecules from tumors in the form of soluble factors and EVPs reach distant organs even before

S. Acharyya (✉)
Institute for Cancer Genetics and Department of Pathology and Cell Biology, Columbia University, New York, NY, USA

Herbert Irving Comprehensive Cancer Center, New York, NY, USA
e-mail: sa3141@cumc.columbia.edu

© Springer Nature Switzerland AG 2022
S. Acharyya (ed.), *The Systemic Effects of Advanced Cancer*,
https://doi.org/10.1007/978-3-031-09518-4_1

the cancer cells arrive, a phenomenon known as pre-metastatic niche (PMN) conditioning [5]. PMN conditioning has been observed in the lung, liver, brain, and bone, which are common sites of metastatic colonization from epithelial tumors [6, 7]. By remodeling the vascular permeability, immune milieu, and extracellular matrix composition, PMN formation can prime distant organs and increase the chances of metastatic success. *PMN formation is therefore an example of a systemic change that is induced remotely by tumors in distant organs without the presence of cancer cells in those organs. The first part of this textbook highlights tumor-initiated systemic changes in distant organs that can influence the metastatic process, patient prognosis, and treatment outcome.*

The systemic changes that occur during cancer progression are not limited to future sites of metastatic growth. Tumor-derived soluble factors and EVPs also induce systemic changes in tissues and organs that are almost never inhabited by cancer cells, such as skeletal muscle and adipose tissue. These systemic effects negatively impact the host and often result in a debilitating muscle-wasting syndrome known as cachexia. Cachexia is present in 80% of cancer patients and is associated with their decreased tolerance to anti-neoplastic therapy, poor prognosis, and accelerated death. While it is not clear whether cachexia benefits tumor growth or merely represents an unintended consequence of advanced cancer, the combined targeting of cancer and cachexia is expected to improve the quality of life and survival for cancer patients. *The second and third parts of this textbook, therefore, highlight the impact of systemic changes induced by advanced cancer on skeletal muscle and how these changes negatively impact muscle signaling, size, and function. Additional topics of discussion include how anti-cancer therapies can further exacerbate muscle wasting and how new strategies propose to overcome both cancer- and therapy-induced muscle wasting.*

1 Author and Text Introductions

The overall objective of this textbook is to provide a comprehensive view of the systemic effects of cancer with examples of inter-organ crosstalk and signaling.

Part I of this textbook (Chapters "A Systemic View of Metastatic Disease: Inter-Organ Crosstalk and Therapeutic Implications; Systemic Regulation of Metastatic Disease by Extracellular Vesicles and Particles; Bone Metastasis: Systemic Regulation and Impact on Host; Targeting Metastatic Disease: Challenges and New Opportunities") includes discussions by lead experts in the field of metastasis that highlight how metastatic disease successfully develops through a web of inter-organ crosstalk. The following questions are addressed in Part I: *Is inter-organ crosstalk a determinant of metastatic success? Which organs participate in inter-organ crosstalk and how is it regulated? Does inter-organ crosstalk have implications for diagnosis or treatment of metastatic disease? Is targeting the metastasis-promoting inter-organ crosstalk a viable option to effectively eliminate metastatic disease?* The lead author of the second chapter is Dr. David Lyden, a physician-scientist and Professor of Pediatrics and Cell and Developmental Biology at Weill Cornell Medicine. Dr. Lyden's pioneering work studying the systemic effects of cancer led to the discovery of the "pre-metastatic niche (PMN)" as a key determinant of metastatic success and helped elucidate the importance of extracellular vesicles and particles (EVPs) during metastatic development. Both phenomena are now widely viewed as classic examples of systemic regulation during metastasis. In chapter "Systemic Regulation of Metastatic Disease by Extracellular Vesicles and Particles", Dr. Lyden and colleagues describe PMN formation, which involves the early molecular and cellular alterations that are induced by primary tumors in distant organs and occur even before tumor cells arrive at these organ sites. PMN formation is associated with vascular leakiness, extracellular matrix remodeling, inflammation, and metabolic reprogramming, all of which can promote successful colonization of incoming cancer cells. Dr. Lyden and colleagues subsequently describe the mechanisms by which pro-metastatic messages are transmitted between organs. One mode of transmission occurs through EVPs, which deliver cancer-cell-derived nucleic acids, proteins, lipids, and metabolites to non-cancerous recipient cells. Tumor-derived EVPs can thereby reprogram the immune system and aid in PMN formation. Lastly, Chapter "A Systemic View of Metastatic Disease: Inter-Organ Crosstalk and Therapeutic Implications" describes the biology of EVPs and discusses their utility as biomarkers for the diagnosis of metastatic disease, and as therapeutic targets and deliverables for metastasis treatment. Beyond their role in cancer progression, EVPs can also contribute to the pathophysiology of cachexia and other paraneoplastic syndromes, a topic that has also been summarized in this chapter.

Chapter "Bone Metastasis: Systemic Regulation and Impact on Host" highlights the multifaceted role of TGF-beta, a cytokine that promotes both metastasis and muscle weakness through tumor-bone-muscle crosstalk. Chapter "Bone Metastasis: Systemic Regulation and Impact on Host" is written by Dr. Theresa Guise, who is a physician-scientist and Professor in the Division of Internal Medicine at the University of Texas MD Anderson Cancer Center, and Co-Director of the Bone Disease Program. Dr. Guise has made seminal contributions to our understanding of bone metastasis and its effects on the musculoskeletal system. In chapter "Bone Metastasis: Systemic Regulation and Impact on Host", co-authors Drs. Sukanya Suresh and Guise describe how cancer cells that have colonized the bone initiate and propagate a vicious cycle that leads to increased bone resorption and destruction through the TGF-beta signaling pathway. The authors discuss therapeutic strategies that can interrupt this vicious cycle and reduce bone metastasis. This chapter also highlights the ability of TGF-beta to promote muscle dysfunction during bone metastasis and outlines the design of novel strategies for TGF-beta pathway inhibition to block both bone metastasis and muscle weakness for enhanced patient survival.

In the final chapter of Part I (Chapter "Targeting Metastatic Disease: Challenges and New Opportunities"), Dr. Hanqiu Zheng and colleagues discuss current opportunities for targeting metastatic disease. Dr. Zheng is an Associate Professor in the Department of Basic Medical Sciences in the School of Medicine at Tsinghua University in Beijing, China. He studies drug resistance and metastasis in breast cancer. Dr. Zheng and colleagues highlight several promising anti-metastatic strategies designed to target the different steps of the metastatic cascade. In this chapter, the authors discuss the effectiveness of immunotherapy, anti-HER2 therapy and bone-metastasis-targeting approaches in the treatment of advanced cancer. In this context, the authors discuss the systemic nature of metastasis and the crosstalk between tumors and organs that harbor metastatic cells, which presents further therapeutic challenges. Of relevance here, several metabolic changes are associated with cancer progression and metastasis, in both the tumor and the host. Some of these metabolic changes represent vulnerabilities that provide new opportunities for targeting metastatic disease, which are also highlighted in this chapter.

Part II of this textbook is focused on describing cachexia, a common systemic effect of advanced cancer that is linked to poor prognosis and reduced patient survival. Cachexia is a complex metabolic syndrome that is associated with the involuntary loss of skeletal muscle and an active catabolic drive that cannot be reversed solely by nutritional support. Patients with cachexia experience impaired daily functioning, decreased tolerance to anti-cancer therapies, and a drastic reduction in quality of life and survival. There are currently no effective FDA-approved therapies to prevent or reverse cachexia.

Chapters "Signaling Pathways That Promote Muscle Catabolism in Cachexia; The Role of Interleukin-6 Cytokines in Cancer Cachexia; NF-kB Signaling in the Macroenvironment of Cancer Cachexia" provide the readers with insights into what drives cachexia. Dr. Shenhav Cohen is an Associate Professor in the Technion Israel Institute of Technology who studies the molecular mechanisms that regulate muscle size and proteolysis during muscle wasting. In chapter "Signaling Pathways That Promote Muscle Catabolism in Cachexia", co-authors Drs. Jennifer Gilda and Cohen highlight the proteolytic pathways that activate the degradation of soluble and contractile myofibrillar proteins and describe the order of the molecular and cellular events, as well as the key regulators, driving this catabolic process.

How do you define cachexia? What are symptoms of cachexia? Is chronic inflammation linked to cachexia? Dr. Teresa Zimmers, Founding Director of the Indiana University (IU) Simon Cancer Center Cachexia Working Group, and Professor of Surgery at the Indiana University School of Medicine, addresses these questions in chapter "The Role of Interleukin-6 Cytokines in Cancer Cachexia". Dr. Zimmers directs the Cachexia Working Group at the IU Simon Comprehensive Cancer Center and has made seminal contributions to the field of cancer cachexia. In chapter "The Role of Interleukin-6 Cytokines in Cancer Cachexia", co-authors Drs. Daenique Jengelley and Zimmers highlight the role of the interleukin-6 (IL-6) family of cytokines in driving cachexia by activating common downstream signaling pathways involving JAK/STAT, MAPK, and AKT. The authors discuss (a) which of the IL-6 family members induce cachexia in preclinical models, (b) which IL-6 family members are associated with clinical cancer cachexia, (c) how IL-6 family members trigger cachexia, and (d) the current progress in targeting cachexia using anti-IL-6 or anti-IL-6R antibodies in preclinical and clinical studies. Finally, the authors discuss the strengths of such anti-cytokine therapies that can reduce both tumor burden and cachexia simultaneously, thereby improving the quality of life and prolonging the survival of cancer patients.

Among inflammation-associated pathways, the nuclear factor kappa B (NF-κB) signaling pathway plays a key role in both tumorigenesis and cachexia. Dr. Denis Guttridge is a leader in the field of cancer cachexia. He is a Professor and the Director of the Darby Children's Research Institute and the Associate Director of

Translational Sciences for the Hollings Cancer Center at the Medical University of South Carolina. Dr. Guttridge's long-term research focuses on the biology of NF-κB and his group has made seminal contributions in demonstrating how the NF-κB signaling pathway regulates tumorigenesis, muscle differentiation, and cachexia. In chapter "NF-kB Signaling in the Macroenvironment of Cancer Cachexia", co-authors Drs. Benjamin Pryce and Guttridge highlight how the NF-κB pathway promotes both cancer development and muscle wasting. This chapter introduces the readers to the canonical and non-canonical NF-κB pathway, how NF-κB signaling contributes to normal muscle development, and how it is perturbed in cachexia. NF-κB signaling is then discussed in the context of pancreatic cancer, in which NF-κB signaling drives tumorigenesis as well as cachexia development. Finally, the authors describe the role of NF-κB signaling in blocking muscle regeneration in response to tumor-factor-induced muscle damage, highlighting how NF-κB signaling influences multiple nodes of cancer progression and cachexia in the tumor macroenvironment.

For patients with metastatic cancer, localized surgical options are limited, and systemic treatments are needed for targeting the disseminated disease. In this regard, chemotherapeutic agents are commonly used for treating metastatic disease across multiple cancer types. Chemotherapies kill cancer cells but also induce damage to normal tissues. In Part III of this textbook, the authors discuss how chemotherapy adversely affects the musculoskeletal system, causing bone loss and exacerbating cachexia.

In chapter "Therapy-Induced Toxicities Associated with the Onset of Cachexia", Dr. Andrea Bonetto, an Associate Professor of Surgery at the Indiana School of Medicine and an expert in therapy-induced cachexia, discusses chemotherapy-induced toxicities. In particular, Dr. Bonetto and colleagues discuss which chemotherapeutic agents cause muscle wasting and/or bone loss, which pathways are activated in muscle and bone in response to treatment with chemotherapeutic agents, and strategies that have been designed to counteract these therapy-induced side effects. Finally, Dr. Bonetto and colleagues discuss some of the promising combination treatments tested in preclinical studies that could reduce chemotherapy-induced toxicities and improve outcomes and survival in cancer patients.

In a complementary chapter on therapy-induced toxicities (Chapter "Bone-muscle Crosstalk in Advanced Cancer and Chemotherapy"), Dr. David Waning focuses on host-specific consequences of bone metastasis, which is common in patients with breast, lung, and prostate cancer, and how some anti-cancer treatments affect bone-muscle crosstalk and exacerbate wasting. Dr. David Waning is an Associate Professor in the Department of Molecular Physiology at Penn State Cancer Institute. Dr. Waning heads the Musculoskeletal Research Laboratory, and his group aims to develop therapies that can reduce metastasis without inducing significant bone-muscle toxicities. In this chapter, Dr. Waning describes how musculoskeletal damage occurs either directly by tumor cells that have colonized the bone, indirectly through paracrine and endocrine signaling between tumor, bone, and muscle, or by anti-cancer therapies that negatively impact bone and muscle. Dr. Waning outlines how a new generation of therapeutics that are being tested in clinical trials can reduce bone metastasis as well as musculoskeletal damage in preclinical models.

The experimental findings described in Parts I-III provide the readers with a background on the systemic effects of advanced cancer and describe the biological problem of cachexia. Part IV of this textbook presents the current strategies for the prevention and treatment of cachexia in cancer patients.

The author of chapter "New Developments in Targeting Cancer Cachexia", Dr. Richard Skipworth, is a physician-scientist and surgeon in Clinical Surgery at the Royal Infirmary in Edinburgh, UK. Dr. Skipworth cares for patients with oesophagogastric malignancy and cachexia and is actively involved in clinical cachexia research. In chapter "New Developments in Targeting Cancer Cachexia", Dr. Skipworth and colleagues discuss the diagnosis of cachexia, the stages of cachexia, and some of the promising strategies to treat cachexia. This chapter also highlights findings from recent cachexia clinical trials as well as new multimodal strategies for managing cachexia in cancer patients.

A key component of cachexia is the associated loss of muscle function in advanced cancer patients, which compromises their daily functioning, prognosis, and treatment outcome. In this context, exercise physiology studies have demonstrated an inverse correlation between physical activity and the initiation of cancer, its recurrence, and the development of cachexia [8, 9]. In the final chapter (Chapter "Exercise: A Critical Component of Cachexia Prevention and Therapy in Cancer"), exercise physiologist Dr. Emidio Pistilli discusses the importance of physical activity and exercise in preventing and treating cancer cachexia. Dr. Pistilli is an Associate Professor in the Department of Human Performance and Exercise Physiology in the School of Medicine at West Virginia University. His work focuses on understanding mechanisms that contribute to cancer-associated fatigue. His group conducts mechanistic studies to inform the design of supportive therapies that alleviate fatigue and improve the quality of life of cancer patients. In this chapter, Dr. Pistilli and colleagues discuss the different exercise guidelines that are prescribed for cancer patients, such as aerobic exercise and resistance training. The authors discuss how structured exercise programs in patients with cancer can prevent or treat cachexia by increasing lean body mass and improving muscle functional capacity.

Acknowledgments We would like to thank our Cancer Biology students and patient advocates who inspired us to compile this textbook on the systemic effects of cancer. We also wish to acknowledge our funding sources: NCI RO1 CA231239, Pershing Square Sohn Prize, Irving Scholar Award, Interdisciplinary Research Initiatives Seed (IRIS) program, The Irma T. Hirschl Monique Weill-Caulier Trust Award, the American Cancer Society (ACS), Phi Beta Psi Sorority and Schaefer Award to S.A.

References

1. Lambert, A.W., Pattabiraman, D.R., Weinberg, R.A.: Emerging biological principles of metastasis. Cell.

168(4), 670–691 (2017). https://doi.org/10.1016/j.cell.
2016.11.037

2. Obenauf, A.C., Massague, J.: Surviving at a distance: organ-specific metastasis. Trends Cancer. **1**(1), 76–91 (2015). https://doi.org/10.1016/j.trecan.2015.07.009

3. Hanahan, D., Coussens, L.M.: Accessories to the crime: functions of cells recruited to the tumor microenvironment. Cancer Cell. **21**(3), 309–322 (2012). https://doi.org/10.1016/j.ccr.2012.02.022

4. Pelissier Vatter, F.A., Cioffi, M., Hanna, S.J., Castarede, I., Caielli, S., Pascual, V., et al.: Extracellular vesicle- and particle-mediated communication shapes innate and adaptive immune responses. J. Exp. Mcd. **218**(8) (2021). https://doi.org/10.1084/jem.20202579

5. Peinado, H., Zhang, H., Matei, I.R., Costa-Silva, B., Hoshino, A., Rodrigues, G., et al.: Pre-metastatic niches: organ-specific homes for metastases. Nat. Rev. Cancer. **17**(5), 302–317 (2017). https://doi.org/10.1038/nrc.2017.6

6. Psaila, B., Kaplan, R.N., Port, E.R., Lyden, D.: Priming the 'soil' for breast cancer metastasis: the pre-metastatic niche. Breast Dis. **26**, 65–74 (2006). https://doi.org/10.3233/bd-2007-26106

7. Psaila, B., Lyden, D.: The metastatic niche: adapting the foreign soil. Nat. Rev. Cancer. **9**(4), 285–293 (2009). https://doi.org/10.1038/nrc2621

8. Jones, L.W.: Precision oncology framework for investigation of exercise as treatment for cancer. J. Clin. Oncol. **33**(35), 4134–4137 (2015). https://doi.org/10.1200/JCO.2015.62.7687

9. Solheim, T.S., Laird, B.J.A., Balstad, T.R., Bye, A., Stene, G., Baracos, V., et al.: Cancer cachexia: rationale for the MENAC (Multimodal-Exercise, Nutrition and Anti-inflammatory medication for Cachexia) trial. BMJ Support. Palliat. Care. **8**(3), 258–265 (2018). https://doi.org/10.1136/bmjspcare-2017-001440

Part I
Systemic Regulation in Metastatic Disease

Systemic Regulation of Metastatic Disease by Extracellular Vesicles and Particles

Gang Wang, Candia M. Kenific, Grace Lieberman, Haiying Zhang, and David Lyden

Abstract

Cancer is a systemic disease that induces functional dysregulation in multiple tissues and organs. The great majority of cancer-related deaths are due to metastasis, a multistep process involving crosstalk between cancer cells and stromal cells and immune cells in pre-metastatic niches (PMNs) at distant organs. As message carriers, extracellular vesicles and particles (EVPs) derived from cancer cells can modulate the phenotype of recipient cells by transferring various biomolecules, including nucleic acids, proteins, metabolites and lipids, thereby contributing to PMN formation and metastasis. In this chapter, we highlight the roles of cancer cell-derived EVPs in metastasis progression, with a focus on PMN establishment.

Learning Objectives

1. Distinguish different types of EVPs, including apoptotic bodies, microvesicles, exosomes, and exomeres.
2. Gain an appreciation for tumor-derived EVPs in remodeling the tissue architecture of the tumor microenvironment in primary tumor, premetastatic niche, and metastatic site.
3. Understand how tumor-derived EVPs modulate innate and adaptive immune systems.
4. Recognize the functional roles of tumor-derived EVPs in angiogenesis and vasculature remodeling.
5. Diagram tumor-derived EVP-mediated pre-metastatic niche formation and metastatic organotropism.
6. Recognize the translational potential of tumor-derived EVPs as biomarkers for cancer diagnosis and prognosis, therapeutic targets and deliverables.

G. Wang · C. M. Kenific · G. Lieberman · H. Zhang · D. Lyden (✉)
Children's Cancer and Blood Foundation Laboratories, Departments of Pediatrics, and Cell and Developmental Biology, Drukier Institute for Children's Health, Meyer Cancer Center, Weill Cornell Medicine, New York, NY, USA
e-mail: haz2005@med.cornell.edu; dcl2001@med.cornell.edu

© Springer Nature Switzerland AG 2022
S. Acharyya (ed.), *The Systemic Effects of Advanced Cancer*,
https://doi.org/10.1007/978-3-031-09518-4_2

1 Introduction

Cancer is the second leading cause of death worldwide [1, 2]. The morbidity and mortality of cancer can mainly be attributed to metastasis, the critical stage of cancer progression resulting in colonization and expansion of cancer cells disseminated from primary tumor into distant organs [3]. Although steps of the metastatic cascade involving tumor intrinsic properties have been well established (i.e., migration, invasion, intravasation, extravasation, colonization, and expansion) [3–6], increasing evidence underscores the importance of extrinsic effects in metastasis, such as the influence of tumor-secreted factors that alter local and distal microenvironments [7–10].

In support of Stephen Paget's "seed and soil" hypothesis [11], the organs of future metastasis are selectively and actively reprogrammed by primary tumors prior to the onset of metastasis [12, 13]. Primary tumor-derived factors, such as soluble proteins, metabolites, and EVPs, can modulate the biophysiological functions of distant organs thereby providing supportive microenvironments known as PMNs, for the survival and colonization of disseminated tumor cells before their arrival [9, 13–16]. Since its inception by David Lyden and colleagues in 2005, PMNs have been characterized by molecular and cellular alterations that result in vascular leakiness, extracellular matrix remodeling, inflammation, hypoxia, metabolic reprogramming, and organ dysfunction to promote metastatic colonization [7–9, 12, 13, 17, 18]. It has been demonstrated that EVPs shed by cancer cells, as well as by tumor-associated stromal cells and infiltrating immune cells, are critical mediators for the establishment of PMNs [12, 19].

EVPs represent a heterogeneous population of membranous extracellular vesicles (EVs), and non-membranous, non-vesicular nanoparticles shed by different cell types, which are secreted into the extracellular space and function as important mediators of cell–cell communication both locally and distally [20, 21]. Based on their origin, there are four major subgroups of EVPs including apoptotic bodies, microvesicles, exosomes, and exomeres [14, 20–22] (Fig. 1). The apoptotic bodies can be generated either by budding from the plasma membrane or by

fragmentation of the apoptotic cells. Therefore, the apoptotic bodies have an expansive size range from 100 nm up to 5 μm in dimension [22]. Microvesicles are generated directly by outward budding of the plasma membrane, typically ranging in size from about 100 nm to 1000 nm. A new class of large microvesicles (1–10 μm) that are shed from non-apoptotic plasma membrane blebs, namely oncosomes, has also been characterized as carriers for intercellular transfer of oncogenic signalings [23, 24]. Exosomes are nanometer-sized vesicles with a diameter of about 50 nm to 150 nm that originate in endosomes through the invagination of endosome membranes. This inward budding forms intraluminal vesicles within the endosome that contain nucleic acids, proteins, metabolites, and lipids. The resulting multivescular endosome can either fuse with lysosomes for content degradation or fuse with the plasma membrane to release the intraluminal vesicles as exosomes [19, 22]. Exosomes also present with heterogenous subpopulations. Because of their overlap in size, density, morphology and even the membrane-associated proteins, it can be challenging to distinguish exosome subpopulations, microvesicles and apoptotic bodies by technologies currently used, such as differential ultracentrifugation, sucrose gradient purification, and immunoisolation technologies [22]. With recent technical advances by asymmetric flow field-flow fractionation (AF4), exosome-large (Exo-L; 90–120 nm) and exosome-small (Exo-S; 60–80 nm) were identified as two subpopulations of exosomes that package distinct molecular contents [21]. In addition, a new subset of non-membranous, non-vesicular nanoparticles, termed exomeres (~35 nm), was identified and shown to primarily package metabolic enzymes [21, 25]. Efforts are ongoing to determine mechanisms of exomere biogenesis and their potential functions in the cancer setting.

EVPs were initially recognized as carriers for eliminating unnecessary proteins and other compounds from the cells [26–28]. However, recent studies have revealed that EVPs also function as messengers for cell–cell communications

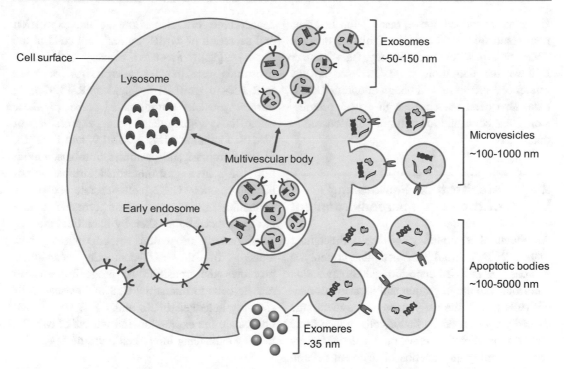

Fig. 1 Schematic illustration depicting the generation of distinct EVP subpopulations. Apoptotic bodies, which have a diameter of about 100 nm to 5 μm, are generated either by budding from the plasma membrane or by fragmentation of the apoptotic cells. Microvesicles can be generated directly by outward budding of the plasma membrane, typically ranging in size from about 100 nm to 1000 nm. Exosomes have a diameter of about 50 nm to 150 nm, and are originated in endosomes through the invagination of endosome membranes. This inward budding forms intraluminal vesicles within the endosome that package a variety of cargos, including nucleic acids, proteins, metabolites, and lipids. The resulting multivesicular body can either fuse with lysosomes for content degradation or fuse with the plasma membrane to release the intraluminal vesicles as exosomes. There is a subset of non-membranous nanoparticles, namely exomeres (~35 nm), secreted from cells, whereas the mechanism of exomeres biogenesis remains unknown

by transferring their bioactive cargos to recipient cells [9, 21, 23, 29–31]. When released extracellularly, EVPs can be uptaken by recipient cells in different ways, such as ligand-receptor mediated interaction, endocytosis, or direct fusion of the EVPs with the plasma membrane of the recipient cells [20, 32]. During cancer progression, cancer cells consistently secrete EVPs that can target neighboring cells in the local microenvironment via autocrine or paracrine signaling, as well as cells located in distant organs via paracrine signaling [19, 33]. Seminal works published since the early 2000s have shown that tumor-derived EVPs play critical roles in all steps of the metastatic cascade including local and distant microenvironment reconstruction, immune system modulation, angiogenesis and vascular leakiness induction, intravasation and extravasation, and survival and proliferation of metastatic cancer cells in distant organs [7, 8, 14, 17, 18, 34, 35]. Moreover, tumor-derived EVPs can carry specific surface cargos and preferentially target resident cells in a specific distant organ, such as lung, liver, or brain, thereby preparing PMNs for organotropic metastasis [8–10, 12]. In this chapter, we summarize the roles of EVPs, with a focus on exosomes and microvesicles, in regulating the metastatic cascade, PMN establishment, and metastatic organotropism. Due to the difficulties in distinguishing the different subpopulations of

EVPs as mentioned above, many published data may result from studies using a mixture of these EVP subtypes. For consistency, this chapter follows the definition of EVPs used in the referenced literatures. Although apoptotic bodies have also been associated with cancer progression, their contributions will not be discussed in detail here.

2 Architectural Remodeling of the Local Microenvironment

Architectural remodeling of the extracellular matrix (ECM) through extracellular proteolysis and changes in ECM deposition and organization contributes to the migration and invasion of cancer cells during the early steps of the metastatic cascade. EVPs shed by tumor cells induce these changes by directly acting on the ECM or by reprogramming the function of recipient cells in the microenvironment which can also alter the ECM (Fig. 2).

2.1 ECM Degradation

The proteolytic activity of matrix metalloproteinases (MMPs) is critical for ECM degradation and cancer development [36]. Recent studies have identified different members of MMPs, including MMP2, MMP9, MMP13, and MMP14 that are enriched in exosomes derived from various types of cancer cells [37–40]. Tumor-derived exosomes can bind to individual components in the ECM through adhesion molecules, such as CD44 and integrin $\alpha_6\beta_4$, and release active MMPs for degradation of collagens, laminin, and fibronectin, leading to the decreased cellular adhesion and increased invasion of cancer cells in extracellular space [41]. Degradation of ECM is also thought to liberate growth factors that may be trapped within the matrix, which can also promote the growth and motility of cancer cells. In addition to directly transferring MMPs, exosomes secreted from

cancer cells can also stimulate the expression and secretion of MMPs by recipient cells in the stroma [42]. Prostate cancer cell-derived exosomes can transfer microRNAs, such as miR-100-5p, miR-21-5p, and miR-139-5p, to normal prostate fibroblasts. Uptake of these microRNAs upregulates the expression of MMPs, including MMP2, MMP9, and MMP13, in the fibroblast and promotes fibroblast migration [43]. In a gastrointestinal stromal tumor (GIST) model, GIST cells secrete exosomes containing oncogenic protein tyrosine kinase (KIT) that can be uptaken by interstitial stromal cells, namely progenitor smooth muscle cells. Uptake of the KIT-containing exosomes promotes the transition of progenitor smooth muscle cells to Interstitial Cell of Caja-like cells showing activated ERKs and AKT, which then upregulate the expression and release of MMP1, thereby enhancing tumor cell invasion [44].

2.2 Fibroblast Reprograming

Cancer-associated fibroblasts (CAFs) are one of the dominant cellular components in the tumor stroma. A subpopulation of these CAFs is activated and exhibits the traits of myofibroblasts, which are characterized by the expression of vimentin and α-smooth actin (α-SMA) [45]. Myofibroblasts have been reported to be involved in maintaining ECM homeostasis by producing growth factors, collagens, laminin, and fibronectin, as well as proteases, including MMPs [46]. Both *in vitro* and *in vivo* evidence demonstrated that cancer cell-derived exosomes trigger the differentiation of fibroblasts into myofibroblasts in support of tumor growth [47, 48] (Fig. 2). Mechanistically, cancer exosomal transforming growth factor-β 1 (TGFβ1) can be delivered to interstitial fibroblasts, and activate fibroblasts through TGFβ1-SMAD3 signaling. Decreased secretion of exosomes by disrupting Rab27a, a member of Rab GTPases [49], attenuated *in vitro* myofibroblastic differentiation and *in vivo* tumor

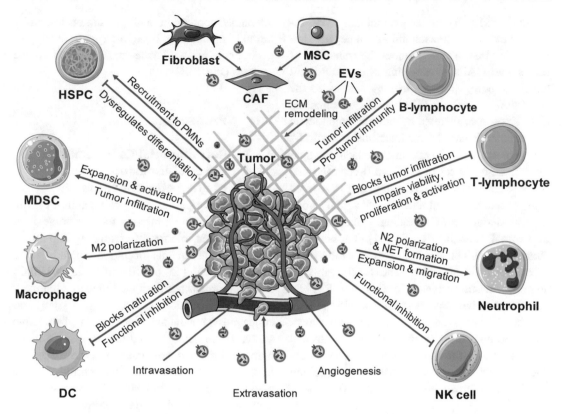

Fig. 2 EV-mediated cell–cell communications in cancer settings. Tumor cells can escape immune attack throughout metastatic progression via secreting EVs, which function in reprogramming the immune system through targeting of diverse cell types, ranging from HSPC to well-differentiated innate (macrophage, DC, neutrophil, and NK cell) and adaptive (T- and B-lymphocyte) immune cells. Tumor-derived EVs can also reconstruct ECM by inducing MMPs-dependent proteolytic degradation and by promoting the differentiation of normal fibroblasts and MSCs into CAFs. Furthermore, tumor cells transfer EVs to endothelial cells to promote angiogenesis and to modulate vascular permeability for intravasation and extravasation

growth [48]. However, only myofibroblasts generated by the stimulation of exosomal TGFβ1, but not soluble TGFβ1 alone, show the capacity of pro-angiogenesis and pro-tumor growth. One explanation is that the intact heparan sulphate proteoglycans on exosome membranes are required for the delivery of functional TGFβ1, as loss of heparan sulphate side chains interrupts TGFβ1-SMAD3 signaling and abrogates myofibroblastic differentiation without affecting the delivered dose of TGFβ1 [48]. Given that there are various cargos carried by the exosomes, other factors such as microRNAs, DNAs,

mRNAs, and cytokines could be co-delivered to the fibroblasts and cooperate to induce the differentiation of fibroblasts into tumor-promoting myofibroblasts.

In addition to the differentiation of resident fibroblasts, mesenchymal stem cells (MSCs) located in the tumor stroma can also function as precursors for tumor-supporting myofibroblasts. Studies from different groups have shown that cancer cell-derived exosomes induce the transition of MSCs into CAFs (Fig. 2), and these functional CAFs then promote cancer progression by releasing inflammatory cytokines and MMPs

[50–53]. Hepatocellular carcinoma (HCC)-derived exosomes transfer miR-21 to hepatic stellate cells (HSCs) and promote the transition of HSCs into CAFs by targeting phosphatase and tensin homolog (PTEN), thereby leading to the activation of PDK1/AKT signaling [50]. Exosomes secreted by chronic lymphocytic leukemia cells can deliver functional microRNA and proteins and activate nuclear factor kappa-light-chain-enhancer of activated B cells (NFκB) and protein kinase B (AKT) signaling in stromal cells, including endothelial cells and bone marrow-derived MSCs, leading to the transition of stromal cells into CAFs [51]. Furthermore, ovarian cancer cell-derived exosomes induce the myofibroblastic phenotype and functionality of adipose tissue-derived MSCs in SMAD-dependent and -independent pathways [53]. Overall, the exosomes derived from different cancer types promote the transition of various stromal cells into CAFs through different ways, probably due to their distinct exosomal cargos eliciting specific downstream signaling pathways.

3 No Immune Cells Are Immune to EV Targeting

The immune system is thought to provide a defense against cancer progression by identifying and destroying nascent tumor cells, a procedure known as immunosurveillance [54]. Multiple immune cells originating from bone marrow-derived progenitor cells, such as macrophages, dendritic cells (DCs), natural killer (NK) cells, and T- and B-lymphocytes, have been found to be responsible for anti-tumor immunity [55]. However, cancer cells can either lose expression of cell surface-bound tumor antigen to avoid recognition by cytotoxic T-lymphocytes, or can instigate an immunosuppressive tumor microenvironment (TME) by secreting tumor-derived factors, such as cytokines and EVPs, that reprogram the functions of immune cells, enabling immune evasion and cancer progression [7, 56–59]. Here, we describe how cancer cell-derived EVs function in immune

tolerance/suppression and pro-tumor immunity through modulating bone marrow-derived progenitor cells and the innate- and adaptive-immune systems (Fig. 2).

3.1 Progenitors

3.1.1 Recruiting HSPCs to Facilitate the Establishment of Pre-metastatic Niche

Hematopoietic stem and progenitor cells (HSPCs) residing in the bone marrow are the progenitors that are responsible for differentiating into all the cellular compartments of blood, including the immune system [55, 60]. Recruitment of HSPCs to distant organs has been identified as a hallmark of PMN formation [13]. David Lyden and colleagues conducted the proof-of-principle study, which demonstrated that metastatic melanoma cell-derived exosomes promote metastasis in the lung by educating and recruiting bone marrow progenitors to PMNs [7]. Briefly, metastatic melanoma cell-derived exosomes can fuse with and transfer the receptor tyrosine kinase MET to the bone marrow progenitors and reprogram them toward a pro-vasculogenic and pro-migratory phenotype, promoting mobilization of these exosome-reprogrammed progenitors from the bone marrow to the lungs for PMN generation [7]. Strikingly, suppression of exosome secretion via knocking down *Rab27a* [7], or inhibition of exosomal uptake using reserpine [61], impairs exosome-induced recruitment of bone marrow-derived cells (BMDCs) to the lungs for PMN formation and subsequent melanoma lung metastasis.

3.1.2 Inducing the Development of MDSCs and Associated Immunosuppression

HSPCs are able to differentiate into two main cell populations, namely common myeloid progenitors (CMPs) and common lymphoid progenitors (CLPs) [60]. CMPs primarily give rise to the components of the innate immune system, such as neutrophils, DCs, and macrophages, while CLPs mainly produce the adaptive immune system, such as T- and B-lymphocytes [60]. CLPs can also generate NKs cells, which show both innate- and adaptive-immune activity [62]. Cancer cell-derived EVs can fuse with and reprogram the function of bone marrow-derived progenitor cells [7, 63], which likely dysregulate the birth of immune cells at the level of bone marrow. For example, in a murine mammary cancer model, tumor exosomes were shown to target myeloid precursor cells and block their differentiation into DCs in the bone marrow, thereby inducing the development of myeloid-derived suppressor cells (MDSCs) [64, 65].

MDSCs have been identified as a population of immature myeloid cells accumulating in humans and mice in pathological conditions, such as chronic inflammation, stress, and cancer [66]. The development of MDSCs and recruitment of these cells to the TME and PMNs plays a critical role in promoting local tumor growth and distant PMN formation by suppressing various types of immune responses. For instance, membrane-associated heat shock protein 72 (Hsp72) is enriched in mouse cancer cell lines (colon cancer CT26, lymphoma EL4 and breast cancer TS/A), and this cancer cell-derived exosomal Hsp72 can bind to TLR2 on MDSCs and induce the expression and production of interleukin-6 (IL-6) in a TLR2/MyD88 dependent manner. Autocrine IL-6 then triggers the activation of STAT3 signaling and the suppressive function of MDSCs [67]. However, this study showed that only tumor-derived exosomes could trigger MDSC immunosuppressive activities while tumor-derived soluble factors that are free of exosomes could induce MDSCs expansion, indicating the functional specificity of the exosomes and other soluble factors derived from cancer cells in shaping the TME. In addition to the exosomal Hsp72, exosomal Hsp86 is another mediator for regulating MDSC immunosuppressive functions [68]. In particular, melanoma cells isolated from *RET* transgenic mice secrete exosomes that are enriched with Hsp86. Exosomal Hsp86 binds to TLR4 on immature myeloid cells, leading to the activation of NFκB signaling and upregulation of programmed cell death ligand 1 (PD-L1) expression. The binding of immature myeloid cells-expressing PD-L1 to CD8+ T cells-expressing PD-1 is sufficient to suppress the activation of anti-tumor CD8+ T cells. Furthermore, miRNAs packaged in the cancer cell-derived exosomes can also induce the immunosuppressive effects of MDSCs [69, 70]. In a murine glioma model, hypoxia promotes the secretion of exosomes from glioma cells. The exosomal miR-10a targets RAR-related orphan receptor alpha (*Rora*), a negative regulator of inflammation responses, to activate NFκB signaling, while the exosomal miR-21 targets *Pten* to induce activation of AKT signaling. The induction of both NFκB and AKT signaling pathways facilitates the differentiation and activation of MDSCs, thereby providing an immunosuppressive TME for tumor growth [70]. In melanoma patients, a set of microRNAs, including miR-146a, miR-155, miR-125b, miR-100, let-7e, miR-125a, miR-146b, and miR-99b, were enriched in melanoma cell-derived EVs and correlated with the infiltration of MDSCs. Moreover, enrichment of these microRNAs in the EVs from plasma of melanoma patients was associated with resistance to immunotherapy [69], pointing to the need for blocking the EV-mediated activation of immunosuppressive MDSCs to revert immunotherapy resistance.

3.2 Innate Immune System

The innate immune system functions in the first line for host defense against pathogens [71], including microbial infection and cancer. Cancer cell-derived EVs can interfere with the innate immune system by targeting and impairing the functions of a variety of innate immune cells, such as macrophages, DCs, NK cells, and neutrophils (Fig. 2).

3.2.1 Promoting Macrophage Polarization Toward an Anti-inflammation and Pro-tumor Phenotype

Macrophages represent a main subpopulation of immune cells in the TME. Two major phenotypes of the macrophages have been described: the type 1 macrophages (M1) that show pro-inflammatory and anti-tumor activities and type 2 macrophages (M2) that exhibit anti-inflammatory and pro-tumor activities [72]. Macrophages exposed to interferon gamma (IFNγ), tumor necrosis factor (TNF), and lipopolysaccharide (LPS) acquire the M1 phenotype, while interleukin-4 (IL-4), IL-10 and TGFβ1 can promote macrophage polarization to the M2 phenotype [72]. Recent studies have demonstrated that cancer cells can also fine-tune macrophagesphenotypes via paracrine transfer of exosomal cargos, including microRNAs and immune-regulatory proteins, to promote tumor growth and metastasis [73–76]. The exosomes isolated from lung cancer cell lines LLC and A549 are enriched with functional microRNAs, including miR-21 and miR-29a. These secreted exosomal miR-21 and miR-29a can bind to and activate murine TLR7 and human TLR8 in macrophages, resulting in the activation of TLR-mediated NFκB signaling and secretion of pro-metastatic inflammatory cytokines [73]. Epithelial ovarian cancer cell-derived exosomes can modulate macrophages by delivering miR-222-3p, which targets and suppresses SOCS3, leading to the upregulation of STAT3 signaling and M2 phenotype polarization [74]. In colon cancer, tumor cells harboring specific missense mutations in *TP53* deliver miR-1246-containing exosomes to neighboring macrophages and facilitate a shift of macrophages toward the M2 state showing enhanced secretion of pro-tumorigenic factors, such as IL-10, TGFβ, and MMPs [75]. During cancer progression, growing tumors are often subject to hypoxia due to insufficient oxygen supply. Hypoxia can stimulate the secretion of exosomes, which contain immuno-regulatory proteins and chemokines, including CSF-1, CCL2, FTH, FTL, and TGFβ, to promote M2-like macrophage polarization of

infiltrated myeloid cells [76]. In addition to the macrophages infiltrated in the TME, tumor EVs can also fuse with the resident macrophages in different organs, such as Kupffer cells in the liver and microglia in the brain [8–10]. How the tumor EVs reprogram the functions of those organ-specific resident macrophages to actively participate in PMNs formation will be discussed below.

3.2.2 Impairing the Priming of Cytotoxic T Cells by Dendritic Cells

DCs are bone marrow-derived antigen-presenting cells (APCs), which can initiate immune responses by capturing, processing, and presenting antigens to T-lymphocytes [77]. DCs can present antigens released from cancer cells to and activate naïve T-lymphocytes, inducing the generation of cytotoxic T-lymphocytes, thereby contributing to anti-tumor activities. However, tumor cells can also impair the functions of DCs and escape anti-tumor responses through transferring exosomal cargos [78, 79]. For example, EGFR mutant (E746-A750 deletion, termed EGFR-19del) lung cancer cells transfer active exosomal EGFR-19del to the surface of DCs and inhibit their ability to prime the activation of cytotoxic CD8+ T cells [78]. Pancreatic cancer cell-derived exosomes contain miR212-3p, targeting the regulatory factor X-associated protein (RFXAP), a transcription factor of the major histocompatibility complex (MHC) class II genes. Uptake of exosomal miR-213-3p induces the downregulation of RFXAP and consequent transcriptional suppression of MHC class II genes in DCs, leading to compromised activation of cytotoxic CD4+ T cells [79].

3.2.3 Escaping from Natural Killer Cell-Mediated Immunosurveillance

NK cells are a population of innate lymphoid cells exerting cytotoxicity on cells that are considered harmful to the host, such as cancer cells and virus-infected cells [80], and thus are considered as a critical cell population for immune surveillance of cancer. NK cells express multiple activating receptors, including NKG2D, NKp30, and

NKp46, and the binding of these receptors by ligands expressed by cancer cells triggers NK cell activation independent of MHC-mediated antigen presentation. To escape from NK cell-mediated immunosurveillance, cancer cells either modulate their cell surface ligands or release immunosuppressive factors to dampen the functions of NK cells [81]. TGFβ is one of the immunosuppressive factors that is enriched in tumor-derived EVs. TGFβ inhibits the activation and function of NK cells by downregulating their activating receptors, such as NKp30, NKp46, NKG2D, and DNAM-1, or by repressing mTOR signaling [82, 83]. As a consequence, EVs derived from many types of cancers, including pancreatic, prostate, and breast cancer, can promote tumor growth via attenuating the functions of NK cells [84–86].

3.2.4 Inducing Neutrophil Polarization and NET Formation to Promote Tumor Growth and Metastasis

Neutrophils are the most abundant immune cells in circulation, and they can infiltrate into the TME and distant organs to support tumor growth and metastasis [87, 88]. Both murine cancer models and human patients with different types of cancers, including melanoma, liver, kidney, and gastric cancers, show increased numbers of neutrophils in the circulation and the tumor, which is associated with poor outcomes [89–93]. Evidence has shown that the primary tumor itself and tumor-induced inflammatory cascade release multiple cytokines and chemokines, such as granulocyte colony-stimulating factor (G-CSF), granulocyte-macrophage colony-stimulating factor (GM-CSF), IL-6, IL-17 and CXCL1, stimulating neutrophil differentiation in the bone marrow and trafficking to the circulation [88, 94–97]. Similar to macrophages, cancer-associated neutrophils (both circulating and tumor-infiltrated) exhibit both anti-tumor and pro-tumor activities, referred to as N1 and N2 types, respectively [87, 98]. For instance, interferon-β (IFNβ) can induce an anti-tumor N1 phenotype, whereas TGFβ can promote a pro-tumor

N2 phenotype [99, 100]. Tumor exosomes have also been reported to prime neutrophils in the bone marrow and induce their N2 polarization to facilitate tumor growth and metastasis [101, 102]. Furthermore, tumor exosome-mediated recruitment of neutrophils to the lung determines metastatic organotropism by inducing PMN formation [16]. Mechanistically, exosomes derived from primary Lewis lung carcinoma (LLC) fuse with alveolar epithelial cells and activate TLR3 via exosomal RNAs, especially the small nuclear RNAs (snRNAs). Activated alveolar epithelial cells secrete chemokines, such as CXCL1, CXCL2, CXCL5, and CXCL12, which mobilize neutrophils to the lung for generating a favorable microenvironment for homing of disseminated cancer cells [16].

Tumor-derived exosomes also promote the formation of neutrophil extracellular traps (NETs), which are extracellular chromatin structures bound with specific granular and cytoplasmic proteins [103, 104]. NETs were first identified as a form of innate immune response to microbe infection, while recent evidence showed that NETs also contribute to cancer-associated thrombosis and cancer metastasis [105–107]. In a murine model of 4T1 breast cancer, tumor-bearing mice showed an increased number of circulating neutrophils and accelerated NET-mediated thrombosis in venous and arterial injury models. Strikingly, 4T1 cell-derived exosomes induced NET formation in neutrophils primed with G-CSF *in vitro* and intravenous injection of 4T1 cell-derived exosomes into G-CSF-treated mice enhanced thrombosis *in vivo*, suggesting exosome-mediated NET formation facilitates the formation of cancer-associated thrombosis [103]. NETs were also found to be able to stimulate 41 cell invasion and migration and trap circulating lung cancer cells (murine H59 and human A549) within chromatin webs, thereby promoting metastasis [106, 107]. Therefore, it is conceivable that tumor-derived exosomes can promote thrombosis and metastatic progression by stimulating NET formation.

3.3 Adaptive Immune System

The adaptive immune system is activated when innate immunity is insufficient to eliminate pathogens or toxins and is induced by antigen processing and presenting cells [108]. The adaptive immune responses are primarily carried out by two major types of immune cells, namely T- and B-lymphocytes, which function through antigen-specific receptors that are generated by gene rearrangements during development. Human adaptive immunity is thought to be able to eliminate cancers, however, as mentioned above, cancers can actively turn off anti-tumor responses by self-remodeling or reprogramming the immune cells (Fig. 2).

3.3.1 Suppressing the Proliferation and Activation of T-lymphocytes

T-lymphocytes, especially CD4+ (helper) and CD8+ (cytotoxic) T cells, are considered as the major immune cells responsible for combating cancer cells. However, the anti-tumor functions of T-lymphocytes can be significantly compromised, which is mediated, at least in part, by tumor-derived EVs. Evidence showed that tumor-derived EVs can suppress T-lymphocyte functions by either inhibiting their infiltration into tumors [109] or by attenuating their proliferation and activation [110, 111]. Head and neck squamous cell carcinoma (HNSCC) cell line-derived exosomes are enriched with immunosuppressive cargos, including FasL, PD-L1 CD39, CD73, and a truncated form of COX-2. After injection of these exosomes into mice bearing premalignant oral/esophageal carcinoma, the numbers of tumor-infiltrating CD4+ and CD8+ T-lymphocytes were significantly reduced, leading to accelerated tumor progression [109]. HCC cells can transfer their functional protein, 14-3-3ζ, to the T-lymphocytes through exosomes. Uptake of 14-3-3ζ by T-lymphocytes impairs their viability, proliferation and activity and induces their exhaustion [110]. In a glioblastoma model, the stem-like brain tumor-initiating cells shed exosomes containing the extracellular protein, tenascin-C (TNC). Exosomal-TNC binds to T-lymphocyte-associated $\alpha5\beta1$ and $\alpha v\beta6$ integrins and attenuates mTOR signaling, resulting in the reduction of T-lymphocyte proliferation, activation and cytokine production [111]. Unfortunately, the specificity of the T-lymphocyte subsets that can take up tumor EVs still remains elusive.

3.3.2 Altering the Development of B-lymphocytes

Unlike T-lymphocytes, which are the key components of cell-mediated responses that can directly destroy target cells, B-lymphocytes mediate the humoral immune response mainly by giving rise to antibodies for recognizing antigens and neutralizing pathogens after activation. It has been shown that B-lymphocytes are present in the TME of different cancer types, such as melanoma, breast, and pancreatic cancer [112–114], whereas both the phenotypes and functions of tumor-associated B-lymphocytes are less characterized than that of T-lymphocytes. There is a subset of tumor-infiltrated B-lymphocytes, namely regulatory B (Breg) cells, which has been reported to exhibit pro-tumor activity by inducing immunosuppression [115, 116]. In human hepatocellular carcinoma (HCC), cancer cell-derived exosomal HMGB1 promotes the expansion of TIM-1$^+$ Breg cells via TLR2/4-MAPK signaling pathways. The accumulation of Breg cells in tumors shows strong suppressive activity against cytotoxic CD8$^+$ T-lymphocytes by expressing and secreting immunosuppressive IL-10, thereby fostering HCC immune evasion [117].

Taken together, cancer cells escape immune attack throughout metastatic progression via reprogramming the immune system through targeting of diverse cell types, ranging from HSPCs to well-differentiated innate and adaptive immune cells. Importantly, this immune modulation is mediated, in part, by cancer cell-derived EVs. However, the factors that determine EV uptake in innate versus adaptive immune systems and the specific subsets of these immune cells that take up EVs are largely unknown. Although the roles of HSPCs and the innate immune system in PMNs formation have been intensively characterized, how the adaptive immune system, especially T- and B-lymphocytes functionally promotes this process still remains elusive. Understanding the functional roles of the adaptive immune system in the development of metastasis may pave the way for providing new anti-tumor and anti-metastasis therapeutic strategies.

4 Angiogenesis

Angiogenesis is a multistep process to generate new vasculature from pre-existing vessels. It involves the breakdown of the vascular membrane and activation, proliferation and migration of endothelial cells to form the lumen and the resultant new blood vessels [118]. Angiogenesis plays a fundamental role in development and growth. However, during cancer progression, angiogenesis is the key event for tumor growth and metastasis by providing essential nutrients and oxygen, as well as allowing trafficking of cells to distant organs via the circulation. The crosstalk between cancer cells and TME-resident stromal cells, mediated by cytokines, growth factors and EVPs, orchestrates the multidimensional process for endothelial cell activation and angiogenic programming of tumor tissues [119]. Here, we describe how tumor-derived EVs function in favoring tumor vascularization (Fig. 2).

One of the major pathways contributing to angiogenesis is vascular endothelial growth factor (VEGF)-mediated signaling. In HNSCC cancer patients, EV-packaged ephrin type B receptor 2 (EPHB2) binds to ephrin-B2 on the surface of endothelial cells and induces ephrin-B2 reverse signaling, which then promotes angiogenesis by activating both VEGF receptor 2 (VEGFR2) and STAT3 signaling pathways [120, 121]. As a result, overexpression of EPHB2 in HNSCC tumors is associated with enhanced angiogenesis and poor prognosis [121]. Hypoxic breast cancer cells secrete exosomes containing miR-210, which targets ephrin-A3 and protein-tyrosine phosphatase 1B (PTP1B) in neighboring cells

[122]. PTP1B is a negative regulator of VEGF [123], and its downregulation leads to increased activation of VEGF, thus promoting VEGF-mediated recruitment and activation of endothelial cells for subsequent angiogenesis [122]. In another hypoxia study, miR-135b was found to be significantly upregulated and enriched in exosomes derived from hypoxia-resistant multiple myeloma cells [124]. The exosomal miR135-b targets and suppresses the expression of factor-inhibiting hypoxia-inducible factor 1 (FIH1) in endothelial cells, thereby releasing the transcriptional activity of hypoxia-inducible factor 1 alpha (HIF1α). HIF1 then transcriptionally upregulates the expression of angiogenic cytokines, such as VEGF, TGFβ3 and angiopoietin, to promote vascularization within the tumor [124]. In addition to miR-135b, another type of RNA, namely piwi-interacting RNA-832 (piRNA-832), was also identified in the EVs derived from multiple myeloma cells [125]. The EVs containing piRNA-832 can be effectively transferred from multiple myeloma cells to endothelial cells where piRNA-832 induces the activation, proliferation, and tube formation of endothelial cells by upregulating VEGF, IL-6, and ICAM1 and downregulating apoptosis. Clinically, the increased expression of piRNA-832 is correlated with advanced stage and poor prognosis of multiple myeloma [125].

Other than delivering EV cargoes that can activate VEGF signaling in endothelial cells, cancer cells can also directly secrete VEGF in EVs to induce angiogenesis. For instance, elevated pro-angiogenic factor VEGF-A was detected in the EVs isolated from glioblastoma patients compared with healthy individuals. A subpopulation of glioblastoma cells, namely glioblastoma stem-like cells, secrete VEGF-A in EVs, and uptake of the VEGF-A-containing EVs by brain endothelial cells increases their capacity for permeability and vascular formation [126]. Recently, new evidence showed that tumor-derived small EVs (EVs with a diameter of <200 nm and soluble proteins of >100 kDa) could also stimulate and activate endothelial cells for migration and tube formation, independent of vesicle uptake [127]. This pro-angiogenic effect is induced by a 189 amino acid isoform of VEGF ($VEGF_{189}$), which preferentially localizes on the surface of small EVs by interacting with heparin. The small EV-bound $VEGF_{189}$ ($sEV\text{-}VEGF_{189}$) is sufficient to activate VEGFR2 signaling via interacting with the extracellular domain of VEGFR2 on target cells. Interestingly, $sEV\text{-}VEGF_{189}$ is insensitive to the therapeutic antibody for VEGF, namely bevacizumab. As a result, $sEV\text{-}VEGF_{189}$ can still stimulate tumor xenograft growth in a mouse model treated with bevacizumab, and elevated level of $sEV\text{-}VEGF_{189}$ is associated with disease progression in renal cell carcinoma patients who received bevacizumab monotherapy, indicating that $sEV\text{-}VEGF_{189}$ contributes to the resistance of tumors for anti-VEGF therapy [127].

In addition to VEGF signaling, exosomal MMP-mediated degradation of the ECM, which has been described above, also functionally promotes tumor angiogenesis [36, 41]. Together, these findings illustrate the functions of EVs and their associated cargos in regulating angiogenesis for tumor growth and metastasis.

5   Vascular Permeability

In order to metastasize to distant organs, disseminated cancer cells need to enter the circulation (termed intravasation), survive in the circulation, and then leave the circulation (termed extravasation) before seeding into the target organs [128, 129]. Here, we present evidence for the role of cancer cell-EVs in contributing to intravasation and extravasation by transferring EV-associated cargos to endothelial cells to modulate vascular permeability (Fig. 2).

A variety of EV-associated cargoes have been identified as modulators of vascular permeability, including proteins and microRNAs. In glioblastoma, glioblastoma stem-like cells transfer pro-permeability factor Semaphorin3A (Sema3A) bound on the EV surface to endothelial cells. Both *in vitro* and *in vivo* evidence showed that EV-bound Sema3A binds to its receptor Neuropilin1 (NRP1) on recipient endothelial

cells, thereby triggering endothelial and vascular permeability dependent on the Sema3A/NRP1 signaling pathway [130]. As mentioned above, VEGF-A packaged in EVs from glioblastoma stem-like cells also induces permeability of brain endothelial cells *in vitro* [126]. MiR-25-3p, targeting Krüppel-like factor 2 (KLF2) and Krüppel-like factor 4 (KLF4), can be transferred from colorectal cancer cells to endothelial cells [131]. As members of the Krüppel-like family transcription factors, KLF2 negatively regulates the expression of the pro-angiogenic factor VEGFR2 [132], while KLF4 positively regulates tight junction-related proteins, including ZO-1, occludin, and claudin-5 [133]. Consequently, uptake of exosomal miR-25-3p by endothelial cells decreases the expression of KLF2 and KLF4, leading to increased vascular permeability and angiogenesis via upregulating VEGFR2 and downregulating tight junctions [131].

Furthermore, microRNAs packaged in cancer cell-derived EVs can also directly target tight junction-related proteins to induce vascular permeability and promote metastasis. For example, exosomal miR-23a derived from lung cancer cells was reported to target and inhibit tight junction protein ZO-1 and prolyl hydroxylase 1 and 2 (PHD1 and 2), leading to increased vascular permeability and metastasis [134]. Additionally, miR-105 and miR-939 are packaged in the exosomes secreted by metastatic breast cancer cells and are transferred to endothelial cells. Exosomal miR-105 and miR-939 destroy the barrier function of endothelial monolayers and induce vascular permeability by downregulating tight junction proteins ZO-1 and vascular endothelial cadherin (VE-cadherin), respectively [18, 135]. Another exosomal microRNA, miR-103 is secreted by HCC cells, can also directly inhibit the expression of VE-cadherin, p120-catenin and ZO-1 in endothelial cells when uptaken, leading to attenuated integrity of tight junctions [136]. Exosomal miR23-a, miR-105, and miR-103 are readily detected in the circulation of lung cancer, breast cancer and HCC patients, respectively, and high levels of these microRNAs are correlated with cancer recurrence and occurrence of metastasis [18, 136], highlighting their potential role as therapeutic targets of metastasis.

Another type of non-coding RNA, namely circular RNAs, is also present in EVs secreted by cancer cells [137]. Abundant evidence demonstrated that circular RNAs function as microRNA sponges competing with microRNAs to regulate downstream gene expression [138]. Pancreatic cancer cells deliver exosomal circular RNA IARS (circ-IARS) to endothelial cells to downregulate the expression of miR-122 in recipient cells, and relieve its inhibition of Ras homolog gene family member A (RhoA) expression. Activated RhoA promotes the formation of actin stress fibers by enhancing contractility of myosin, which induces cell contraction and leads to a breakdown of tight junctions, resulting in elevated tumor invasion and metastasis [137].

6 PMN Formation and Metastatic Organotropism

In 1889, Stephen Paget raised the hypothesis that the communication between 'seeds' (referring to tumor cells) and 'soil' (referring to the host distant microenvironments) provides a supportive microenvironment in distant organs favoring organotropic metastasis [11]. It took nearly one century before Isaiah Fidler proved this hypothesis by providing evidence showing that metastasis is a non-random biological event dependent on both tumor properties and host microenvironments [139, 140]. Indeed, distinct cancer types display significant variations in organ-specific metastasis, termed metastatic organotropism. For instance, prostate cancer typically colonizes the bone and colon cancer most frequently metastasizes to the liver, while breast cancer spontaneously spreads to bone, lung, liver, and brain [4, 141, 142]. During the past few decades, extensive studies have illustrated that the target tissues and organs of future metastasis are not determined by just passive homing of disseminated cancer cells but instead are actively manipulated by the primary tumors prior to initiation of metastasis. Biophysiological manipulation of local lymph nodes and distant organs by tumor-derived factors, including EVPs, provides supportive microenvironments, termed pre-metastatic niches or PMNs, whose establishment is formed prior to the arrival of metastatic tumor cells and whose identity favors, in part, organotropic metastasis by addressing the mystery of why cancer metastasizes to specific organ sites [12, 13]. Here we summarize the functional roles of tumor-derived EVs in preparing PMNs in local lymph nodes and distant organs, including lung, liver, bone, and brain for the development of metastasis (Fig. 3).

6.1 Local Metastasis to Lymph Nodes

Disseminated tumor cells can travel from primary tumors to local lymph nodes through the lymphatic system or to distant organs through the bloodstream. Lymph node metastasis is frequently an early event in tumor dissemination [143, 144]. Tumor-derived EVs are readily detected in the lymph nodes in mouse cancer models and human cancer patients, wherein they are found to be fused with DCs, macrophages, B-lymphocytes, stromal cells and endothelial cells [145–148]. Evidence has shown that tumor-derived EVs promote PMN formation in the lymph nodes by modulating immune responses, remodeling the lymphatic network and reconstructing ECM [145, 149, 150] (Fig. 3).

The major role of the lymphatic system is to coordinate the trafficking of antigens and immune cells [151]. The lymph node-localized immune cells, such as DCs, macrophages and B-lymphocytes are targeted by tumor-derived EVs, suggesting a role for tumor-derived EVs in modulating immune responses within the lymph node to support the colonization of disseminated cancer cells. Indeed, the cargos in human melanoma-derived EVs, including S100A8, S100A9, AnnexinA1, and ICAM1, have been found to significantly compromise the maturation of DCs in the sentinel lymph node (SLN) [146]. Furthermore, breast cancer cell-derived exosomes can fuse with macrophages in the axillary lymph nodes and activate these target

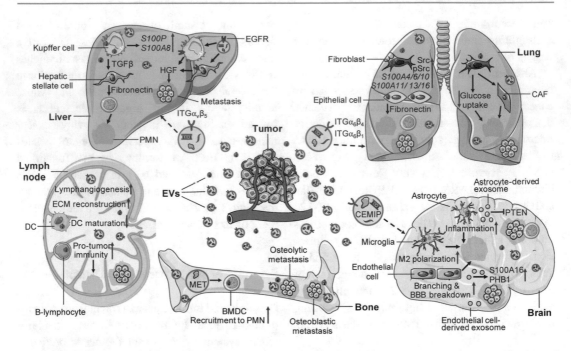

Fig. 3 EVs induce the formation of PMNs in local lymph nodes and distant organs, and mediate organotropic metastasis. During cancer progression, primary tumor-derived EVs, which contain diverse types of cargos, can travel to different tissues and organs for PMN formation and subsequent metastasis. Lymph nodes. Tumor-derived EVs can promote PMN formation by modulating immune responses (for example by blocking DCs maturation and by inducing B-lymphocyte-mediated pro-tumor immunity), remodeling the lymphatic network to promote lymphangiogenesis and reconstructing ECM. Liver. Pancreatic cancer cell-derived exosomes containing ITGα$_v$β$_5$ target to liver and can fuse with Kupffer cells and stimulate their release of TGFβ, which in turn, activate hepatic stellate cells to produce fibronectin. Fibronectin deposition in the liver creates a fibrotic microenvironment for arresting BMDCs. Furthermore, uptake of tumor-derived exosomes can also upregulate *S100P* and *S100A8* expression in Kupffer cells, leading to the formation of a pro-inflammatory milieu for PMN formation and liver-tropic metastasis. Gastric cancer cell-derived exosomes transfer EGFR to Kupffer cells and hepatic stellate cells, leading to the activation and release of HGF into extracellular space to promote colonization and proliferation of cancer cells. Lung. Breast cancer cell-derived exosomes enriched with integrins α6β4 and α6β1 can fuse with fibroblasts and epithelial cells in the lung microenvironment, and activate Src signaling and upregulate pro-inflammatory *S100* gene expression for PMN formation and lung-tropic metastasis. Furthermore, breast cancer cell-derived exosome can also downregulate PKM2 and GLUT1 thereby suppressing glucose uptake by the fibroblasts, leading to enhanced glucose utilization of circulating cancer cells in PMNs. HCC cell-derived exosomes can convert normal fibroblast to CAF, and melanoma-derived exosomes can promote the recruitment of BMDCs to the lung. Brain. CEMIP is specifically packaged into the exosomes derived from brain-tropic breast cancer cells. Exosomal CEMIP induces endothelial cell branching and microglial cell inflammation for PMN formation. Breast cancer cell-derived exosomes can also breach the BBB and induce M1-to-M2 polarization of microglial cells. Melanoma-derived EVs can fuse with astrocyte and promote the expression of multiple cytokines and chemokines to create an inflammatory PMN. Furthermore, brain stromal cell-derived exosomes can also functionally facilitate brain metastasis. For example, microvascular endothelial cell-derived exosomes fuse with SCLC cells and upregulate their expression of S100A16 and PHB1, resulting in enhanced survival of SCLC cells in the brain. Astrocyte-derived exosomes induce loss of PTEN in brain metastatic melanoma and breast cancer cells, leading to recruitment of BMDCs (myeloid cells) to the TME for supporting the outgrowth of brain metastatic lesions. Bone. Melanoma-derived exosomes containing MET can fuse with and reprogram bone marrow progenitors. These reprogrammed progenitors are then recruited to PMNs in different organs as BMDCs. Depending on the different types of cancers, bone metastasis could be osteolytic or osteoblastic. For instance, breast cancer predominantly induces osteolytic metastasis, while prostate cancer bone metastasis usually involves an osteoblastic phenotype

macrophages, triggering them to secrete inflammatory cytokines, such as IL-6, thus contributing to metastatic tumor development in lymph nodes [152]. However, in a mouse model and in patients with melanoma, the subcapsular sinus (SCS) CD169+ macrophage layer in tumor-draining lymph nodes (tdLNs) act as a physical barrier to block the penetration of tumor-derived EVs by fusing with them, thereby suppressing melanoma. During the natural course of tumor progression, continuous challenge of tumor induces tdLNs to enlarge without expanding their SCS CD169+ macrophage pool, leading to decreased density of CD169+ macrophages and resultant breakdown of the CD169+ macrophage barrier. The disrupted CD169+ macrophage barrier enables tumor-derived EVs to penetrate tdLN cortex and interact with B-lymphocytes, which foster tumor progression by producing autoantibodies [145]. Another study pinpointed that SCS CD169+ macrophages are dispensable for melanoma growth and its lymph node metastasis, since subcutaneously injected melanoma showed similar metastatic lymph node lesions in both wild type and CD169$^{-/-}$ mice [153]. Hence, although tumor-derived EVs can fuse with different immune cells in the lymph nodes, their functional roles in establishing PMNs for metastatic progression need further investigation.

Lymphatic network remodeling, mediated by the growth of new lymphatic vessels through lymphangiogenesis and remodeling of pre-existing lymphatics has been reported to functionally promote cancer metastasis to regional lymph nodes and distant organs [154–156]. The factors, including lymphangiogenic growth factors (such as VEGF-C and VEGF-D that are secreted by tumor cells or immune cells) [154, 155] and tumor-derived EVs [149, 150, 157, 158], are important inducers of lymphatic network remodeling. For patients with cervical squamous cell carcinoma (CSCC), primary tumor-derived exosomes can transfer miR-221-3p to lymphatic endothelial cells. Exosomal miR-221-3p promotes the migration and tube formation of lymphatic endothelial cells via targeting vasohibin-1 and activating ERK/AKT signaling, thereby inducing peritumoral lymphangiogenesis and lymph node metastasis [158]. In colorectal cancer patients, primary tumor-derived exosomes enriched with interferon regulatory factor 2 (IRF-2) are preferentially uptaken by macrophages in SLN. Uptake of exosomal IRF-2 induces the release of VEGF-C by macrophages, which then remodel the lymphatic network by promoting the proliferation of lymphatic endothelial cells for the formation of PMNs and subsequent development of sentinel lymph node metastasis [150]. Bladder cancer patients with lymph node metastasis have poor prognosis [159]. In addition to VEGF-C, which has been demonstrated to promote bladder cancer lymph node metastasis, exosomes are another mediator linking bladder cancer and the lymphatic system independent of VEGF-C [149, 157]. Bladder cancer cell-derived exosomes containing a complex consisting of lymph node metastasis-associated transcript 2 (*LNMAT2*, a long noncoding RNA) and heterogeneous nuclear ribonucleoprotein A2B1 (hnRNPA2B1) can be internalized by lymphatic endothelial cells. Then, LNMAT2 forms a triplex with the prospero homeobox 1 (*PROX1*) promoter and enhances its transcription by increasing the hnRAPA2B1-mediated H3K4 trimethylation, resulting in lymphangiogenesis and lymphatic metastasis [149].

ECM reconstruction occurs not only in the TME but also in local lymph nodes and distant organs to facilitate PMN formation and resultant cancer metastasis. Evidence showed that homing of melanoma-derived exosomes to SLNs leads to the induction of molecular signals involved in melanoma cell recruitment (Stabilin 1, Ephrin R β4, and Integrin αv), extracellular matrix deposition (Mapk14, uPA, Laminin 5, Collagen 18α-1 and G-α13) and vascular endothelial cell proliferation (TNFα, VEGF-B, HIF1-α, and Thbs1) within the lymph nodes [143]. In gastric carcinoma, cancer cell-derived exosomes containing CD97 were shown to promote the PMN formation in draining lymph nodes by inducing the expression of pro-metastatic factors, including CD55, CD44v6, integrin α5β1, CD31, and EpCam [160]. Furthermore, in murine pancreatic cancer, CD44v6 positive cancer cell-derived exosomes can transfer microRNAs (such as miR-494 and miR-542-3p) to lymph node stroma cells, leading to the downregulation of cadherin-17 and upregulation of metalloproteinases (MMP2, MMP3, and MMP14) [147]. Thus, exosome-induced dysregulation of adhesion molecules and proteases modulates ECM within the lymph nodes, leading to the preparation of PMNs for tumor cell hosting.

Lymph node metastasis has staging and prognostic values for disease progression and outcome prediction in different cancer types, such as melanoma and prostate cancer [161, 162]. The findings of EV-mediated communications between cancer cells and lymph nodes provide potential non-invasive diagnostic approaches and therapeutic targets for cancer patients in early stages with lymph node metastasis. Intervening therapeutically with lymph node metastasis by targeting EV-dependent pathways should be considered as a promising strategy for preventing metastatic disease progression.

6.2 Distant Organotropic Metastasis for Lung, Liver, Bone, and Brain

Tumor-derived EVs have been proven to be critical regulators of PMN establishment for organotropic metastasis by transferring bioactive molecules, including proteins, metabolites, lipids, and nucleic acids, to recipient cells through circulation. Critical studies provide evidence that tumor-derived EVs containing unique surface molecules can target specific organs via ligand-receptor means to prepare the PMNs for organotropic metastasis [9]. Consistently, downregulation of EV secretion from cancer cells by genetic engineering or pharmacological inhibition reduces the formation of PMNs and resultant spontaneous metastasis in different mouse cancer models [7, 163].

6.2.1 Lung Metastasis

Lung is a frequent site of metastasis formation from various malignant lesions, such as melanoma, liver cancer, and breast cancer [141, 142, 164]. Research in the past few years has greatly

improved our understanding of the mechanisms underlying lung-tropic metastasis mediated by tumor-derived EVs (Fig. 3).

As mentioned above, melanoma cell-derived exosomes can promote metastasis to the lung by inducing vascular leakiness, and educating bone marrow progenitors toward a pro-metastatic phenotype. The educated bone marrow progenitors are then recruited to the lung for PMN preparation [7]. Furthermore, in a spontaneous metastasis model achieved by tumor inoculation-resection-relapse approach, primary LLC tumor-derived exosomes was shown to activate TLR3 in lung epithelial cells through exosomal RNAs, inducing chemokines induction and consequent neutrophil recruitment to the lung for PMN formation [16]. These activated lung epithelial cells were also shown to secrete fibronectin, a protein that facilitates the adhesion of BMDCs in PMNs [13, 16, 165].

In addition to recruiting immune cells to the lungs, tumor-derived exosomes can also directly fuse with lung resident cells, including fibroblasts and epithelial cells [9, 166]. In a breast cancer model, cancer cell-derived exosomes enriched with integrins $\alpha 6\beta 4$ and $\alpha 6\beta 1$ can fuse with fibroblasts and epithelial cells in the laminin-rich lung microenvironment, and activate Src signaling and upregulate pro-inflammatory $S100$ gene expression for PMN formation and lung-tropic metastasis [9]. Furthermore, metastatic HCC cells exhibit a high capacity for lung metastasis by converting normal fibroblasts to CAFs in the lung via exosomal miR-1247-3p [166]. Mechanistically, exosomal miR-1247-3p downregulates β-1,4-galactosyltransferases III (B4GALT3) in normal fibroblasts, inducing the activation of $\beta 1$-integrin-NFκB signaling. These activated CAFs promote the establishment of inflammatory PMNs by secreting cytokines, including IL-6 and IL-8, to facilitate lung metastasis [166].

Tumor-derived exosomes can also metabolically reprogram the microenvironment in lungs for generating PMNs [17]. In a breast cancer model, cancer cell-derived exosome-encapsulated miR-122 can be transferred to lung fibroblasts. MiR-122 mediated downregulation of PKM2 and GLUT1 then suppresses glucose uptake by the fibroblasts, promoting enhanced glucose utilization and consequent growth of circulating cancer cells in PMNs [17]. Taken together, these studies demonstrated that tumor-derived exosomes promote PMN formation via providing microenvironmental immune permissiveness, stromal cell remodeling and metabolic functional reprogramming for lung metastasis.

6.2.2 Liver Metastasis

In addition to the lung, liver is another favored distant organ for metastasis from gastrointestinal malignancies, including pancreatic, gastric, and colorectal cancer. Cancer patients with liver metastasis often show poor prognosis, due to the fact that liver is the major organ for nutrient metabolism, detoxification, and bile production. The liver is composed of five major cell types including hepatocytes, Kupffer cells, hepatic stellate cells, sinusoid endothelial cells, and biliary epithelial cells [167]. EVs secreted from different types of primary tumors might communicate with different cell types in the liver to make it hospitable for metastasis (Fig. 3).

Exosomes derived from pancreatic cancer cells can be specifically uptaken by Kupffer cells in the liver [8, 9]. Notably, exosome-bound integrins $\alpha_V\beta5$ promote the binding of exosomes to Kupffer cells, and exosomal migratory inhibitory factor (MIF) then stimulates the release of TGFβ by Kupffer cells, which, in turn, activate hepatic stellate cells to produce fibronectin. Deposition of fibronectin in the liver creates a fibrotic microenvironment for arresting BMDCs, such as macrophages and neutrophils. Furthermore, uptake of tumor-derived exosomes can also upregulate S100P and S100A8 expression in Kupffer cells, contributing to the formation of a pro-inflammatory milieu for establishing PMNs and liver-tropic metastasis [8, 9]. Clinically, integrin α_V was increased in plasma-derived exosomes isolated from pancreatic cancer patients with liver metastasis compared with patients with no metastasis or control subjects [9], indicating that specific exosomal integrins can serve as biomarkers for organotropic metastasis.

Gastric cancer cell-derived exosomes deliver epidermal growth factor receptor (EGFR) to the liver, wherein EGFR integrates into the plasma membrane of stromal cells, including Kupffer cells and hepatic stellate cells [168]. Integration of EGFR into these liver stromal cells can suppress the expression of miR-26a/b, which target hepatocyte growth factor (HGF), leading to the upregulation and activation of HGF. Activated HGF can be secreted to the extracellular space in liver, and binds to its receptor c-MET on the cancer cells, thereby inducing the colonization and proliferation of metastatic gastric cancer cells [168].

As mentioned above, colorectal cancer cell-derived exosomal miR-25-3p reprograms the endothelial cells to induce vascular permeability and angiogenesis in the liver microenvironment, leading to the formation of PMNs for liver metastasis [131]. Another microRNA, namely miR-21-5p, enriched in exosomes derived from colorectal cancer cells also functionally promotes liver-specific metastasis [169]. Mechanistically, miR-21-5p binds to TLR7 in Kupffer cells and induces Kupffer cell polarization toward a pro-inflammatory phenotype that produces IL-6. This miR-21-5p-TLR7-IL-6 axis creates permissive inflammatory PMNs in the liver for homing of the circulating colorectal cancer cells [169].

6.2.3 Bone Metastasis

Many types of cancers, such as breast and prostate cancer, frequently metastasize to the bone. Bone and bone marrow consist of different cell types, including osteoblasts (synthesize bone), osteoclasts (resorb bone), and osteocytes [141, 170]. Due to the unique features of these cell types, different patterns of bone effects have been observed in cancer patients with bone metastasis, including osteolytic and osteoblastic [170]. Indeed, tumor-derived EVs can alter the bone microenvironment in both ways to support osteolytic or osteoblastic metastasis (Fig. 3).

Breast cancer predominantly induces osteolytic metastasis, which requires the activity of bone-resorbing osteoclasts [170]. Receptor activator of NFκB ligand (RANKL) and its receptor (RANK) are crucial factors for inducing osteoclasts differentiation and bone formation [171]. Evidence showed that breast cancer cell-derived exosomal L-plastin, a member of the actin-binding proteins family, could induce Ca^{2+}/NFATc1-mediated osteoclast formation from RANKL-primed precursors, thereby facilitating metastatic bone osteolysis [172]. Furthermore, breast cancer cells can also transfer exosomal miR-20a-5p, which targets *SRCIN1* (SRC kinase signaling inhibitor 1), to the pre-osteoclasts, namely bone marrow macrophages, resulting in osteoclastogenesis and osteolytic lesions [173].

In contrast to breast cancer, the progression of prostate cancer to metastatic bone lesions involves an osteoblastic phenotype [170]. EVs derived from prostate cancer cells were shown to enhance osteoblast viability and produce a growth supportive microenvironment for circulating tumor cells via transferring micro- and messenger RNA cargos, which are related to the genes involved in protein translation, cell–cell interaction, and cell surface signaling [174]. For example, prostate cancer cells promote osteoblast activity via transferring exosomal miR-141-3p to osteoblasts [175]. Briefly, miR-141-3p targets and downregulates *DLC1* (deleted in liver cancer 1), a gene encoding a Rho GTPase-activating protein in osteoblasts. DLC1 downregulation leads to the activation of p38MAPK and resultant upregulation of osteoblast osteoprotegerin (OPG), which serves as a decoy receptor of RANKL and competes for binding of RANKL to RANK [176]. As mentioned above, RANKL/RANK signaling functions in promoting osteo-clast differentiation but inhibits the differentiation of osteoblasts [171]. OPG-mediated suppression of RANKL/RANK signaling thereby promotes osteoblasts activity, which reconstructs the bone microenvironment favoring metastasis [175]. Instead of directly modifying osteoblasts activity, prostate cancer cells can also promote the osteogenic differentiation of MSCs via transferring exosomal has-miR-940, targeting *ARHGAP1* (Rho GTPase activating protein 1) and *FAM134A* (family with sequence similarity 134, member A) to MSCs [177], although the mechanism remains

unclear. Interestingly, overexpression of has-miR-940 in MDA-MB-231 breast cancer cells, which preferentially induce osteolytic lesions, leads to the formation of osteoblastic metastasis through promoting osteogenic differentiation of MSCs. Taken together, these findings suggest that cancer cells can remotely modify the bone microenvironment to support the osteolytic or osteoblastic metastatic lesions via transferring various exosomal cargos.

6.2.4 Brain Metastasis

Brain metastasis, another important factor contributing to cancer-associated deaths, mainly comes from breast cancer, lung cancer, and melanoma [141]. Emerging evidence shows that these primary tumors are able to remotely manipulate the brain microenvironment to support colonization, survival, and outgrowth of circulating cancer cells via secreted EVs [9, 10, 178–180] (Fig. 3).

A key event for successful brain metastasis is to break the blood-brain barrier (BBB), a selective semipermeable structure formed by endothelial cells with tight junction [181]. The BBB regulates the transport of molecules between the central nervous system and the circulation, and prevents neurotoxic blood components, pathogens, and circulating tumor cells from invading the brain [141, 181]. However, tumor cells are able to destroy the integrity of the BBB via secreted EVs, to facilitate brain metastasis [182]. For example, EVs derived from brain-tropic breast cancer cells can downregulate *PDPK1* (3-phosphoinositide-dependent protein kinase-1) in brain endothelial cells via transfer of miR-181c [178]. Downregulation of PDPK1 reduces the phosphorylation of cofilin, and the resultant activated cofilin dysregulates the localization of tight junction proteins, N-cadherin and actin, leading to the breakdown of BBB to enhance brain metastasis [178].

In a mouse model, exosomes derived from brain-tropic, but not lung- or bone-tropic, breast cancer cells predominantly distribute to the brain and treating mice with brain-tropic exosomes facilitates metastatic lesions in the brain [9, 10]. One important finding made by Lyden and colleagues is that cell migration-inducing and hyaluronan-binding protein (CEMIP) is specifically packaged in brain-tropic exosomes secreted by brain-tropic breast cancer cells and can be uptaken by brain endothelial and microglial cells. Exosomal CEMIP induces endothelial cell branching and inflammation in the perivascular niche by upregulating prostaglandin-endoperoxide synthase 2 (Ptgd2), Tnf, and Ccl/Cxcl, thereby conditioning PMNs for brain metastasis [10]. Uptake of brain-tropic exosomes by microglial cells can also trigger their M1-to-M2 polarization, and this conversion induces a local immune suppressive microenvironment for the colonization of circulating breast cancer cells [179]. Furthermore, in a murine melanoma model, cancer cell-derived EVs are able to transfer inflammation-inducing mRNA cargos, including high-mobility group box 1 (Hmgb1), thymic stromal lymphopoietin (Tslp), and interferon regulatory factor 1 (Irf1), to the astrocytes to promote the expression of various cytokines and chemokines, thereby creating a hospitable inflammatory PMN favoring brain metastasis [180].

Studies discussed above indicate that cancer cell-derived EVs can promote brain metastatic lesions by communicating with brain stromal cells. On the other hand, brain stromal cells-derived EVs can also functionally facilitate cancer cell survival and colonization in the brain. For example, human brain microvascular endothelial cell-derived exosomes can fuse with small cell lung cancer (SCLC) cells and upregulate their expression of S100A16. Elevated S100A16 expression in the SCLC cells functions in maintaining mitochondrial membrane potential by upregulating prohibitin 1 (PHB1), leading to enhanced survival of SCLC cells under stressful conditions in the brain [183]. Furthermore, astrocyte-derived exosomes induce loss of PTEN in multiple brain metastatic cancer cells, including breast cancer and melanoma, by transferring PTEN-targeting microRNAs, especially miR-19a [184]. This adaptive PTEN loss increases the secretion of chemokine CCL2 from brain metastatic cancer cells, leading to the recruitment of IBA1-positive myeloid cells,

which then reciprocally promote the outgrowth of brain metastatic lesions [184]. Taken together, these findings indicate that the reciprocal crosstalk between cancer cells and microenvironment via EVs in PMNs determines the tropism of circulating cancer cells that metastasize to specific organs, including the brain.

7 Concluding Remarks and Future Directions

7.1 Summary

In summary, we have discussed how tumor-derived EVPs can promote the systemic effects of metastatic cancers in various aspects. Here, we highlight two categories of systemic effects caused by tumor-derived EVPs. First, tumor-derived EVPs can influence the immune system via different mechanisms (affecting both progenitors and more mature immune lineages; innate and adaptive immune systems) to facilitate the escape of tumor cells from immune surveillance [7, 63]. What determines the selective uptake of tumor-derived EVPs by immune cells is an area of active investigation. Immune checkpoint inhibitors have resulted in promising treatments for patients with widespread metastatic disease. However, the malfunction of immune cells targeted by tumor-derived EVPs can jeopardize the full benefits of these therapies [185]. Adjuvant therapy to intervene with the adverse effects exerted by tumor-derived EVPs upon the immune system needs to be developed for efficient immunotherapy.

Secondly, tumor-derived EVPs circulate and fuse with cells at future sites of metastasis to promote the establishment of PMN [9]. Specific molecules carried by EVPs on their surface, such as integrins, can dictate their unique biodistribution and aid the formation of PMN for organotropic metastasis. These studies help to answer, in part, the 130-year-old mystery for Stephen Paget's "seed and soil" hypothesis.

7.2 Future Directions

7.2.1 Comprehensive Functional Characterization of the EVP Cargos

EVPs can transfer various cargos, including proteins, nucleic acids, metabolites, lipids, and probably a variety of other biomolecules, to recipient cells. However, most of the efforts have focused on determining the functions of EVP proteins and nucleic acids, especially the RNAs. Limited is known about the functions of DNAs, metabolites, or lipids. It has been reported that tumor-derived exosomes carry both single-stranded and double-stranded DNAs, as well as mitochondrial DNAs [21, 186–188]. Although some pilot studies have pointed out that exosomal DNAs can modulate tumor immunity [189–191], its functional effects and mechanisms in recipient cells are far from being fully elucidated. Successful metabolomics and lipidomics mass spectrometry have been performed to reveal the metabolites and lipids composition of tumor-derived exosomes [21, 192], whereas their physiological functions in the context of cancer need to be further investigated.

7.2.2 Tracking of Tumor-Derived EVPs In Vivo

Although it is clear that tumor-derived EVPs contribute to the PMN formation, how EVPs travel from the tumor to local and distant organs through circulation and how they are internalized by specific target cells remain elusive. In mouse models, the dissemination of tumor-derived EVPs to multiple tissues and organs through different approaches has been sparsely characterized [8–10, 21, 193–195]. For example, purified exosomes can be labeled with PKH fluorescent dye to label exosomal membrane and then injected into the circulation. Biodistribution of these labeled exosomes in different tissues and organs can be visualized by *ex vivo* imaging [8–10, 21]. EV transfer can also be visualized *in vivo* through high-resolution intravital imaging using Cre-loxP-mediated expression of fluorescent reporters [194, 195]. In this case, malignant tumor cells (donor) release EVs carrying Cre mRNA, which can induce fluorescence in recipient cells, thus marking and facilitating visualization of cells that have uptaken EVs [194, 195]. However, due to the technical limitation of detection sensitivity, these approaches are not able to accurately track EVPs in the blood circulation, and mouse models are not ideal for real-time monitoring of the behavior of exosomes *in vivo*.

Recently, a genetically engineered zebrafish embryo model has been shown to be a powerful tool to track tumor-derived EVs in the blood circulation *in vivo* [196]. The transparent zebrafish embryo has a well-characterized vasculature, which makes it suitable for *in vivo* tracking of tumor-derived EVs at high spatiotemporal resolution. After intravascular injection of melanoma EVs or melanoma cells labeled with MemBright, a membrane dye, or injection of melanoma cells expressing Syntenin2-GFP, which marks secreted EVs, the behavior of injected EVs or tumor-derived EVs in the blood circulation can be tracked using high-resolution intravital imaging [196]. This real-time *in vivo* imaging demonstrated that tumor-secreted EVs are mostly taken up by endothelial cells and patrolling macrophages, and the pre-treatment of zebrafish with melanoma EVs enhances metastasis, which is in consistent with studies in mouse models. The technical advances of this study shed light on the *in vivo* biology of EVPs.

7.2.3 Potential Role of Tumor-Derived EVPs in Other Systemic Complications of Cancer

Cancer is a systemic disease. Besides metastasis, other systemic complications include cachexia and paraneoplastic syndrome. Whether tumor-derived EVPs functionally contribute to these systemic complications, and if so, what are the underlying mechanisms, are attracting more research interest.

About 80% of cancer patients with metastatic malignancies experience skeletal muscle wasting with or without fat loss, a metabolic syndrome known as cachexia, which contributes to over

20% of cancer deaths [197, 198]. Cachectic cancer patients are too weak to tolerate standard dose of therapy and often die prematurely of respiratory dysfunction and cardiac failure. Although the mechanisms underlying metastasis-induced cachexia have been extensively investigated [199, 200], the functional role of tumor EVPs in cachexia progression, including muscle atrophy and fat loss, has just emerged.

Recent studies have shown that the EV concentration in serum is correlated with cachexia progression of cancer patients and tumor EVs are able to induce muscle cell catabolism and apoptosis, thereby inducing muscle atrophy [201–203]. In murine models of diverse types of cancers, including lung, colon, and pancreatic cancer, tumor cells secrete exosomal Hsp70 and Hsp90, which stimulate muscle catabolism and wasting by activating TLR4-p38β MAPK-C/EBPβ catabolic signaling [202]. In a murine model of lung cancer and in patients with pancreatic cancer, cancer cell-derived microvesicles containing miR-21 can promote cachexia by inducing muscle cell death. Particularly, microvesicle-carried miR-21 activates TLR-7 receptor on myoblasts and induces their apoptosis via c-Jun N-terminal kinase (JNK) signaling [203]. Furthermore, murine colon cancer exosome-packaged miR-195a-5p and miR-125b-1-3p can target and downregulate the expression of Bcl-2, an apoptosis inhibitor protein, in muscle cells, thereby activating muscle cell apoptosis [201].

In addition to muscle wasting, tumor-derived exosomes also contribute to cachexia development by inducing fat loss, including increased lipolysis, reduced adipogenesis, and white adipose tissue (WAT) browning. In this scenario, exosomal transfer of adrenomedullin (AM) from pancreatic cancer to adipose tissues stimulates lipolysis by interacting with adrenomedullin receptor (ADMR), which then induces the activation of hormone-sensitive lipase dependent on ERK1/2 and p38 MAPKs [204]. Evidence showed that lung cancer exosomes can be uptaken by human adipose tissue-derived mesenchymal stem cells (hAD-MSCs), and inhibit hAD-MSC adipogenesis by transferring TGFβ

signaling [205]. The phenotypic switch from WAT to brown fat, i.e. WAT browning, increases energy expenditure in cancer cachexia [206]. Recent studies using murine breast and colorectal cancer models suggested that cancer cells trigger WAT browning by transferring exosomal miR-155 and miR-146-5p, which targets and downregulates the expression of peroxisome proliferator-activated receptor-gamma (PPARγ) and homeobox C10 (HOXC10), respectively, in adipocytes [207, 208]. Clearly, understanding the role of tumor EVPs in the development of cancer-associated cachexia may provide new diagnostic biomarkers and potential therapeutic targets of cachexia in cancer patients.

Paraneoplastic syndromes comprise a set of symptoms, including endocrine, neurological, mucocutaneous, and hematological dysfunction, that are not induced directly by primary or metastatic tumors but are the consequence of cancer-associated soluble factors, such as cytokines, hormones, and EVPs, or result from immune responses against the tumor [209]. Therefore, paraneoplastic syndromes are frequently considered as indirect effects of underlying malignancies. A pilot study demonstrated that pancreatic cancer-derived exosomes containing AM and CA19–9 can fuse with β cells via caveolin-mediated endocytosis or micropinocytosis, and exosomal AM induces increased endoplasmic reticulum (ER) stress and unfolded protein response failure and suppresses insulin secretion, ultimately promoting β-cell death [210]. This study provided novel insights into how pancreatic cancer induces paraneoplastic β-cell dysfunction and diabetes via transferring exosomes. Furthermore, tumor-derived exosomes can also induce thrombosis by delivering procoagulant components, such as tissue factor, resulting in a poor prognosis of cancer patients [211, 212]. Increasing evidence shows that cancer patients frequently develop cognitive impairments, such as declined memory, learning, attention, and executive capability, which occurs before cancer diagnosis or after anti-cancer therapy [213, 214]. EVs could contribute to the cancer-associated cognitive impairments, since EVs released by the central nervous system

(CNS)-originated cancer cells are able to reprogram the brain environment, while non-CNS cancer cell-derived EVs can destruct the BBB to induce brain metastasis [18, 178, 182, 214]. Moreover, anti-cancer therapy, including irradiation and chemotherapy, can also enhance the EV secretion or modulate the composition of EV cargos from cancer cells, which have been found to be correlated with peripheral neuropathy or cognitive dysfunction [215–217]. Undoubtedly, more in-depth investigations into how EVPs can promote paraneoplastic syndromes will broaden our understanding of the systemic effects of cancer.

7.2.4 EVPs as Biomarkers for Diagnosis

Late diagnosis of cancer is one of the major reasons for poor prognosis of cancer patients. Detection of cancer at early stages increases the chance of successful treatment, which improves patient survival rate and quality of life. EVPs in the circulation may serve as critical biomarkers, especially as a non-invasive liquid biopsy tool for early cancer detection when circulating cancer cells are scarce and undetectable, since EVP cargos can partially reflect the genetic and molecular complexity of the cells from which they originated [218]. For example, exosomal miR-210 and miR-1233 expression levels were found to be able to differentiate clear-cell renal cell carcinoma patients from healthy individuals [219]. Moreover, a signature of five plasma EV proteins, including EGFR, EPCAM, MUC1, GPC1, and WNT2, has been identified as a promising biomarker for detecting pancreatic ductal adenocarcinoma (PDAC) early when local surgery or radiation can be performed and perhaps eradicate this disease more [220].

Tumor-derived exosomes have been shown to determine organotropic metastasis dependent on the specificity of exosomal integrins [9]. For example, exosomal integrin $\alpha_6\beta_4$ and $\alpha_6\beta_1$ are linked to lung metastasis, while $\alpha_v\beta_5$ is associated with liver metastasis in mouse cancer models [9]. The association of these integrins with metastatic organotropism has also been validated in

cancer patients. Integrin β_4 and α_v expression level was found increased in plasma-derived exosomes from patients with lung metastasis (regardless of the origin of primary tumors) and liver metastasis (originated from PDAC), respectively, compared with patients with no metastasis. Importantly, exosomal integrin α_v level at diagnosis was higher in PDAC patients who developed liver metastasis than in those who did not develop liver metastasis within three years of diagnosis, or in healthy donors [9]. Therefore, analysis of these and other cargos of tumor-derived exosomes could provide new diagnostic strategy to predict organ-specific metastasis.

Although EVPs, particularly exosomes, are promising biomarkers for early detection of cancer and prediction of organotropic metastasis, there are still many challenges ahead for clinical application. For instance, in the plasma, it remains unclear what percent of tumor-associated exosomes represent exosomes specifically derived from primary tumors and tumor microenvironments. In addition, it is plausible that the contribution of altered stromal cell-derived exosomes in the cancer setting could also be a contributory factor that makes up tumor-associated exosomes in circulation. Through the collective detection of tumor cell-derived or stromal cell-derived EVPs, systemic, cancer-specific signatures of circulating exosomes can be established for liquid biopsy. Furthermore, due to overlapping features of EVP subpopulations and technical limitations for exosome isolation, it remains challenging to distinguish exosomes from other subsets of EVPs. Indeed, all methods currently used for exosome isolation, including ultracentrifugation, size-based isolation (ultrafiltration, size exclusion chromatography, and AF4), immunoaffinity capture-based isolation, microfluidic technology have their own shortcomings [221] that could compromise the consistency of clinical practice. Therefore, the development of a robust and reproducible exosome-specific isolation method is expected for high-throughput clinical application.

7.2.5 EVPs as Therapeutic Targets and Deliverables

Metastasis is the leading cause of cancer-related deaths; however, there is still no effective strategy for clinical treatment of metastasis. The emerging roles of EVPs in cancer metastasis suggest that blockade of EVP uptake in local lymph nodes and distant organs could be a promising therapeutic strategy for the prevention and treatment of metastasis. A pilot study identified the anti-hypertensive drug reserpine as a suppressor of tumor-derived exosome uptake, leading to impaired PMN formation and melanoma lung metastasis [61]. However, caution should be considered before adjuvant treatment with reserpine, since chemotherapy could induce hypotension in cancer patients [222]. Moreover, the discovery of the role of exosomal CEMIP in preconditioning PMNs in the brain microenvironment for metastasis could lead to the development of strategies to block CEMIP for preventing and treating brain metastasis [10], which are currently tested in pre-clinical models of brain metastasis (unpublished data).

Cancer cells can evade immune surveillance by expressing cell surface programmed death-ligand 1 (PD-L1), which binds to programmed death-1 (PD-1) on T-lymphocytes to suppress their anti-tumor activity [223, 224]. Therefore, immunotherapies that block PD-1 and PD-L1 have shown remarkable promise in treating a variety of cancers, including melanoma, renal cancer, and non-small cell lung cancer [224]. However, only 10%–30% of patients with these cancers respond to anti-PD-L1/PD-1 therapies [225]. Evidence showed that tumor-secreted exosomal PD-L1 is resistant to anti-PD-L1 antibody blockade, thereby promoting cancer progression through suppression of T-lymphocytes activity in the draining lymph nodes [226]. This discovery pinpoints that the combination of anti-exosomal PD-L1 strategy with current therapeutics might benefit cancer patients that are not responding to anti-PD-L1/PD-1 therapies.

EVPs function as mediators of intercellular communication. The native structure and characteristics of EVPs enable their passage through physiological barriers, which are generally impermeable to synthetic drug-delivery vehicles. Therefore, EVPs are ideal vehicles for delivering nano-sized drugs to distant sites for clinical application [227]. For instance, the protective structure of the brain, the BBB, contributes to chemotherapeutic resistance of brain metastasis by restricting the penetration of large chemical molecules [228]. The capability of EVs to breach the intact BBB via transcytosis shed light on their potential application as vehicles for delivering drugs targeting brain metastasis [182]. A number of studies have been conducted to investigate the promising application of exosomes for delivering anti-tumor agents. For example, exosomes can deliver doxorubicin to tumor tissues in mice with breast cancer to inhibit tumor growth *in vivo* [229]. The exosomes used for packaging doxorubicin were purified from immature mouse DCs to reduce immunogenicity and toxicity and were engineered to express a tumor-targeting motif for enhanced specificity [229]. Furthermore, exosomes that were engineered to carry short interfering RNA or short hairpin RNA targeting $KRAS^{G12D}$ showed high efficiency for suppressing cancer progression in both orthotopic and genetically engineered mouse models of pancreatic cancer [230]. In a murine lung cancer model, anthocyanidins encapsulated in bovine milk-derived exosomes showed significantly enhanced anti-tumor response without signs of toxicity compared with free compound [231].

Nevertheless, challenges still remain for the clinical application of EVP-based drug delivery for anti-cancer therapy. First, the yield of EVPs might not be sufficient for long-term treatment in cancer patients. Second, the heterogeneity of EVPs might compromise the therapeutic efficiency, since the EVP cargos are still not fully characterized. Third, to avoid triggering unexpected immune responses, the appropriate donor cells should be chosen for producing EVPs. Thus, further investigation needs to be conducted to increase the yield and purity, examine the biodistribution, as well as characterize the molecular composition of EVPs to make it applicable for EVP utilization as drug-delivery vehicles.

Acknowledgments The authors gratefully acknowledge support from the following funding sources: the National Cancer Institute CA224175 (to D.L.), CA210240 (to D. L.), CA232093 (to D.L.), CA163117 (to D.L.), CA207983 (to D.L.), CA163120 (to D.L.), CA169416 (to D.L.), CA169538 (to D.L.), CA218513 (to D.L. and H.Z.) and AI144301 (to D.L.), the United States Department of Defense (W81XWH-13-1-0425, W81XWH-13-1-0427, W81XWH-13-1-0249 and W81XWH-14-1-0199 (to D. L.)), the Malcolm Hewitt Weiner Foundation (to D.L.), the Hartwell Foundation (to D.L.), the Manning Foundation (to D.L.), the Thompson Family Foundation (to D.L.), The Daniel P. and Nancy C. Paduano Family Foundation (to D.L.), the James Paduano Foundation (to D.L.), the Sohn Foundation (to D.L.), the STARR Consortium I9-A9-056 (to D.L. and H.Z.), the STARR Consortium I8-A8-123 (to D.L.), the AHEPA Vth District Cancer Research Foundation (to D.L.), the Pediatric Oncology Experimental Therapeutics Investigator's Consortium (to D.L.), the Alex's Lemonade Stand Foundation (to D. L.), the Breast Cancer Research Foundation (to D.L.), the Daedalus Fund (Weill Cornell Medicine, to D.L.), the Feldstein Medical Foundation (to D.L.), the Tortolani Foundation (to D.L.), the Children's Cancer and Blood Foundation (to D.L.), the Clinical & Translational Science Center (to D.L. and H.Z.), the Mary K. Ash Charitable Foundation (to D.L.), the Selma and Lawrence Ruben Science to Industry Bridge Fund (to D.L.). Figures were created using the templates from Servier Medical Art, licensed under a Creative Commons Attribution 3.0 Unported License; https://smart.servier.com.

References

1. David, A.R., Zimmerman, M.R.: Cancer: an old disease, a new disease or something in between? Nat. Rev. Cancer. **10**, 728–733 (2010)
2. Siegel, R.L., Miller, K.D., Jemal, A.: Cancer statistics, 2020. CA Cancer J. Clin. **70**, 7–30 (2020)
3. Lambert, A.W., Pattabiraman, D.R., Weinberg, R.A.: Emerging biological principles of metastasis. Cell. **168**, 670–691 (2017)
4. Obenauf, A.C., Massague, J.: Surviving at a distance: organ-specific metastasis. Trends Cancer. **1**, 76–91 (2015)
5. Massague, J., Obenauf, A.C.: Metastatic colonization by circulating tumour cells. Nature. **529**, 298–306 (2016)
6. Valastyan, S., Weinberg, R.A.: Tumor metastasis: molecular insights and evolving paradigms. Cell. **147**, 275–292 (2011)
7. Peinado, H., et al.: Melanoma exosomes educate bone marrow progenitor cells toward a pro-metastatic phenotype through MET. Nat. Med. **18**, 883–891 (2012)
8. Costa-Silva, B., et al.: Pancreatic cancer exosomes initiate pre-metastatic niche formation in the liver. Nat. Cell Biol. **17**, 816–826 (2015)
9. Hoshino, A., et al.: Tumour exosome integrins determine organotropic metastasis. Nature. **527**, 329–335 (2015)
10. Rodrigues, G., et al.: Tumour exosomal CEMIP protein promotes cancer cell colonization in brain metastasis. Nat. Cell Biol. **21**, 1403–1412 (2019)
11. Paget, S.: The distribution of secondary growths in cancer of the breast. Lancet. **133**, 571–573 (1889)
12. Peinado, H., et al.: Pre-metastatic niches: organ-specific homes for metastases. Nat. Rev. Cancer. **17**, 302–317 (2017)
13. Kaplan, R.N., et al.: VEGFR1-positive haematopoietic bone marrow progenitors initiate the pre-metastatic niche. Nature. **438**, 820–827 (2005)
14. Wortzel, I., Dror, S., Kenific, C.M., Lyden, D.: Exosome-mediated metastasis: communication from a distance. Dev. Cell. **49**, 347–360 (2019)
15. Elia, I., et al.: Breast cancer cells rely on environmental pyruvate to shape the metastatic niche. Nature. **568**, 117–121 (2019)
16. Liu, Y., et al.: Tumor exosomal RNAs promote lung pre-metastatic niche formation by activating Alveolar epithelial TLR3 to recruit neutrophils. Cancer Cell. **30**, 243–256 (2016)
17. Fong, M.Y., et al.: Breast-cancer-secreted miR-122 reprograms glucose metabolism in premetastatic niche to promote metastasis. Nat. Cell Biol. **17**, 183–194 (2015)
18. Zhou, W., et al.: Cancer-secreted miR-105 destroys vascular endothelial barriers to promote metastasis. Cancer Cell. **25**, 501–515 (2014)
19. Wang, S.E.: Extracellular vesicles and metastasis. Cold Spring Harb. Perspect. Med. (2019)
20. Mathieu, M., Martin-Jaular, L., Lavieu, G., Thery, C.: Specificities of secretion and uptake of exosomes and other extracellular vesicles for cell-to-cell communication. Nat. Cell Biol. **21**, 9–17 (2019)
21. Zhang, H., et al.: Identification of distinct nanoparticles and subsets of extracellular vesicles by asymmetric flow field-flow fractionation. Nat. Cell Biol. **20**, 332–343 (2018)
22. Colombo, M., Raposo, G., Thery, C.: Biogenesis, secretion, and intercellular interactions of exosomes and other extracellular vesicles. Annu. Rev. Cell Dev. Biol. **30**, 255–289 (2014)
23. Al-Nedawi, K., et al.: Intercellular transfer of the oncogenic receptor EGFRvIII by microvesicles derived from tumour cells. Nat. Cell Biol. **10**, 619–624 (2008)
24. Di Vizio, D., et al.: Oncosome formation in prostate cancer: association with a region of frequent

chromosomal deletion in metastatic disease. Cancer Res. **69**, 5601–5609 (2009)

25. Zhang, Q., et al.: Transfer of functional cargo in exomeres. Cell Rep. **27**, 940–954 e946 (2019)
26. Johnstone, R.M.: The Jeanne Manery-Fisher Memorial Lecture 1991. Maturation of reticulocytes: formation of exosomes as a mechanism for shedding membrane proteins. Biochem. Cell Biol. **70**, 179–190 (1992)
27. Harding, C., Heuser, J., Stahl, P.: Receptor-mediated endocytosis of transferrin and recycling of the transferrin receptor in rat reticulocytes. J. Cell Biol. **97**, 329–339 (1983)
28. Pan, B.T., Johnstone, R.M.: Fate of the transferrin receptor during maturation of sheep reticulocytes in vitro: selective externalization of the receptor. Cell. **33**, 967–978 (1983)
29. Haraszti, R.A., et al.: High-resolution proteomic and lipidomic analysis of exosomes and microvesicles from different cell sources. J Extracell Vesicles. **5**, 32570 (2016)
30. Zakharova, L., Svetlova, M., Fomina, A.F.: T cell exosomes induce cholesterol accumulation in human monocytes via phosphatidylserine receptor. J. Cell. Physiol. **212**, 174–181 (2007)
31. Clement, E., et al.: Adipocyte extracellular vesicles carry enzymes and fatty acids that stimulate mitochondrial metabolism and remodeling in tumor cells. EMBO J. **39**, e102525 (2020)
32. Steinbichler, T.B., Dudas, J., Riechelmann, H., Skvortsova, I.I.: The role of exosomes in cancer metastasis. Semin. Cancer Biol. **44**, 170–181 (2017)
33. Tomasetti, M., Lee, W., Santarelli, L., Neuzil, J.: Exosome-derived microRNAs in cancer metabolism: possible implications in cancer diagnostics and therapy. Exp. Mol. Med. **49**, e285 (2017)
34. Adem, B., Vieira, P.F., Melo, S.A.: Decoding the biology of exosomes in metastasis. Trends Cancer. **6**, 20–30 (2020)
35. Daassi, D., Mahoney, K.M., Freeman, G.J.: The importance of exosomal PDL1 in tumour immune evasion. Nat. Rev. Immunol. (2020)
36. Kessenbrock, K., Plaks, V., Werb, Z.: Matrix metalloproteinases: regulators of the tumor microenvironment. Cell. **141**, 52–67 (2010)
37. Ginestra, A., et al.: Urokinase plasminogen activator and gelatinases are associated with membrane vesicles shed by human HT1080 fibrosarcoma cells. J. Biol. Chem. **272**, 17216–17222 (1997)
38. Dolo, V., et al.: Selective localization of matrix metalloproteinase 9, beta1 integrins, and human lymphocyte antigen class I molecules on membrane vesicles shed by 8701-BC breast carcinoma cells. Cancer Res. **58**, 4468–4474 (1998)
39. Shan, Y., et al.: Hypoxia-induced matrix metalloproteinase-13 expression in exosomes from nasopharyngeal carcinoma enhances metastases. Cell Death Dis. **9**, 382 (2018)
40. Hakulinen, J., Sankkila, L., Sugiyama, N., Lehti, K., Keski-Oja, J.: Secretion of active membrane type 1 matrix metalloproteinase (MMP-14) into extracellular space in microvesicular exosomes. J. Cell. Biochem. **105**, 1211–1218 (2008)
41. Mu, W., Rana, S., Zoller, M.: Host matrix modulation by tumor exosomes promotes motility and invasiveness. Neoplasia. **15**, 875–887 (2013)
42. Redzic, J.S., et al.: Extracellular vesicles secreted from cancer cell lines stimulate secretion of MMP-9, IL-6, TGF-beta1 and EMMPRIN. PLoS One. **8**, e71225 (2013)
43. Sanchez, C.A., et al.: Exosomes from bulk and stem cells from human prostate cancer have a differential microRNA content that contributes cooperatively over local and pre-metastatic niche. Oncotarget. **7**, 3993–4008 (2016)
44. Atay, S., et al.: Oncogenic KIT-containing exosomes increase gastrointestinal stromal tumor cell invasion. Proc. Natl. Acad. Sci. USA. **111**, 711–716 (2014)
45. Hinz, B., Celetta, G., Tomasek, J.J., Gabbiani, G., Chaponnier, C.: Alpha-smooth muscle actin expression upregulates fibroblast contractile activity. Mol. Biol. Cell. **12**, 2730–2741 (2001)
46. Kalluri, R.: The biology and function of fibroblasts in cancer. Nat. Rev. Cancer. **16**, 582–598 (2016)
47. Webber, J., Steadman, R., Mason, M.D., Tabi, Z., Clayton, A.: Cancer exosomes trigger fibroblast to myofibroblast differentiation. Cancer Res. **70**, 9621–9630 (2010)
48. Webber, J.P., et al.: Differentiation of tumour-promoting stromal myofibroblasts by cancer exosomes. Oncogene. **34**, 290–302 (2015)
49. Ostrowski, M., et al.: Rab27a and Rab27b control different steps of the exosome secretion pathway. Nat. Cell Biol. **12**, 19–30.; sup pp 11-13 (2010)
50. Zhou, Y., et al.: Hepatocellular carcinoma-derived exosomal miRNA-21 contributes to tumor progression by converting hepatocyte stellate cells to cancer-associated fibroblasts. J. Exp. Clin. Cancer Res. **37**, 324 (2018)
51. Paggetti, J., et al.: Exosomes released by chronic lymphocytic leukemia cells induce the transition of stromal cells into cancer-associated fibroblasts. Blood. **126**, 1106–1117 (2015)
52. Chowdhury, R., et al.: Cancer exosomes trigger mesenchymal stem cell differentiation into pro-angiogenic and pro-invasive myofibroblasts. Oncotarget. **6**, 715–731 (2015)
53. Cho, J.A., et al.: Exosomes from ovarian cancer cells induce adipose tissue-derived mesenchymal stem cells to acquire the physical and functional characteristics of tumor-supporting myofibroblasts. Gynecol. Oncol. **123**, 379–386 (2011)
54. Vesely, M.D., Kershaw, M.H., Schreiber, R.D., Smyth, M.J.: Natural innate and adaptive immunity to cancer. Annu. Rev. Immunol. **29**, 235–271 (2011)
55. Wildes, T.J., Flores, C.T., Mitchell, D.A.: Concise review: modulating cancer immunity with

hematopoietic stem and progenitor cells. Stem Cells. **37**, 166–175 (2019)

56. Gonzalez, H., Hagerling, C., Werb, Z.: Roles of the immune system in cancer: from tumor initiation to metastatic progression. Genes Dev. **32**, 1267–1284 (2018)

57. Wellenstein, M.D., de Visser, K.E.: Cancer-cell-intrinsic mechanisms shaping the tumor immune landscape. Immunity. **48**, 399–416 (2018)

58. McGranahan, N., et al.: Allele-specific HLA loss and immune escape in lung cancer evolution. Cell. **171**, 1259–1271 e1211 (2017)

59. Tauriello, D.V.F., et al.: TGFbeta drives immune evasion in genetically reconstituted colon cancer metastasis. Nature. **554**, 538–543 (2018)

60. King, K.Y., Goodell, M.A.: Inflammatory modulation of HSCs: viewing the HSC as a foundation for the immune response. Nat. Rev. Immunol. **11**, 685–692 (2011)

61. Ortiz, A., et al.: An interferon-driven oxysterol-based defense against tumor-derived extracellular vesicles. Cancer Cell. **35**, 33–45 e36 (2019)

62. Vivier, E., et al.: Innate or adaptive immunity? The example of natural killer cells. Science. **331**, 44–49 (2011)

63. Hornick, N.I., et al.: AML suppresses hematopoiesis by releasing exosomes that contain microRNAs targeting c-MYB. Sci. Signal. **9**, ra88 (2016)

64. Yu, S., et al.: Tumor exosomes inhibit differentiation of bone marrow dendritic cells. J. Immunol. **178**, 6867–6875 (2007)

65. Xiang, X., et al.: Induction of myeloid-derived suppressor cells by tumor exosomes. Int. J. Cancer. **124**, 2621–2633 (2009)

66. Gabrilovich, D.I.: Myeloid-derived suppressor cells. Cancer Immunol. Res. **5**, 3–8 (2017)

67. Chalmin, F., et al.: Membrane-associated Hsp72 from tumor-derived exosomes mediates STAT3-dependent immunosuppressive function of mouse and human myeloid-derived suppressor cells. J. Clin. Invest. **120**, 457–471 (2010)

68. Fleming, V., et al.: Melanoma extracellular vesicles generate immunosuppressive myeloid cells by upregulating PD-L1 via TLR4 signaling. Cancer Res. **79**, 4715–4728 (2019)

69. Huber, V., et al.: Tumor-derived microRNAs induce myeloid suppressor cells and predict immunotherapy resistance in melanoma. J. Clin. Invest. **128**, 5505–5516 (2018)

70. Guo, X., et al.: Immunosuppressive effects of hypoxia-induced glioma exosomes through myeloid-derived suppressor cells via the miR-10a/Rora and miR-21/Pten Pathways. Oncogene. **37**, 4239–4259 (2018)

71. Akira, S., Uematsu, S., Takeuchi, O.: Pathogen recognition and innate immunity. Cell. **124**, 783–801 (2006)

72. Vitale, I., Manic, G., Coussens, L.M., Kroemer, G., Galluzzi, L.: Macrophages and metabolism in the tumor microenvironment. Cell Metab. **30**, 36–50 (2019)

73. Fabbri, M., et al.: MicroRNAs bind to toll-like receptors to induce prometastatic inflammatory response. Proc. Natl. Acad. Sci. USA. **109**, E2110–E2116 (2012)

74. Ying, X., et al.: Epithelial ovarian cancer-secreted exosomal miR-222-3p induces polarization of tumor-associated macrophages. Oncotarget. **7**, 43076–43087 (2016)

75. Cooks, T., et al.: Mutant p53 cancers reprogram macrophages to tumor supporting macrophages via exosomal miR-1246. Nat. Commun. **9**, 771 (2018)

76. Park, J.E., et al.: Hypoxia-induced tumor exosomes promote M2-like macrophage polarization of infiltrating myeloid cells and microRNA-mediated metabolic shift. Oncogene. **38**, 5158–5173 (2019)

77. Palucka, K., Banchereau, J.: Cancer immunotherapy via dendritic cells. Nat. Rev. Cancer. **12**, 265–277 (2012)

78. Yu, S., et al.: EGFR E746-A750 deletion in lung cancer represses antitumor immunity through the exosome-mediated inhibition of dendritic cells. Oncogene. (2020)

79. Ding, G., et al.: Pancreatic cancer-derived exosomes transfer miRNAs to dendritic cells and inhibit RFXAP expression via miR-212-3p. Oncotarget. **6**, 29877–29888 (2015)

80. Guillerey, C., Huntington, N.D., Smyth, M.J.: Targeting natural killer cells in cancer immunotherapy. Nat. Immunol. **17**, 1025–1036 (2016)

81. Baginska, J., et al.: The critical role of the tumor microenvironment in shaping natural killer cell-mediated anti-tumor immunity. Front. Immunol. **4**, 490 (2013)

82. Han, B., et al.: Altered NKp30, NKp46, NKG2D, and DNAM-1 expression on circulating NK cells is associated with tumor progression in human gastric cancer. J Immunol Res. **2018**, 6248590 (2018)

83. Viel, S., et al.: TGF-beta inhibits the activation and functions of NK cells by repressing the mTOR pathway. Sci. Signal. **9**, ra19 (2016)

84. Zhao, J.G., et al.: Tumor-derived extracellular vesicles inhibit natural killer cell function in pancreatic cancer. Cancers. **11** (2019)

85. Lundholm, M., et al.: Prostate tumor-derived exosomes down-regulate NKG2D expression on natural killer cells and CD8+ T cells: mechanism of immune evasion. PLoS One. **9**, e108925 (2014)

86. Liu, C., et al.: Murine mammary carcinoma exosomes promote tumor growth by suppression of NK cell function. J. Immunol. **176**, 1375–1385 (2006)

87. Coffelt, S.B., Wellenstein, M.D., de Visser, K.E.: Neutrophils in cancer: neutral no more. Nat. Rev. Cancer. **16**, 431–446 (2016)

88. Kowanetz, M., et al.: Granulocyte-colony stimulating factor promotes lung metastasis through mobilization

of Ly6G+Ly6C+ granulocytes. Proc. Natl. Acad. Sci. USA. **107**, 21248–21255 (2010)

89. Granot, Z., et al.: Tumor entrained neutrophils inhibit seeding in the premetastatic lung. Cancer Cell. **20**, 300–314 (2011)

90. Li, Y.W., et al.: Intratumoral neutrophils: a poor prognostic factor for hepatocellular carcinoma following resection. J. Hepatol. **54**, 497–505 (2011)

91. Zhao, J.J., et al.: The prognostic value of tumor-infiltrating neutrophils in gastric adenocarcinoma after resection. PLoS One. **7**, e33655 (2012)

92. Jensen, H.K., et al.: Presence of intratumoral neutrophils is an independent prognostic factor in localized renal cell carcinoma. J. Clin. Oncol. **27**, 4709–4717 (2009)

93. Schmidt, H., et al.: Elevated neutrophil and monocyte counts in peripheral blood are associated with poor survival in patients with metastatic melanoma: a prognostic model. Br. J. Cancer. **93**, 273–278 (2005)

94. Casbon, A.J., et al.: Invasive breast cancer reprograms early myeloid differentiation in the bone marrow to generate immunosuppressive neutrophils. Proc. Natl. Acad. Sci. USA. **112**, E566–E575 (2015)

95. Bayne, L.J., et al.: Tumor-derived granulocyte-macrophage colony-stimulating factor regulates myeloid inflammation and T cell immunity in pancreatic cancer. Cancer Cell. **21**, 822–835 (2012)

96. Coffelt, S.B., et al.: IL-17-producing gammadelta T cells and neutrophils conspire to promote breast cancer metastasis. Nature. **522**, 345–348 (2015)

97. Wang, D., Sun, H., Wei, J., Cen, B., DuBois, R.N.: CXCL1 is critical for premetastatic niche formation and metastasis in colorectal cancer. Cancer Res. **77**, 3655–3665 (2017)

98. Shaul, M.E., Fridlender, Z.G.: Tumour-associated neutrophils in patients with cancer. Nat. Rev. Clin. Oncol. **16**, 601–620 (2019)

99. Andzinski, L., et al.: Type I IFNs induce anti-tumor polarization of tumor associated neutrophils in mice and human. Int. J. Cancer. **138**, 1982–1993 (2016)

100. Fridlender, Z.G., et al.: Polarization of tumor-associated neutrophil phenotype by TGF-beta: "N1" versus "N2" TAN. Cancer Cell. **16**, 183–194 (2009)

101. Zhang, X., et al.: Tumor-derived exosomes induce N2 polarization of neutrophils to promote gastric cancer cell migration. Mol. Cancer. **17**, 146 (2018)

102. Hwang, W.L., Lan, H.Y., Cheng, W.C., Huang, S.C., Yang, M.H.: Tumor stem-like cell-derived exosomal RNAs prime neutrophils for facilitating tumorigenesis of colon cancer. J. Hematol. Oncol. **12**, 10 (2019)

103. Leal, A.C., et al.: Tumor-derived exosomes induce the formation of neutrophil extracellular traps: implications for the establishment of cancer-associated thrombosis. Sci. Rep. **7**, 6438 (2017)

104. Brinkmann, V., et al.: Neutrophil extracellular traps kill bacteria. Science. **303**, 1532–1535 (2004)

105. Demers, M., et al.: Cancers predispose neutrophils to release extracellular DNA traps that contribute to cancer-associated thrombosis. Proc. Natl. Acad. Sci. USA. **109**, 13076–13081 (2012)

106. Park, J., et al.: Cancer cells induce metastasis-supporting neutrophil extracellular DNA traps. Sci. Transl. Med. **8**, 361ra138 (2016)

107. Cools-Lartigue, J., et al.: Neutrophil extracellular traps sequester circulating tumor cells and promote metastasis. J. Clin. Invest. (2013)

108. Litman, G.W., Rast, J.P., Fugmann, S.D.: The origins of vertebrate adaptive immunity. Nat. Rev. Immunol. **10**, 543–553 (2010)

109. Razzo, B.M., et al.: Tumor-derived exosomes promote carcinogenesis of murine oral squamous cell carcinoma. Carcinogenesis. (2019)

110. Wang, X., et al.: 14-3-3zeta delivered by hepatocellular carcinoma-derived exosomes impaired antitumor function of tumor-infiltrating T lymphocytes. Cell Death Dis. **9**, 159 (2018)

111. Mirzaei, R., et al.: Brain tumor-initiating cells export tenascin-C associated with exosomes to suppress T cell activity. Onco. Targets. Ther. **7**, e1478647 (2018)

112. Griss, J., et al.: B cells sustain inflammation and predict response to immune checkpoint blockade in human melanoma. Nat. Commun. **10**, 4186 (2019)

113. Garaud, S., et al.: Tumor infiltrating B-cells signal functional humoral immune responses in breast cancer. JCI Insight. **5** (2019)

114. Meng, Q., Valentini, D., Rao, M., Maeurer, M.: KRAS renaissance(s) in tumor infiltrating B cells in pancreatic cancer. Front. Oncol. **8**, 384 (2018)

115. Pylayeva-Gupta, Y., et al.: IL35-producing B cells promote the development of pancreatic neoplasia. Cancer Discov. **6**, 247–255 (2016)

116. Xiao, X., et al.: PD-1hi identifies a novel regulatory B-cell population in human hepatoma that promotes disease progression. Cancer Discov. **6**, 546–559 (2016)

117. Ye, L., et al.: Tumor-derived exosomal HMGB1 fosters hepatocellular carcinoma immune evasion by promoting TIM-1(+) regulatory B cell expansion. J. Immunother. Cancer. **6**, 145 (2018)

118. Katoh, M.: Therapeutics targeting angiogenesis: genetics and epigenetics, extracellular miRNAs and signaling networks (Review). Int. J. Mol. Med. **32**, 763–767 (2013)

119. De Palma, M., Biziato, D., Petrova, T.V.: Microenvironmental regulation of tumour angiogenesis. Nat. Rev. Cancer. **17**, 457–474 (2017)

120. Sawamiphak, S., et al.: Ephrin-B2 regulates VEGFR2 function in developmental and tumour angiogenesis. Nature. **465**, 487–491 (2010)

121. Sato, S., et al.: EPHB2 carried on small extracellular vesicles induces tumor angiogenesis via activation of ephrin reverse signaling. JCI Insight. **4** (2019)

122. Jung, K.O., Youn, H., Lee, C.H., Kang, K.W., Chung, J.K.: Visualization of exosome-mediated miR-210 transfer from hypoxic tumor cells. Oncotarget. **8**, 9899–9910 (2017)

123. Lanahan, A.A., et al.: PTP1b is a physiologic regulator of vascular endothelial growth factor signaling in endothelial cells. Circulation. **130**, 902–909 (2014)

124. Umezu, T., et al.: Exosomal miR-135b shed from hypoxic multiple myeloma cells enhances angiogenesis by targeting factor-inhibiting HIF-1. Blood. **124**, 3748–3757 (2014)

125. Li, B.B., et al.: piRNA-823 delivered by multiple myeloma-derived extracellular vesicles promoted tumorigenesis through re-educating endothelial cells in the tumor environment. Oncogene. **38**, 5227–5238 (2019)

126. Treps, L., Perret, R., Edmond, S., Ricard, D., Gavard, J.: Glioblastoma stem-like cells secrete the pro-angiogenic VEGF-A factor in extracellular vesicles. J. Extracell. Vesicles. **6**, 1359479 (2017)

127. Ko, S.Y., et al.: Cancer-derived small extracellular vesicles promote angiogenesis by heparin-bound, bevacizumab-insensitive VEGF, independent of vesicle uptake. Commun. Biol. **2**, 386 (2019)

128. Strilic, B., Offermanns, S.: Intravascular survival and extravasation of tumor cells. Cancer Cell. **32**, 282–293 (2017)

129. Mohme, M., Riethdorf, S., Pantel, K.: Circulating and disseminated tumour cells – mechanisms of immune surveillance and escape. Nat. Rev. Clin. Oncol. **14**, 155–167 (2017)

130. Treps, L., et al.: Extracellular vesicle-transported Semaphorin3A promotes vascular permeability in glioblastoma. Oncogene. **35**, 2615–2623 (2016)

131. Zeng, Z., et al.: Cancer-derived exosomal miR-25-3p promotes pre-metastatic niche formation by inducing vascular permeability and angiogenesis. Nat. Commun. **9**, 5395 (2018)

132. Bhattacharya, R., et al.: Inhibition of vascular permeability factor/vascular endothelial growth factor-mediated angiogenesis by the Kruppel-like factor KLF2. J. Biol. Chem. **280**, 28848–28851 (2005)

133. Ma, J., et al.: Kruppel-like factor 4 regulates blood-tumor barrier permeability via ZO-1, occludin and claudin-5. J. Cell. Physiol. **229**, 916–926 (2014)

134. Hsu, Y.L., et al.: Hypoxic lung cancer-secreted exosomal miR-23a increased angiogenesis and vascular permeability by targeting prolyl hydroxylase and tight junction protein ZO-1. Oncogene. **36**, 4929–4942 (2017)

135. Di Modica, M., et al.: Breast cancer-secreted miR-939 downregulates VE-cadherin and destroys the barrier function of endothelial monolayers. Cancer Lett. **384**, 94–100 (2017)

136. Fang, J.H., et al.: Hepatoma cell-secreted exosomal microRNA-103 increases vascular permeability and promotes metastasis by targeting junction proteins. Hepatology. **68**, 1459–1475 (2018)

137. Li, J., et al.: Circular RNA IARS (circ-IARS) secreted by pancreatic cancer cells and located within exosomes regulates endothelial monolayer permeability to promote tumor metastasis. J. Exp. Clin. Cancer Res. **37** (2018)

138. Hansen, T.B., et al.: Natural RNA circles function as efficient microRNA sponges. Nature. **495**, 384–388 (2013)

139. Fidler, I.J., Nicolson, G.L.: Organ selectivity for implantation survival and growth of B16 melanoma variant tumor lines. J. Natl. Cancer Inst. **57**, 1199–1202 (1976)

140. Hart, I.R., Fidler, I.J.: Role of organ selectivity in the determination of metastatic patterns of B16 melanoma. Cancer Res. **40**, 2281–2287 (1980)

141. Gao, Y., et al.: Metastasis organotropism: redefining the congenial soil. Dev. Cell. **49**, 375–391 (2019)

142. Nguyen, D.X., Bos, P.D., Massague, J.: Metastasis: from dissemination to organ-specific colonization. Nat. Rev. Cancer. **9**, 274–U265 (2009)

143. Hood, J.L., San, R.S., Wickline, S.A.: Exosomes released by melanoma cells prepare sentinel lymph nodes for tumor metastasis. Cancer Res. **71**, 3792–3801 (2011)

144. Xiao, Z., et al.: Molecular mechanism underlying lymphatic metastasis in pancreatic cancer. Biomed. Res. Int. **2014**, 925845 (2014)

145. Pucci, F., et al.: SCS macrophages suppress melanoma by restricting tumor-derived vesicle-B cell interactions. Science. **352**, 242–246 (2016)

146. Maus, R.L.G., et al.: Human melanoma-derived extracellular vesicles regulate dendritic cell maturation. Front. Immunol. **8**, 358 (2017)

147. Rana, S., Malinowska, K., Zoller, M.: Exosomal tumor microRNA modulates premetastatic organ cells. Neoplasia. **15**, 281–295 (2013)

148. Srinivasan, S., Vannberg, F.O., Dixon, J.B.: Lymphatic transport of exosomes as a rapid route of information dissemination to the lymph node. Sci. Rep. **6**, 24436 (2016)

149. Chen, C., et al.: Exosomal long noncoding RNA LNMAT2 promotes lymphatic metastasis in bladder cancer. J. Clin. Invest. **130**, 404–421 (2020)

150. Sun, B., et al.: Colorectal cancer exosomes induce lymphatic network remodeling in lymph nodes. Int. J. Cancer. **145**, 1648–1659 (2019)

151. Randolph, G.J., Ivanov, S., Zinselmeyer, B.H., Scallan, J.P.: The lymphatic system: integral roles in immunity. Annu. Rev. Immunol. **35**, 31–52 (2017)

152. Chow, A., et al.: Macrophage immunomodulation by breast cancer-derived exosomes requires toll-like receptor 2-mediated activation of NF-kappaB. Sci. Rep. **4**, 5750 (2014)

153. Muhsin-Sharafaldine, M.R., Saunderson, S.C., Dunn, A.C., McLellan, A.D.: Melanoma growth and lymph node metastasis is independent of host CD169 expression. Biochem. Biophys. Res. Commun. **486**, 965–970 (2017)

154. Skobe, M., et al.: Induction of tumor lymphangiogenesis by VEGF-C promotes breast cancer metastasis. Nat. Med. **7**, 192–198 (2001)

155. Stacker, S.A., et al.: VEGF-D promotes the metastatic spread of tumor cells via the lymphatics. Nat. Med. **7**, 186–191 (2001)

156. Stacker, S.A., et al.: Lymphangiogenesis and lymphatic vessel remodelling in cancer. Nat. Rev. Cancer. **14**, 159–172 (2014)

157. Park, R.J., Hong, Y.J., Wu, Y., Kim, P.M., Hong, Y. K.: Exosomes as a communication tool between the lymphatic system and bladder cancer. Int. Neurourol. J. **22**, 220–224 (2018)

158. Zhou, C.F., et al.: Cervical squamous cell carcinoma-secreted exosomal miR-221-3p promotes lymphangiogenesis and lymphatic metastasis by targeting VASH1. Oncogene. **38**, 1256–1268 (2019)

159. Hautmann, R.E., de Petriconi, R.C., Pfeiffer, C., Volkmer, B.G.: Radical cystectomy for urothelial carcinoma of the bladder without neoadjuvant or adjuvant therapy: long-term results in 1100 patients. Eur. Urol. **61**, 1039–1047 (2012)

160. Liu, D., et al.: CD97 promotion of gastric carcinoma lymphatic metastasis is exosome dependent. Gastric Cancer. **19**, 754–766 (2016)

161. Morton, D.L., et al.: Sentinel-node biopsy or nodal observation in melanoma. N. Engl. J. Med. **355**, 1307–1317 (2006)

162. Cheng, L., et al.: Cancer volume of lymph node metastasis predicts progression in prostate cancer. Am. J. Surg. Pathol. **22**, 1491–1500 (1998)

163. Im, E.J., et al.: Sulfisoxazole inhibits the secretion of small extracellular vesicles by targeting the endothelin receptor A. Nat. Commun. **10**, 1387 (2019)

164. Altorki, N.K., et al.: The lung microenvironment: an important regulator of tumour growth and metastasis. Nat. Rev. Cancer. **19**, 9–31 (2019)

165. Erler, J.T., et al.: Hypoxia-induced lysyl oxidase is a critical mediator of bone marrow cell recruitment to form the premetastatic niche. Cancer Cell. **15**, 35–44 (2009)

166. Fang, T., et al.: Tumor-derived exosomal miR-1247-3p induces cancer-associated fibroblast activation to foster lung metastasis of liver cancer. Nat. Commun. **9**, 191 (2018)

167. Trefts, E., Gannon, M., Wasserman, D.H.: The liver. Curr. Biol. **27**, R1147–R1151 (2017)

168. Zhang, H., et al.: Exosome-delivered EGFR regulates liver microenvironment to promote gastric cancer liver metastasis. Nat. Commun. **8**, 15016 (2017)

169. Shao, Y., et al.: Colorectal cancer-derived small extracellular vesicles establish an inflammatory premetastatic niche in liver metastasis. Carcinogenesis. **39**, 1368–1379 (2018)

170. Mundy, G.R.: Metastasis to bone: causes, consequences and therapeutic opportunities. Nat. Rev. Cancer. **2**, 584–593 (2002)

171. Cao, X.: RANKL-RANK signaling regulates osteoblast differentiation and bone formation. Bone Res. **6**, 35 (2018)

172. Tiedemann, K., et al.: Exosomal release of L-plastin by breast cancer cells facilitates metastatic bone osteolysis. Transl. Oncol. **12**, 462–474 (2019)

173. Guo, L., et al.: Breast cancer cell-derived exosomal miR-20a-5p promotes the proliferation and differentiation of osteoclasts by targeting SRCIN1. Cancer Med. **8**, 5687–5701 (2019)

174. Probert, C., et al.: Communication of prostate cancer cells with bone cells via extracellular vesicle RNA; a potential mechanism of metastasis. Oncogene. **38**, 1751–1763 (2019)

175. Ye, Y., et al.: Exosomal miR-141-3p regulates osteoblast activity to promote the osteoblastic metastasis of prostate cancer. Oncotarget. **8**, 94834–94849 (2017)

176. Lacey, D.L., et al.: Bench to bedside: elucidation of the OPG-RANK-RANKL pathway and the development of denosumab. Nat. Rev. Drug Discov. **11**, 401–419 (2012)

177. Hashimoto, K., et al.: Cancer-secreted hsa-miR-940 induces an osteoblastic phenotype in the bone metastatic microenvironment via targeting ARHGAP1 and FAM134A. Proc. Natl. Acad. Sci. USA. **115**, 2204–2209 (2018)

178. Tominaga, N., et al.: Brain metastatic cancer cells release microRNA-181c-containing extracellular vesicles capable of destructing blood-brain barrier. Nat. Commun. **6**, 6716 (2015)

179. Xing, F., et al.: Loss of XIST in breast cancer activates MSN-c-Met and reprograms microglia via exosomal miRNA to promote brain metastasis. Cancer Res. **78**, 4316–4330 (2018)

180. Gener Lahav, T., et al.: Melanoma-derived extracellular vesicles instigate proinflammatory signaling in the metastatic microenvironment. Int. J. Cancer. **145**, 2521–2534 (2019)

181. Sweeney, M.D., Zhao, Z., Montagne, A., Nelson, A. R., Zlokovic, B.V.: Blood-brain barrier: from physiology to disease and back. Physiol. Rev. **99**, 21–78 (2019)

182. Morad, G., et al.: Tumor-derived extracellular vesicles breach the intact blood-brain barrier via transcytosis. ACS Nano. **13**, 13853–13865 (2019)

183. Xu, Z.H., et al.: Brain microvascular endothelial cell exosome-mediated S100A16 up-regulation confers small-cell lung cancer cell survival in brain. FASEB J. **33**, 1742–1757 (2019)

184. Zhang, L., et al.: Microenvironment-induced PTEN loss by exosomal microRNA primes brain metastasis outgrowth. Nature. **527**, 100–104 (2015)

185. Chen, G., et al.: Exosomal PD-L1 contributes to immunosuppression and is associated with anti-PD-1 response. Nature. **560**, 382–386 (2018)

186. Balaj, L., et al.: Tumour microvesicles contain retrotransposon elements and amplified oncogene sequences. Nat. Commun. **2**, 180 (2011)

187. Guescini, M., Genedani, S., Stocchi, V., Agnati, L.F.: Astrocytes and Glioblastoma cells release exosomes carrying mtDNA. J. Neural Transm. (Vienna). **117**, 1–4 (2010)

188. Thakur, B.K., et al.: Double-stranded DNA in exosomes: a novel biomarker in cancer detection. Cell Res. **24**, 766–769 (2014)

189. Sharma, A., Johnson, A.: Exosome DNA: critical regulator of tumor immunity and a diagnostic biomarker. J. Cell. Physiol. **235**, 1921–1932 (2020)

190. Kitai, Y., et al.: DNA-containing exosomes derived from cancer cells treated with topotecan activate a STING-dependent pathway and reinforce antitumor immunity. J. Immunol. **198**, 1649–1659 (2017)

191. Xu, M.M., et al.: Dendritic cells but not macrophages sense tumor mitochondrial DNA for cross-priming through signal regulatory protein alpha signaling. Immunity. **47**, 363-+ (2017)

192. Puhka, M., et al.: Metabolomic profiling of extracellular vesicles and alternative normalization methods reveal enriched metabolites and strategies to study prostate cancer-related changes. Theranostics. **7**, 3824–3841 (2017)

193. Lai, C.P., et al.: Visualization and tracking of tumour extracellular vesicle delivery and RNA translation using multiplexed reporters. Nat. Commun. **6**, 7029 (2015)

194. Zomer, A., et al.: In vivo imaging reveals extracellular vesicle-mediated phenocopying of metastatic behavior. Cell. **161**, 1046–1057 (2015)

195. Zomer, A., Steenbeek, S.C., Maynard, C., van Rheenen, J.: Studying extracellular vesicle transfer by a Cre-loxP method. Nat. Protoc. **11**, 87–101 (2016)

196. Hyenne, V., et al.: Studying the fate of tumor extracellular vesicles at high spatiotemporal resolution using the zebrafish embryo. Dev. Cell. **48**, 554–572 e557 (2019)

197. Fearon, K.C., Glass, D.J., Guttridge, D.C.: Cancer cachexia: mediators, signaling, and metabolic pathways. Cell Metab. **16**, 153–166 (2012)

198. Porporato, P.E.: Understanding cachexia as a cancer metabolism syndrome. Oncogenesis. **5** (2016)

199. Wang, G., et al.: Metastatic cancers promote cachexia through ZIP14 upregulation in skeletal muscle. Nat. Med. **24**, 770–781 (2018)

200. Baracos, V.E., Martin, L., Korc, M., Guttridge, D.C., Fearon, K.C.H.: Cancer-associated cachexia. Nat. Rev. Dis. Primers. **4**, 17105 (2018)

201. Miao, C., et al.: Cancer-derived exosome miRNAs induce skeletal muscle wasting by Bcl-2-mediated apoptosis in colon cancer cachexia. Mol. Ther. Nucleic Acids. **24**, 923–938 (2021)

202. Zhang, G., et al.: Tumor induces muscle wasting in mice through releasing extracellular Hsp70 and Hsp90. Nat. Commun. **8**, 589 (2017)

203. He, W.A., et al.: Microvesicles containing miRNAs promote muscle cell death in cancer cachexia via TLR7. Proc. Natl. Acad. Sci. USA. **111**, 4525–4529 (2014)

204. Sagar, G., et al.: Pathogenesis of pancreatic cancer exosome-induced lipolysis in adipose tissue. Gut. **65**, 1165–1174 (2016)

205. Wang, S., Li, X., Xu, M., Wang, J., Zhao, R.C.: Reduced adipogenesis after lung tumor exosomes priming in human mesenchymal stem cells via TGFbeta signaling pathway. Mol. Cell. Biochem. **435**, 59–66 (2017)

206. Petruzzelli, M., et al.: A switch from white to brown fat increases energy expenditure in cancer-associated cachexia. Cell Metab. **20**, 433–447 (2014)

207. Wu, Q., et al.: Tumour-originated exosomal miR-155 triggers cancer-associated cachexia to promote tumour progression. Mol. Cancer. **17**, 155 (2018)

208. Di, W.J., et al.: Colorectal cancer prompted adipose tissue browning and cancer cachexia through transferring exosomal miR-146b-5p. J. Cell. Physiol. **236**, 5399–5410 (2021)

209. Manger, B., Schett, G.: Paraneoplastic syndromes in rheumatology. Nat. Rev. Rheumatol. **10**, 662–670 (2014)

210. Javeed, N., et al.: Pancreatic cancer-derived exosomes cause paraneoplastic beta-cell dysfunction. Clin. Cancer Res. **21**, 1722–1733 (2015)

211. Lima, L.G., Leal, A.C., Vargas, G., Porto-Carreiro, I., Monteiro, R.Q.: Intercellular transfer of tissue factor via the uptake of tumor-derived microvesicles. Thromb. Res. **132**, 450–456 (2013)

212. Gomes, F.G., et al.: Breast-cancer extracellular vesicles induce platelet activation and aggregation by tissue factor-independent and -dependent mechanisms. Thromb. Res. **159**, 24–32 (2017)

213. Wefel, J.S., Vardy, J., Ahles, T., Schagen, S.B.: International Cognition and Cancer Task Force recommendations to harmonise studies of cognitive function in patients with cancer. Lancet Oncol. **12**, 703–708 (2011)

214. Koh, Y.Q., et al.: Role of exosomes in cancer-related cognitive impairment. Int. J. Mol. Sci. **21** (2020)

215. Jabbari, N., Nawaz, M., Rezaie, J.: Ionizing radiation increases the activity of exosomal secretory pathway in MCF-7 human breast cancer cells: a possible way to communicate resistance against radiotherapy. Int. J. Mol. Sci. **20** (2019)

216. Bandari, S.K., et al.: Chemotherapy induces secretion of exosomes loaded with heparanase that degrades extracellular matrix and impacts tumor and host cell behavior. Matrix Biol. **65**, 104–118 (2018)

217. Chen, E.I., et al.: Identifying predictors of taxane-induced peripheral neuropathy using mass spectrometry-based proteomics technology. PLoS One. **10** (2015)

218. LeBleu, V.S., Kalluri, R.: Exosomes as a multicomponent biomarker platform in cancer. Trends Cancer. (2020)

219. Zhang, W., et al.: MicroRNAs in serum exosomes as potential biomarkers in clear-cell renal cell carcinoma. Eur. Urol. Focus. **4**, 412–419 (2018)

220. Yang, K.S., et al.: Multiparametric plasma EV profiling facilitates diagnosis of pancreatic malignancy. Sci. Transl. Med. **9** (2017)

221. Li, P., Kaslan, M., Lee, S.H., Yao, J., Gao, Z.: Progress in exosome isolation techniques. Theranostics. **7**, 789–804 (2017)

222. Senkus, E., Jassem, J.: Cardiovascular effects of systemic cancer treatment. Cancer Treat. Rev. **37**, 300–311 (2011)

223. Sanmamed, M.F., Chen, L.: A paradigm shift in cancer immunotherapy: from enhancement to normalization. Cell. **175**, 313–326 (2018)

224. Chen, L., Han, X.: Anti-PD-1/PD-L1 therapy of human cancer: past, present, and future. J. Clin. Invest. **125**, 3384–3391 (2015)

225. Page, D.B., Postow, M.A., Callahan, M.K., Allison, J.P., Wolchok, J.D.: Immune modulation in cancer with antibodies. Annu. Rev. Med. **65**, 185–202 (2014)

226. Poggio, M., et al.: Suppression of exosomal PD-L1 induces systemic anti-tumor immunity and memory. Cell. **177**, 414–427 e413 (2019)

227. Luan, X., et al.: Engineering exosomes as refined biological nanoplatforms for drug delivery. Acta Pharmacol. Sin. **38**, 754–763 (2017)

228. Lin, X., DeAngelis, L.M.: Treatment of brain metastases. J. Clin. Oncol. **33**, 3475–3484 (2015)

229. Tian, Y., et al.: A doxorubicin delivery platform using engineered natural membrane vesicle exosomes for targeted tumor therapy. Biomaterials. **35**, 2383–2390 (2014)

230. Kamerkar, S., et al.: Exosomes facilitate therapeutic targeting of oncogenic KRAS in pancreatic cancer. Nature. **546**, 498–503 (2017)

231. Munagala, R., et al.: Exosomal formulation of anthocyanidins against multiple cancer types. Cancer Lett. **393**, 94–102 (2017)

Bone Metastases: Systemic Regulation and Impact on Host

Sukanya Suresh and Theresa A. Guise

Abstract

Advanced cancers frequently metastasize to bone and disrupt normal bone remodeling to cause bone destruction. Bone metastases are incurable and result in significant morbidity: pain, fracture, hypercalcemia, spinal cord compression, and muscle weakness. The associated muscle weakness may occur prior to or in the absence of cachexia. Tumor-induced bone destruction causes the release of growth factors that can fuel tumor progression in bone. These factors can also act systemically to cause muscle weakness. The resulting muscle dysfunction adds to cancer-associated morbidity and can increase the risk of falls and fractures in patients with bone metastases. In this review, we discuss the mechanisms by which bone metastases can induce muscle weakness and potential ways to identify and treat cancer-induced systemic dysfunction.

Learning Objectives

- To identify bone as a preferred site for certain cancer metastases.
- To recognize that a feed-forward cycle exists in the bone microenvironment in which tumors hijack the bone cells to disrupt normal bone remodeling.
- To understand how tumor cell–bone interactions result in systemic muscle weakness.
- To recognize that muscle dysfunction can occur prior to or in the absence of loss of muscle mass.
- To consider that cancer therapy can result in muscle dysfunction.

1 Introduction

In the United States, cancer is the second leading cause of death with approximately 4950 new cases diagnosed each day in 2020. In men, prostate and lung cancers account for 43% of cancers, and in women, breast cancer alone makes up 30% of new diagnoses. In 2020, the estimated deaths from cancer is projected to be 606,250. The cancer death rates peaked in 1991 and now have an overall decline of 29%. This decline is driven by the reduced mortality rates for lung, breast, prostate, and colorectal cancers, largely due to

S. Suresh · T. A. Guise (✉)
Department of Endocrine Neoplasia and Hormonal Disorders, The University of Texas MD Anderson Cancer Center, Houston, TX, USA
e-mail: TAGuise@mdanderson.org

S. Acharyya (ed.), *The Systemic Effects of Advanced Cancer*,
https://doi.org/10.1007/978-3-031-09518-4_3

lifestyle changes, such as reduced smoking rates, as well as due to the significant breakthroughs in treatments and advanced diagnostic tools that detect cancer at an early stage [1].

Cancer is a complex and heterogenous disease with systemic effects in the host including bone metastases resulting in fractures, body weight loss, muscle weakness, muscle wasting, loss of fat reserves, and dysfunction of metabolism, immune system and cognition. These secondary complications arise due to the deregulation of signaling pathways mediated by cytokines such as interleukins, tumor necrosis factor (TNF), interferons and growth factors such as transforming growth factor-beta (TGF-β), platelet-derived growth factor (PDGF), insulin-like growth factors (IGF) that can be either tumor-derived or induced by the host as a response to the tumor. Thus, even cancers that are not metastatic can have systemic effects due to the paracrine nature of cytokine and growth factor signaling [2]. In this chapter, we discuss the systemic effects of cancer due to excess bone destruction, observed in cancer patients, studies in animal models that elucidate the molecular mechanisms involved in these effects and the state of clinical care in managing cancer-associated systemic dysfunction.

2   Bone Metastases

In patients with advanced cancers, bone is the most common site for metastasis [3]. The incidence of bone metastases varies with the type of cancer. For example, bone metastases occur in 70% of patients with advanced breast cancer and 85% of patients with prostate cancer [3]. Incidence of bone metastases increases with the advancement in disease stages and is also reported in advanced lung, renal, colorectal and gastrointestinal cancers and melanoma [4]. In an autopsy-based study, 30–40% of patients with thyroid, kidney, and bronchus carcinomas had bone metastases [5]. In hematological cancers, multiple myeloma has a high rate of bone metastases with over 85% of patients reporting bone disease, making it the most common cancer

that metastasizes to bone [6]. Bone metastases are incurable; resulting in significant pain, hypercalcemia, fractures, spinal cord compression, and bone marrow failure and cause considerable morbidity [7].

How the tumor cells survive in the bone marrow may be best explained by Sir James Paget's seed and soil theory in which bone is described as the fertile soil for the growth and colonization of cancer cells, the seed [8]. Bone metastases are predominantly found in trabecular bone with red marrow including the vertebrae, ribs, pelvis, and in the extremities of long bones. The higher prevalence of metastases in these sites suggest that the higher rate of bone remodeling seen in trabecular bone and the rich vasculature of the red marrow provide a suitable seeding ground for the tumor. The major component of the bone microenvironment is the mineralized bone matrix, which is the largest storehouse of growth factors, such as TGF-β, deposited by bone-forming osteoblasts and released by bone-destroying osteoclasts. These growth factors can be released into the bone microenvironment by excessive tumor stimulation of osteoclast and osteoblast activity to fuel tumor growth [9]. Further, the bone microenvironment consists of hematopoietic cells, stromal cells, bone marrow adipocytes, and endothelial cells which contribute to tumor progression in bone [10]. Collectively, these facts support the fertile soil theory. Bone is a dynamic organ with constant remodeling, the result of a coordinated activity of osteoclast-induced bone destruction and osteoblast-mediated new bone formation. Bone strength and function is maintained by the coupling process of osteoclast-osteoblast activity [11]. During bone metastases, the normal bone remodeling process is severely disrupted, leading to bone metastases that can be classified as osteolytic (bone destructive) or osteoblastic (bone forming) or both.

2.1   Osteolytic Bone Metastases

Breast cancer is the most common cancer with osteolytic metastases characterized by increased osteoclast activity. Excessive osteoclast activity occurs due to osteoclast interaction with tumor-derived factors, such as parathyroid hormone-related protein (PTHrP), interleukin (IL)-11, IL-6, and IL-8. Tumor-derived PTHrP upregulates receptor activator of nuclear factor KB ligand (RANKL) production by osteoblasts. The RANKL upon binding to the RANK receptor on osteoclasts increases osteoclast activity and bone resorption [12]. Jagged1, another tumor-derived osteolytic factor, is known to promote osteoclast differentiation by directly binding to the monocyte precursors [13]. The excessive bone resorption observed during bone metastases releases growth factors, such as TGF-β from the bone matrix, promoting tumor cell growth and invasion [14]. This cascade results in a vicious cycle of increasing tumor burden in the bone and bone destruction [9, 15]. Considering the central role of TGF-β in promoting bone metastases, several TGF-β inhibitors have been developed to block the TGF-β signaling pathway [16]. The mainstay of treatment for bone metastases, outside of treating cancer with radiation and chemotherapy is to use bone modifying agents, such as those which inhibit osteoclastic bone resorption: bisphosphonates and inhibitors of RANKL [3]. Amino bisphosphonates, such as pamidronate and more recently, zoledronate, induce osteoclast apoptosis [3] while denosumab, a monoclonal antibody targeting RANKL prevents osteoclast formation and activation [17]. All are highly effective to treat osteolytic bone metastases due to solid tumors and myeloma [3, 18, 19].

2.2 Osteoblastic Bone Metastases

Predominantly seen in patients with prostate cancer, osteoblastic metastases are characterized by abnormal new bone formation resulting in irregular trabecular bone which can be painful and susceptible to fracture [20]. Despite the osteoblastic nature, as classified by radiographic appearance, these metastases also have dysregulated osteoclast function. In prostate cancer patients with bone metastases, high concentrations of bone resorption marker N-telopeptide predicts poor clinical outcome [21]. Bisphosphonates

and denosumab inhibitors of bone destruction are effective to reduce skeletal morbidity due to osteoblastic bone metastases, but Radium 223 dichloride (radium-223), is the only bone-modifying agent to improve survival. This alpha-emitting radionuclide was approved by the FDA for the treatment of bone metastases in castration-resistant prostate cancer [3]. Radium 223 is a calcium mimic that complexes with hydroxyapatite at areas of increased bone remodeling, such as sites of bone metastases, and induces double-strand DNA breaks [22]. It is also under investigation for other solid tumor bone metastases and myeloma, which are osteolytic in nature.

Osteoblastic metastases arise due to tumor-derived growth factors that aid osteoblastogenesis. For example, prostate cancer cells secrete bone morphogenetic proteins (BMPs) that are detected in bone lesions in patients with prostate cancer [23] as well as endothelin-1 [24, 25]. Excess TGF-β is observed in the serum of prostate cancer patients with bone metastases compared to those without bone metastases [26]. In recent studies in a metastatic-castration resistant prostate cancer model, elevated TGF-β in the tumor-bearing bone marrow promoted T_h17 lineage of CD4 T cells and reduced the differentiation of T_h1 effector cells. In this model, blocking of TGF-β in addition to the checkpoint therapy using cytotoxic T-lymphocyte-associated protein 4 (CTLA-4) inhibitor restricted bone metastases compared with CTLA-4 therapy alone [27]. Other tumor-derived growth factors implicated in osteoblastic bone metastases include IGF, PDGF, vascular endothelial growth factor, endothelin-1, and Wnt [28]. Similar to the osteolytic lesions in breast cancer, the osteoblastic lesions in prostate cancer express high levels of PTHrP [29]. However, while PTHrP activates osteoclasts in breast cancer, it is postulated that prostate-specific antigen cleaved PTHrP fragments stimulate osteoblastic bone formation by activating endothelin-A receptors in prostate cancer [30].

Bone thus represents a fertile environment for the cancer cells to colonize. Studies from pre-clinical models have shown that the postmenopausal state favors bone metastases. In one study, 83% of ovariectomized mice had breast cancer bone metastases compared to 17% of mice in the sham controls. The increased bone metastases with ovariectomy were attributed to enhanced osteoclast activity, which was blocked with bisphosphonate treatment [31]. Complete estrogen deprivation using aromatase inhibitors in ovariectomized mice also resulted in increased bone resorption and bone loss which was also blocked by bisphosphonate treatment [32]. Similarly, androgen deprivation in prostate cancer models promoted bone metastases [33]. These studies show that the hormone status of the host plays an important role in the propensity of cancer cells to metastasize bone. Understanding the mechanistic pathways of bone metastases has enabled the development of therapies curbing tumor progression in bone; however, these therapies do not cure bone metastases. Studies in bone metastases are needed in models that reflect the age and associated factors, such as hormonal deficiency and increased adiposity commonly seen in patients. Animal models should also accurately reflect the sequential development of bone metastases to understand the early to late molecular signals that could be targeted for effective bone metastases management.

3   Cancer and Muscle Dysfunction

Nearly 80% of patients with advanced cancers experience progressive loss of muscle mass, strength and function defined as cachexia [34]. Cachexia arises due to a negative protein and energy balance along with reduced food intake, anorexia, and impaired metabolism [35]. Such extreme progressive loss of muscle mass cannot be reversed by nutritional support, rendering treatments ineffective and results in significant impairment in survival, with one-year mortality of patients with cachexia ranging from 20% to 60% [36]. The criteria for cachexia are weight loss of more than 5% of their body weight or weight loss of more than 2% of their body weight in individuals with a body mass index less than 20 kg/m^2, or reduction in muscle mass

[35]. However, patients with a normal body mass index or obesity can also experience cachexia, which underestimates the true prevalence of this condition [37]. Many types of cancer are associated with cachexia, with pancreatic, gastrointestinal and lung cancer showing the most extreme examples. However, it is now recognized that most patients with cancer show muscle dysfunction, even in the absence of frank cachexia [38].

In older individuals, the decrease in muscle mass results in physical deterioration that limits daily functions and often results in the development of a disability [39]. Studies in older men and women have shown that reduction in quadriceps and grip strength, not the loss of muscle mass were strong independent predictors of mortality [40]. Men lose muscle strength almost twice as women, and gain of muscle mass in both men and women was not associated with improvement in muscle strength and function [41]. Loss of muscle strength has also been reported in breast cancer patients, and patients who have undergone radical mastectomy have lower shoulder flexibility. Chemotherapy also induced severe muscle fatigue in patients with breast cancer [42]. Women with breast cancer that metastasized to bone and other visceral organs had less fitness and reduced physical activity compared to healthy women [43]. Assessment of functional capacity using a stationary bike protocol in newly diagnosed breast cancer patients who were receiving endocrine therapy found that loss of muscle function can occur without loss of muscle mass [44]. Lack of association between muscle mass loss and muscle strength loss was also reported in patients with digestive cancer undergoing chemotherapy [45]. Similar to patients with breast cancer, patients with prostate cancer undergoing androgen deprivation therapy had a significant reduction in both upper and lower muscle strength along with reduced bone mineral density and increased body fat percentage compared to age matched healthy men [46]. Muscle wasting was reported in patients with foregut cancers who were undergoing neoadjuvant therapy; these patients had reduced survival compared to those in palliative care [47].

Besides cancer, chemotherapy can also have profound effects in deteriorating muscle health. In patients with metastatic renal cancer, treatment with sorafenib resulted in progressive muscle loss and was reported as an occult condition in patients with a normal or high BMI [48]. Similarly in animal models, the use of tyrosine kinase inhibitors regorafenib and sorafenib resulted in loss of both skeletal and cardiac muscle mass and function [49]. In mice, carboplatin administration resulted in loss of bone, muscle and muscle weakness. Inhibition of TGF-β signaling using neutralizing antibody in these mice, prevented bone loss and muscle weakness but not muscle atrophy [50].

3.1 Bone-Derived Factors Impacting Muscle

Bone and muscle have close developmental and physical association and their interaction is important for mobility and normal development [51]. Bone acts as an endocrine organ and growth factors released from the bone can have systemic effects. For example, osteocalcin primarily secreted by osteoblasts has emerged as an important mediator of glucose metabolism [52]. In patients with metabolic syndrome, undercarboxylated osteocalcin has been inversely linked to cardiovascular diseases [53]. In older women, levels of undercarboxylated osteocalcin were correlated with lower limb strength [54]. Similarly other bone-derived factors including IGF1 [55] and BMP2 [56] have been implicated in myogenic development.

TGF-β and members of the TGF-β pathway superfamily also have major roles in muscle homeostasis during health and disease states. In skeletal muscle, TGF-β inhibits muscle differentiation by repressing MyoD and Myogenin through the Smad3 signaling pathway [57]. In mice, overexpression of TGF-β in muscle resulted in muscle weakness [58], and treating mouse muscles with TGF-β inhibited satellite cell fusion and reduced the size of muscle fibers [59]. In muscles, TGF-β also regulates extracellular matrix remodeling and is activated by matrix

metalloproteases (MMP)-2 and MMP-9. TGF-β inhibits extracellular matrix degradation by activating these proteases by increasing TIMP and plasminogen activator, both of which inhibit matrix breakdown while simultaneously increasing extracellular matrix protein production. However, excessive production of TGF-β results in increased accumulation of extracellular matrix, resulting in tissue fibrosis [60]. Another member of the TGF-β superfamily is Myostatin, which is a negative regulator of skeletal muscle growth. Myostatin binds to the activin type II receptors ACVR2 and ACVR2B, activating the SMAD2/3 signaling pathway [61]. Activin A and activin B proteins, which are also members of TGF-β superfamily, function synergistically with myostatin to suppress muscle growth. The use of activin A and activin B antagonists increased muscle mass by 20% and 45%, respectively, while the addition of myostatin inhibitors increased muscle mass by 150% by completely abrogating Smad2/3 signaling and activating the muscle mass, promoting the BMP/Smad1/5 pathway [62].

3.2 Bone Metastases-Induced Muscle Dysfunction

In bone, osteoblasts deposit TGF-β in the bone matrix, making bone the largest source of TGF-β in the body [63]. TGF-β deposited in the bone matrix is released and activated during bone resorption by osteoclasts. The released TGF-β induces migration of bone marrow stromal cells and coordinates bone formation, thus coupling both bone resorption and formation [64]. In mice, osteoblast-specific overexpression of TGF-β increased matrix deposition and bone resorption, causing bone loss and resulting in an osteoporosis-like phenotype [65]. Consistent with these findings, the pharmacologic inhibition of TGF-β in mice using the TβRI/ALK5 kinase inhibitor, SD-208 resulted in increased trabecular bone volume [66]. In bone metastases, TGF-β has a central role in driving osteolysis, resulting in elevated levels of TGF-β being released from bone during the increased resorption [67]. TGF-β also induces tumor cells to secrete several osteolytic factors such as IL-11 and connective tissue growth factor further driving bone breakdown [68, 69]. Disruption of the TGF-β-Smad signaling pathway early in the metastatic process reduced the tumor burden in bone; however, inhibiting the signaling process was ineffective at reducing bone metastases once the metastasis is established [69]. Studies in mice models of bone metastases from multiple myeloma and breast, prostate, and lung cancers have shown that excess TGF-β released from the bone can result in reduced skeletal muscle function. Reduced forelimb grip strength and specific force of the extensor digitorum longus muscle was observed in the mice with metastases suggesting muscle contractility deficit [70]. The increased TGF-β released from the bone during bone metastases upregulates NADPH oxidase leading to the oxidation and nitrosylation of the calcium release channel ryanodine receptor 1 (RyR1) and reduction of its stabilization unit, calstabin1. This process results in a leaky RyR1 channel, which causes a calcium leak in the sarcoplasmic reticulum, leading to less force produced by the muscle, ultimately resulting in muscle weakness. These features were present only in mice with bone metastases and not in mice with primary tumors. Consistent with the animal models, leaky RyR1 channels were also observed in the skeletal muscles of breast cancer patients with bone metastases [70]. The RyR1 calcium release channel stabilizer Rycal S107 stabilizes the leaky channels by inhibiting the reduction of calstabin1 from the RyR1 complex [71]. In mice with bone metastases administration of Rycal S107 improved forelimb grip strength and muscle function without reducing tumor progression or bone metastases [70].

Recently, a role for zinc binding and zinc transport was reported to contribute to muscle dysfunction. In metastatic breast and colon cancer models, upregulation of zinc-transport encoding gene ZIP14 in multiple muscle types was associated with a marked reduction in grip strength and muscle fiber diameter. Increased ZIP14 resulted in excessive zinc accumulation in the muscle fibers, causing muscle wasting. A similar expression of ZIP14 was also observed in the muscles of cancer patients with muscle weakness and cachectic phenotype. In myoblast cultures, the addition of TGF-β and TNF-α induced ZIP14 expression suggesting the elevated levels of these cytokines in cancer could be causing muscle dysfunction by upregulating ZIP14 expression [72].

4   Conclusion

The systemic effects of cancer on bone and muscle result in poor quality of life, undermine responses to cancer therapies, and further complicate the care of patients who are already undergoing extreme physical and emotional stress. It is clear from recent studies that the effects of cancer and cancer treatment to induce pathologic bone destruction compound these systemic effects to induce muscle weakness. Currently, there are no effective treatments for curing bone metastases and improving muscle strength in patients with cancer. Better animal models and clinical research are needed to accurately model the sequential stages of systemic muscle dysfunction in cancer and determine if other systemic effects of pathologic bone destruction co-exist. Treatments that can prevent the negative systemic effects of cancer and cancer treatment should be developed and implemented with the cancer treatment. In many patients with cancer, muscle weakness can occur during the early stages of cancer, and measures to detect these syndromes need to be routinely employed in the clinics. Including patients with early-stage cancer in clinical trials is necessary to understand the effectiveness of treating systemic dysfunction in cancer in improving the outcome of cancer therapies and in providing a better quality of life for the patients. Awareness and understanding of these secondary complications in cancer patients by clinicians and caregivers are important for diagnosing patients at an early stage to provide supportive care.

Acknowledgments The authors are grateful for support from The University of Texas MD Anderson Cancer

Center Department of Endocrine Neoplasia and Hormonal Disorders, The Rolanette and Berdon Bone Disease Research Program of Texas, the Cancer Prevention Research institute of Texas (CPRIT) and Dive into The Pink. Dr. Guise is supported by grants from CPRIT (RR190108) the NIH (R01CA206025), the Department of Defense (BC171929), The Rolanette and Berdon Bone Disease Research Program of Texas and Dive into The Pink. Dr. Guise is a CPRIT Scholar in Cancer Research. We thank Ashli Nguyen-Villarreal, Associate Scientific Editor, in the Research Medical Library at The University of Texas MD Anderson Cancer Center, for assistance with editing.

References

1. Siegel, R.L., Miller, K.D., Jemal, A.: Cancer statistics, 2020. CA Cancer J. Clin. **70**(1), 7–30 (2020)
2. Dunlop, R.J., Campbell, C.W.: Cytokines and advanced cancer. J. Pain Symptom Manag. **20**(3), 214–232 (2000)
3. Coleman, R.E., et al.: Bone metastases. Nat. Rev. Dis. Primers. **6**(1), 83 (2020)
4. Hernandez, R.K., et al.: Incidence of bone metastases in patients with solid tumors: analysis of oncology electronic medical records in the United States. BMC Cancer. **18**(1), 44 (2018)
5. Coleman, R.E.: Clinical features of metastatic bone disease and risk of skeletal morbidity. Clin. Cancer Res. **12**(20 Pt 2), 6243s–6249s (2006)
6. Roodman, G.D.: Pathogenesis of myeloma bone disease. J. Cell. Biochem. **109**(2), 283–291 (2010)
7. Coleman, R.E.: Skeletal complications of malignancy. Cancer. **80**(8 Suppl), 1588–1594 (1997)
8. Paget, S.: The distribution of secondary growths in cancer of the breast. 1889. Cancer Metastasis Rev. **8**(2), 98–101 (1989)
9. Weilbaecher, K.N., Guise, T.A., McCauley, L.K.: Cancer to bone: a fatal attraction. Nat. Rev. Cancer. **11**(6), 411–425 (2011)
10. Fornetti, J., Welm, A.L., Stewart, S.A.: Understanding the bone in cancer metastasis. J. Bone Miner. Res. **33**(12), 2099–2113 (2018)
11. Raggatt, L.J., Partridge, N.C.: Cellular and molecular mechanisms of bone remodeling. J. Biol. Chem. **285**(33), 25103–25108 (2010)
12. Thomas, R.J., et al.: Breast cancer cells interact with osteoblasts to support osteoclast formation. Endocrinology. **140**(10), 4451–4458 (1999)
13. Sethi, N., et al.: Tumor-derived JAGGED1 promotes osteolytic bone metastasis of breast cancer by engaging notch signaling in bone cells. Cancer Cell. **19**(2), 192–205 (2011)
14. Yin, J.J., et al.: TGF-beta signaling blockade inhibits PTHrP secretion by breast cancer cells and bone metastases development. J. Clin. Invest. **103**(2), 197–206 (1999)
15. Mundy, G.R.: Metastasis to bone: causes, consequences and therapeutic opportunities. Nat. Rev. Cancer. **2**(8), 584–593 (2002)
16. Buijs, J.T., Juarez, P., Guise, T.A.: Therapeutic strategies to target TGF-beta in the treatment of bone metastases. Curr. Pharm. Biotechnol. **12**(12), 2121–2137 (2011)
17. Stopeck, A.T., et al.: Denosumab compared with zoledronic acid for the treatment of bone metastases in patients with advanced breast cancer: a randomized, double-blind study. J. Clin. Oncol. **28**(35), 5132–5139 (2010)
18. Hortobagyi, G.N., et al.: Efficacy of pamidronate in reducing skeletal complications in patients with breast cancer and lytic bone metastases. Protocol 19 Aredia Breast Cancer Study Group. N. Engl. J. Med. **335**(24), 1785–1791 (1996)
19. Berenson, J.R., et al.: Efficacy of pamidronate in reducing skeletal events in patients with advanced multiple myeloma. Myeloma Aredia Study Group. N. Engl. J. Med. **334**(8), 488–493 (1996)
20. Reddington, J.A., et al.: Imaging characteristic analysis of metastatic spine lesions from breast, prostate, lung, and renal cell carcinomas for surgical planning: osteolytic versus osteoblastic. Surg. Neurol. Int. **7**(Suppl 13), S361–S365 (2016)
21. Brown, J.E., et al.: Bone turnover markers as predictors of skeletal complications in prostate cancer, lung cancer, and other solid tumors. J. Natl. Cancer Inst. **97**(1), 59–69 (2005)
22. Gupta, N., et al.: Usefulness of radium-223 in patients with bone metastases. Proc. (Bayl. Univ. Med. Cent.). **30**(4), 424–426 (2017)
23. Autzen, P., et al.: Bone morphogenetic protein 6 in skeletal metastases from prostate cancer and other common human malignancies. Br. J. Cancer. **78**(9), 1219–1223 (1998)
24. Yin, J.J., et al.: A causal role for endothelin-1 in the pathogenesis of osteoblastic bone metastases. Proc. Natl. Acad. Sci. USA. **100**(19), 10954–10959 (2003)
25. Nelson, J.B., et al.: Identification of endothelin-1 in the pathophysiology of metastatic adenocarcinoma of the prostate. Nat. Med. **1**(9), 944–949 (1995)
26. Shariat, S.F., et al.: Preoperative plasma levels of transforming growth factor beta(1) (TGF-beta(1)) strongly predict progression in patients undergoing radical prostatectomy. J. Clin. Oncol. **19**(11), 2856–2864 (2001)
27. Jiao, S., et al.: Differences in tumor microenvironment dictate T helper lineage polarization and response to immune checkpoint therapy. Cell. **179**(5), 1177–1190 e13 (2019)
28. Logothetis, C.J., Lin, S.H.: Osteoblasts in prostate cancer metastasis to bone. Nat. Rev. Cancer. **5**(1), 21–28 (2005)
29. Asadi, F., et al.: Enhanced expression of parathyroid hormone-related protein in prostate cancer as compared with benign prostatic hyperplasia. Hum. Pathol. **27**(12), 1319–1323 (1996)

30. Schluter, K.D., Katzer, C., Piper, H.M.: A N-terminal PTHrP peptide fragment void of a PTH/PTHrP-receptor binding domain activates cardiac ET (A) receptors. Br. J. Pharmacol. **132**(2), 427–432 (2001)

31. Ottewell, P.D., et al.: Zoledronic acid has differential antitumor activity in the pre- and postmenopausal bone microenvironment in vivo. Clin. Cancer Res. **20**(11), 2922–2932 (2014)

32. Wright, L.E., et al.: Aromatase inhibitor-induced bone loss increases the progression of estrogen receptor-negative breast cancer in bone and exacerbates muscle weakness in vivo. Oncotarget. **8**(5), 8406–8419 (2017)

33. Ottewell, P.D., et al.: Castration-induced bone loss triggers growth of disseminated prostate cancer cells in bone. Endocr. Relat. Cancer. **21**(5), 769–781 (2014)

34. Biswas, A.K., Acharyya, S.: Understanding cachexia in the context of metastatic progression. Nat. Rev. Cancer. **20**(5), 274–284 (2020)

35. Fearon, K., et al.: Definition and classification of cancer cachexia: an international consensus. Lancet Oncol. **12**(5), 489–495 (2011)

36. von Haehling, S., Anker, S.D.: Prevalence, incidence and clinical impact of cachexia: facts and numbers-update 2014. J. Cachexia. Sarcopenia Muscle. **5**(4), 261–263 (2014)

37. Sun, L., Quan, X.Q., Yu, S.: An epidemiological survey of cachexia in advanced cancer patients and analysis on its diagnostic and treatment status. Nutr. Cancer. **67**(7), 1056–1062 (2015)

38. Baracos, V.E., et al.: Cancer-associated cachexia. Nat. Rev. Dis. Primers. **4**, 17105 (2018)

39. Visser, M., et al.: Skeletal muscle mass and muscle strength in relation to lower-extremity performance in older men and women. J. Am. Geriatr. Soc. **48**(4), 381–386 (2000)

40. Newman, A.B., et al.: Strength, but not muscle mass, is associated with mortality in the health, aging and body composition study cohort. J. Gerontol. A Biol. Sci. Med. Sci. **61**(1), 72–77 (2006)

41. Goodpaster, B.H., et al.: The loss of skeletal muscle strength, mass, and quality in older adults: the health, aging and body composition study. J. Gerontol. A Biol. Sci. Med. Sci. **61**(10), 1059–1064 (2006)

42. Klassen, O., et al.: Muscle strength in breast cancer patients receiving different treatment regimes. J. Cachexia. Sarcopenia Muscle. **8**(2), 305–316 (2017)

43. Yee, J., et al.: Physical activity and fitness in women with metastatic breast cancer. J. Cancer Surviv. **8**(4), 647–656 (2014)

44. Ballinger, T.J., et al.: Impact of primary breast cancer therapy on energetic capacity and body composition. Breast Cancer Res. Treat. **172**(2), 445–452 (2018)

45. Moreau, J., et al.: Correlation between muscle mass and handgrip strength in digestive cancer patients undergoing chemotherapy. Cancer Med. **8**(8), 3677–3684 (2019)

46. Galvao, D.A., et al.: Reduced muscle strength and functional performance in men with prostate cancer undergoing androgen suppression: a comprehensive cross-sectional investigation. Prostate Cancer Prostatic Dis. **12**(2), 198–203 (2009)

47. Daly, L.E., et al.: Loss of skeletal muscle during systemic chemotherapy is prognostic of poor survival in patients with foregut cancer. J. Cachexia. Sarcopenia Muscle. **9**(2), 315–325 (2018)

48. Antoun, S., et al.: Association of skeletal muscle wasting with treatment with sorafenib in patients with advanced renal cell carcinoma: results from a placebo-controlled study. J. Clin. Oncol. **28**(6), 1054–1060 (2010)

49. Huot, J.R., et al., Chronic treatment with multi-kinase inhibitors causes differential toxicities on skeletal and cardiac muscles. Cancers (Basel), 2019. 11(4)

50. Hain, B.A., et al.: Zoledronic acid improves muscle function in healthy mice treated with chemotherapy. J. Bone Miner. Res. **35**(2), 368–381 (2020)

51. Regan, J.N., Waning, D.L., Guise, T.A.: Skeletal muscle Ca(2+) mishandling: another effect of bone-to-muscle signaling. Semin. Cell Dev. Biol. **49**, 24–29 (2016)

52. Lee, N.K., et al.: Endocrine regulation of energy metabolism by the skeleton. Cell. **130**(3), 456–469 (2007)

53. Riquelme-Gallego, B., et al.: Circulating undercarboxylated osteocalcin as estimator of cardiovascular and type 2 diabetes risk in metabolic syndrome patients. Sci. Rep. **10**(1), 1840 (2020)

54. Levinger, I., et al.: Undercarboxylated osteocalcin, muscle strength and indices of bone health in older women. Bone. **64**, 8–12 (2014)

55. Yoshida, T., Delafontaine, P.: Mechanisms of IGF-1-mediated regulation of skeletal muscle hypertrophy and atrophy. Cell. **9**(9) (2020)

56. Sartori, R., et al.: BMP signaling controls muscle mass. Nat. Genet. **45**(11), 1309 (2013)

57. Liu, D., Black, B.L., Derynck, R.: TGF-beta inhibits muscle differentiation through functional repression of myogenic transcription factors by Smad3. Genes Dev. **15**(22), 2950–2966 (2001)

58. Narola, J., et al.: Conditional expression of TGF-beta1 in skeletal muscles causes endomysial fibrosis and myofibers atrophy. PLoS One. **8**(11), e79356 (2013)

59. Mendias, C.L., et al.: Transforming growth factor-beta induces skeletal muscle atrophy and fibrosis through the induction of atrogin-1 and scleraxis. Muscle Nerve. **45**(1), 55–59 (2012)

60. Mann, C.J., et al.: Aberrant repair and fibrosis development in skeletal muscle. Skelet. Muscle. **1**(1), 21 (2011)

61. Benny Klimek, M.E., et al.: Acute inhibition of myostatin-family proteins preserves skeletal muscle in mouse models of cancer cachexia. Biochem. Biophys. Res. Commun. **391**(3), 1548–1554 (2010)

62. Chen, J.L., et al.: Specific targeting of TGF-beta family ligands demonstrates distinct roles in the regulation of muscle mass in health and disease. Proc. Natl. Acad. Sci. USA. **114**(26), E5266–E5275 (2017)

63. Dallas, S.L., et al.: Proteolysis of latent transforming growth factor-beta (TGF-beta)-binding protein-1 by osteoclasts. A cellular mechanism for release of TGF-beta from bone matrix. J. Biol. Chem. **277**(24), 21352–21360 (2002)

64. Tang, Y., et al.: TGF-beta1-induced migration of bone mesenchymal stem cells couples bone resorption with formation. Nat. Med. **15**(7), 757–765 (2009)

65. Erlebacher, A., Derynck, R.: Increased expression of TGF-beta 2 in osteoblasts results in an osteoporosis-like phenotype. J. Cell Biol. **132**(1–2), 195–210 (1996)

66. Mohammad, K.S., et al.: Pharmacologic inhibition of the TGF-beta type I receptor kinase has anabolic and anti-catabolic effects on bone. PLoS One. **4**(4), e5275 (2009)

67. Kang, Y., et al.: Breast cancer bone metastasis mediated by the Smad tumor suppressor pathway. Proc. Natl. Acad. Sci. USA. **102**(39), 13909–13914 (2005)

68. Kang, Y., et al.: A multigenic program mediating breast cancer metastasis to bone. Cancer Cell. **3**(6), 537–549 (2003)

69. Korpal, M., et al.: Imaging transforming growth factor-beta signaling dynamics and therapeutic response in breast cancer bone metastasis. Nat. Med. **15**(8), 960–966 (2009)

70. Waning, D.L., et al.: Excess TGF-beta mediates muscle weakness associated with bone metastases in mice. Nat. Med. **21**(11), 1262–1271 (2015)

71. Andersson, D.C., et al.: Ryanodine receptor oxidation causes intracellular calcium leak and muscle weakness in aging. Cell Metab. **14**(2), 196–207 (2011)

72. Wang, G., et al.: Metastatic cancers promote cachexia through ZIP14 upregulation in skeletal muscle. Nat. Med. **24**(6), 770–781 (2018)

Targeting Metastatic Disease: Challenges and New Opportunities

Haitian Hu, Zeping Hu, and Hanqiu Zheng

Abstract

Metastasis is the major cause of cancer-associated deaths in most cancers. Research in the past two decades revealed the complexity of metastasis process. Metastasis involved not only tumor-intrinsic signaling, but often their interactions with tumor microenvironment composed of extracellular matrix and stromal cells including fibroblast, endothelial cells, macrophages etc. Tumor immune environment and cancer metabolism shunting also contribute to metastasis cascade. Lagging this mechanistic understanding of metastatic cascade is the scarce of effective therapeutic options specifically designed for metastatic patients. A few successful treatment examples include bisphosphonate and denosumab against bone metastasis, immune checkpoint inhibitors for certain metastatic cancer patients. In this chapter, we will discuss the approved therapeutic agents and procedures for metastatic patients and discuss the current progress on metastasis research that could shed light upon future drug development.

H. Hu · H. Zheng (✉)
Department of Basic Medical Sciences, School of Medicine, Tsinghua University, Beijing, China
e-mail: hanzheng@tsinghua.edu.cn

Z. Hu
School of Pharmaceutical Sciences, Tsinghua University, Beijing, China

Learning Objectives

Cancer represents a major health burden worldwide. It is the leading cause of premature death among most countries with high or very high Human Development Index (HDI) and is one of the top causes among many other countries [1]. In 2018 alone, it is estimated that there were 17 million new cancer diagnosed and 9.5 million cancer deaths. By 2040, the worldwide cancer burden is expected to be 27.5 million new cases and 16.3 million deaths, largely due to population growth and aging.

Cancer metastasis is a multi-step process. As primary tumor grows, it remodels the extracellular microenvironment including the resident stromal cells, immune infiltrated cells, and extracellular matrix, invade the basement membrane. A small number of tumor cells will be able to intravasate into blood vessels, travel by the blood circulation before arriving at distal organ sites. These disseminated tumor cells (DTCs) then extravasate through the micro-vessels into parenchymal tissues. It is estimated that only a very small portion of tumor cells will survive and adapt this new environment and eventually grow into clinically detectable metastatic colonies. [2]. Cancer metastasis is the major cause of cancer related death. Patients with

(continued)

localized disease, without signs of local invasion or distal metastasis have 5-year survival rate of usually more than 80%. However, patients with metastatic disease will have dismal prognosis with 5-year survival rate of less than 20% [3, 4]. Here we review the successful examples for treating metastatic disease and discuss the future endeavors of developing efficient therapies.

1 Successful Examples for Targeting Metastasis

1.1 Bisphosphonates and Denosumab

Since the 1970s, modern therapy strategies have been developed for cancer patients. For example, chemotherapy, radiation therapy, targeted therapy, and immune therapy have all been developed for treating cancer patients. Little strategies are considering how to target metastasis and thus fail to curb this deadly form of cancer. Perhaps the most notable successful therapeutic agent specifically designed for metastatic patients comes from understanding the bone metastasis biology. Bone is a common affected site during metastasis. For example, more than 70% patients develop skeletal complications in breast, prostate and lung cancer patients [5, 6]. Symptoms include aberrant bone absorption or build-up, hypercalcemia, bone pain, and bone fracture [5–7]. Bone metastasis is a perfect example proving Paget's "seed and soil" hypothesis for metastasis progression [8], where DTCs (seeds) that survive and expand in the bone microenvironment (soil). In healthy bones, two major bone resident cells cooperate with each other to maintain the bone homeostasis: The bone building osteoblasts and the bone degrading osteoclasts. However, when metastasis develops, the balance between these two cells is disrupted. In breast cancer, tumor cells help increase the bone microenvironment RANKL (receptor activator of nuclear factor kappa B (RANK) ligand)

level to activate osteoclastogenesis and bone absorption [9–11]. The osteolytic bone also releases growth factors and cytokines from bone matrix, many of which help tumor cells survive and grow [9, 12, 13]. Two clinically approved therapeutic agents are both designed to block the bone degradation process. Bisphosphonate limits bone degradation by binding to bone surface and inhibiting osteoclast activity [14]. Denosumab, a humanized monoclonal antibody against RANKL, binds to RANKL and blocks its interaction with its receptor on pre-osteoclast cells, thus inhibiting the osteoclast differentiation [15–17]. These two therapeutic agents are specifically designed for bone metastasis and are not suitable for treating patients with localized disease. The success of these two agents set the examples that biological events in specific metastatic organs might differ significantly from the primary site due to the drastic difference in their tumor microenvironment, and could be explored for therapeutic intervention (Fig. 1a).

1.2 Immunotherapy

Another effective treatment for metastatic disease comes from immune therapy. Among the many immune surveillance processes, T lymphocytes mediated tumor killing plays a major role. Through recognizing tumor neoantigens presented on tumor cell surface by major histocompatibility complex (MHC), T cell receptors (TCR) and downstream signaling molecules are activated through clustering at the tumor-T cell interface to transduce the tumor killing effects. This can be achieved through direct releasing of cytotoxic granules containing tumor killing enzymes like perforin and granzyme B into tumor cells or through the pro-apoptotic cytokine secretion [18, 19]. Unfortunately, tumor cells evolve to evade T cell surveillance through multiple mechanisms including utilizing T cells' own co-stimulatory/inhibitory system [20]. Besides its TCR activation, proper co-stimulatory/inhibitory receptor activation statuses are also finely regulated for T cells to be fully activated and releasing its effective attack on target cells.

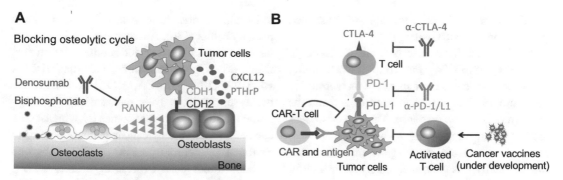

Fig. 1 Successful Examples for Targeting Metastasis. (**a**) Through engaging tumor-osteoblast interactions, including CXCL12-CXCR4, CDH1-CDH2, JAG1-Notch signaling, tumor cells seed and colonize the bone. The proliferating tumor cells and their secreted factors including PTHrP help osteoblasts up-regulate the expression and secretion of RANKL, a major osteoclastogenesis factor. Upon excessive RANKL stimulation, mature multi-nuclei osteoclasts are formed to degrade the bone. Cytokines and growth factors are then released from the bone matrix to stimulate metastatic tumor growth. Denosumab, a RANKK neutralizing antibody, deplete bone osteoclasts by binding to RANKL and blocking its ligand-receptor interaction. Bisphosphonate could bind to bone surface to inhibit osteoclast activity. (**b**) Current Immunotherapy strategies. CTLA-4 antibody and PD-1, PD-L1 antibodies belong to immune checkpoint inhibitors as they block the co-inhibitory receptors on T cells and re-activate T cells for tumor killing. CAR-T cells suppress tumor cells by recognizing tumor associated antigen. Cancer vaccines utilize tumor neoantigens to stimulate the immune system to specifically recognize tumor cells for elimination

PD-1 (Programmed Death 1) and CTLA-4 (Cytotoxic T-lymphocyte antigen 4) are among these important co-inhibitory receptors [21–23]. In melanoma and other solid tumors, ligands of inhibitory receptors like PD-L1 are highly expressed on tumor cells as well as on antigen presenting cells, thus inhibiting cytotoxic T cell activation. Neutralizing antibodies against these molecules, including PD-1 antibodies, PD-L1 antibodies, and CTLA-4 antibodies are able to re-activate cytotoxic T cells and approved for cancer therapy in many cancers including melanoma [24–26]. In certain cases, these immune checkpoint inhibitors (ICIs) display excellent therapeutic efficacy in even metastatic patients, for whom there were no effective therapy previously. For example, in late-stage melanoma patients with metastatic disease, combinational therapy of Nivolumab (PD-1 blocking antibody) and Ipilimumab (CTLA-4 blocking antibody) generates 61% overall response rate and a two-year survival rate of 64%, all of which are significantly improved over previous standard therapies [27] (Fig. 1b).

Combinational therapies of ICIs with radiotherapy, chemotherapy, anti-angiogenic therapy, and targeted therapy are tested in many clinical trials for their efficacies. For example, in a phase Ib clinical trial, a multi-kinase inhibitor regorafenib and a PD-1 inhibitor Nivolumab was tested in combination in advanced metastatic gastric or colorectal cancer patients irrespective of their microsatellite status. The overall responsive rate was 40%. Considering 95% patients were with microsatellite stable (MSS) diseases and expected to be non-responders by Nivolumab treatment alone, this was a significant improvement over previous standard therapy [28]. A phase III trial will confirm the results in the near future.

Other immunotherapy strategies, including CAR-T and cancer vaccines are either approved or under the development, the progress of these therapies is reviewed elsewhere [29–32]. In a few recent reports, personalized cancer vaccine displayed excellent therapeutic effect on patients with late-stage metastatic disease. In two metastatic melanoma patients treated with tumor neoantigen vaccine, T cell infiltration and tumor killing were enhanced in post-vaccination resected metastatic nodules. The metastatic events were reduced, leading to sustained progression-free survival.

Similar beneficial effects were observed in another report [33, 34] (Fig. 1b). It is worth noting that cancer vaccines are dependent on identifying neo-isotopes that can be utilized to activate the anti-tumor immune response. These neo-isotopes are derived from neoantigens which are not necessarily come from driver gene mutations, but often from bypass gene mutations and thus differ from patient to patient. Using next generation sequencing, individual patient's exome mutation landscape can be revealed and the most promising tumor neoantigens are predicted. However, with many neoantigen prediction algorisms developed, it is still very difficult to accurately tell which neoantigen combination will best boost patients' anti-tumor immunity [35]. The application of tumor vaccine in other "immune-cold" tumors, such as most breast cancer and prostate cancer patients are still difficult. Future large-scale clinical trials in broader cancer types will provide clear evidence to support cancer vaccine in treating metastatic disease.

2   Genomic Evolution of Primary Tumor and Metastatic Colonies

Cancer is a genetic disease with genetic mutations that increase the fitness of competing tumor clones. Understanding the genomic differences between primary tumors and metastatic colonies provide critical information to better understand how and when metastases are developed. Potential metastasis specific genes could also be identified by such an approach.

Small-scale genome sequencing and exome sequencing were used in the past to identify such differences. Through years of research, it is generally believed that there are almost no new mutations specific to metastatic tumors. However, some driver gene mutations are indeed presented at much higher frequency at metastatic sites. For example, in a prostate cancer metastasis study, samples from lethal prostate cancer patients were sequenced, TP53 mutation was discovered

to be highly enriched in metastatic clones than in primary tumors [36]. Another analysis in colorectal cancer patients revealed similar pattern that many driver gene mutations, including KRAS, BRAF, PTEN, PI3KCA, KDR, and FLT, are highly enriched in liver metastasis [37]. Additional interesting example comes from breast cancer study. 70% breast cancer patients belong to Estrogen Receptor positive (ER^+) subgroup because tumor cells express high level of ER, one of the major mitogenic signals. Estrogen signaling modulators like tamoxifen and aromatase inhibitors, provide exceptional survival benefits for ER^+ patients. Unfortunately, hormone therapy resistance is developed at metastatic sites by multiple mechanisms. In 2013, two research groups independently discovered that around 25% of therapy-refractory metastatic patients bore mutations on *ESR1* gene, the coding gene for ER. The mutations affect ligand binding domain of the ER protein and generate ligand- independent estrogen signaling activation. Some of the most prevalent mutation sites include p. Tyr537Ser, p.Tyr537Asn and p.Asp538Gly single amino acid substitutions [38, 39]. These mutations are rarely identified in primary cancer tissues and can be considered to be metastatic-specific mutations. Current research is focusing on identifying small molecule inhibitors or degraders that can specifically block the signaling of ER mutations [40].

With the development of next generation sequencing technology, it is now possible to perform large scale pan-cancer analysis on paired primary and metastatic tissues. A recent study sequenced metastatic tumor genomes from 2520 patients, including data from the pairing primary tumors and normal tissues [41]. Similar to previous findings, mutation rates of driver genes are generally higher in metastases than in primary tumors. A few novel gene mutations associated with metastasis were also identified. For example, in pan-cancers, MLK4 (also known as MAP 3K21) mutation is significantly associated with metastasis; in breast cancer, ZFPM1 (also known as FOG1, a zinc finger transcription factor) is a mutated gene in metastatic colonies.

Future studies could determine whether these mutations are metastasis driver mutations with critical functions.

3 Targeting the Metastasis Cascade

Despite the aforementioned successful therapy strategies treating metastasis, a majority of metastatic patients still succumb to cancer due to therapy failure and subsequent disease progression. Most current therapies are approved based on their efficacy to inhibit primary tumor growth. The measurements in clinical trials are the shrinkage of primary tumors and patient survival without considering metastasis status. Sometimes, this design led to controversial results in clinical usage when measured by metastasis events. Bevacizumab (Avastin), a humanized anti-VEGF (Vascular endothelial growth factor) monoclonal antibody was designed to block VEGF's binding to its endothelial receptor VEGFR(1/2) and thus significantly inhibited tumor angiogenesis and extended patient survival [42]. Combinational therapy of Bevacizumab with standard chemotherapy extended progression-free survival time by about 2.1 months in metastatic colorectal cancer patients [43]. It was later discovered that although anti-angiogenic agents inhibited primary tumor growth, it sometimes promoted tumor invasion and metastasis largely due to the side effects from increased hypoxia signaling within tumor tissues and thus limited their usage [44, 45]. It is thus important to consider the candidate drug's therapeutic efficacy on both primary tumor growth and metastasis progression.

3.1 Preventing Metastasis

Metastasis cascade is a series of inter-connected steps that lead to eventual success in tumor outgrowth at distant organs [46, 47]. It is speculated that colonization might be the most rate-limiting step in metastasis cascade. But in theory, blocking any step will likely reduce the overall metastasis events [48, 49]. For patients with early-stage disease without signs of lymph node invasion, metastasis prevention might be the best strategy. Patients with localized disease are expected to have good prognosis after tumor resection [50]. However, tumor cells are already disseminated to distal organ tissues even at very early stage of primary tumor progression [51, 52]. Indeed, blood circulating tumor cells (CTCs) and DTCs in bone marrow could be detected in most cancer patients [53, 54]. Early dissemination theory is also supported by mouse model. Utilizing genetically engineered pancreatic mouse model and crossed with fluorescent labeling strain, labeled tumor cells invaded into surrounding stromal tissue and intravasated into the blood circulation extremely early, well before the pathological determination of primary malignancy [55]. Tumor cells leave the primary site through two different migration mechanism -- collective cell movement or single cell movement generated by a process named epithelial to mesenchymal transition. The reviews for these two types of tumor invasions and underlying signaling mechanisms can be found elsewhere [56, 57]. Even dissemination occurs early, fact in doubt is whether these early DTCs are the "seed" cells that eventually outgrow into full-blown clinically detectable metastasis. Some hints come from adjuvant therapy for patients with localized and operable disease. For example, many early stage cancer patients receive adjuvant chemotherapy or hormone therapies after primary tumor resection, which significantly reduces the risk of future relapse or distal metastasis [58], indicating the early disseminated cells can generate eventual metastases.

For DTCs to successfully lodge to distal site and adapt to the foreign environment, tumor cells need to go through many physical and biological barriers. Tumor cells must escape the immune surveillance and resist sheer stress within the blood circulation. Platelets provide this protective niche during CTCs travelling in the blood vessels by aggregating around the tumor cells [59]. After successfully overcoming these obstacles and arriving at distant organs, only very few cells will survive and generate clinically detectable metastasis, a process named metastasis colonization. It is estimated that only 0.1% of tumor cells entering circulation produce clinically detectable metastases eventually [2]. Many recent studies support Paget's "seed and soil" hypothesis that only matched tumor cells and the suitable fertile foreign tissues will generate the full-blown metastasis. By secreting critical factors or via exosome trafficking, primary tumor cells re-model the foreign tissues before the arrival, a process named "pre-metastatic niche" formation. For example, in breast cancer, primary tumor secretes Angiopoietin-Like 4 (ANGPTL4) to induce vascular permeability of endothelial cells in the lung and facilitate tumor cell extravasate into lung tissues [60]. Exosomes, as carriers of proteins, DNA, mRNA, and miRNAs, are important mediators to remodel pre-metastatic niche and even determine the metastasis organotropism. Exosomes from tumor cells prone to metastasize to lung, liver, and brain re-direct bone-tropic tumor cells to their respective sites. This reprograming of tumor cells is mediated by specific integrins from exosomes. Exosomes of pro-lung metastatic cells carry high level of $\alpha 6\beta 4$ and $\alpha 6\beta 1$ integrins, while exosomes from pro- liver metastatic cells carry high $\alpha V\beta 5$ integrin [61–64]. Blocking the functions of these integrins decreased exosome accumulation in their respective metastatic sites, and inhibited metastasis. Similarly, cell migration-inducing and hyaluronan-binding protein (CEMIP) is a major exosomal protein secreted by brain metastasis cells, which induce endothelial cell branching and inflammation in perivascular niche to promote brain metastasis [65].

Sometimes, DTCs might keep indolent at distant organs for years or even decades before their eventual breakout into clinically detectable metastases [61]. This is especially common in breast cancer and prostate cancer [66, 67]. The phenotype of the delayed onset of clinically detectable metastasis is named "tumor dormancy". The molecular mechanisms underlying dormancy is currently less studied due to lack of pre- clinical animal models. Tumor cells are considered dormant if they generate primary tumor

efficiently, but can only form metastasis in low incidence and long latency after inoculation.

Dormant cancer cell lines are generated by in vivo clonal selection for special capability of developing metastasis after long latency. In breast cancer, 4T07 was developed from in vivo selected mouse mammary tumor cells with weak lung metastatic ability [68]. SCP6 cell line was initially derived from MDA-MB-231 cells with almost no bone metastasis potential [69]. With its xenografting in immune-compromised athymic nude mice, a few mice developed bone metastasis after four months and cells derived were re-selected in vivo before the final generation of PD2D line with much stronger bone metastasis potential [70]. A few other dormancy models and recently developed patient-derived xenograft models are also utilized in research [71]. Mechanistically, dormancy cells survive as stable clusters of very few cells or as micrometastases with balanced cell proliferation and cell death [61, 72]. Bone marrow seems to be a common site for these dormant tumor cells to reside in [51, 73, 74]. These tumor cells occupy a specialized bone microenvironment that generally support hematopoietic stem cells (HSCs), namely the endothelial vascular niche and the osteoblast cell niche [75, 76] (Fig. 2). In a prostate cancer model, tumor cells compete with HSCs for osteoblast niche to generate bone metastasis [77]. More detailed signaling, including heterogenic N-E-cadherin, calcium signaling, JAG1-Notch signaling have been indicated in tumor -- osteoblast niche interaction to support tumor survival in breast cancer models [78–81]. DTCs in bone marrow also utilize the common HSC signaling axil to survive the foreign environment. HSC niche has been well defined and could be lineage tracked by utilizing CXCL12-labeled mice [82]. As perivascular niche cells including endothelial cells and mesenchymal stem cells are the major source for CXCL12, tumor cells express CXCR4 to interact with these niche cells for lodging and surviving the perivascular niche [83–85]. These extravasated tumor cells are expected to be quiescent during dormancy, these tumor cells might also share similar signaling

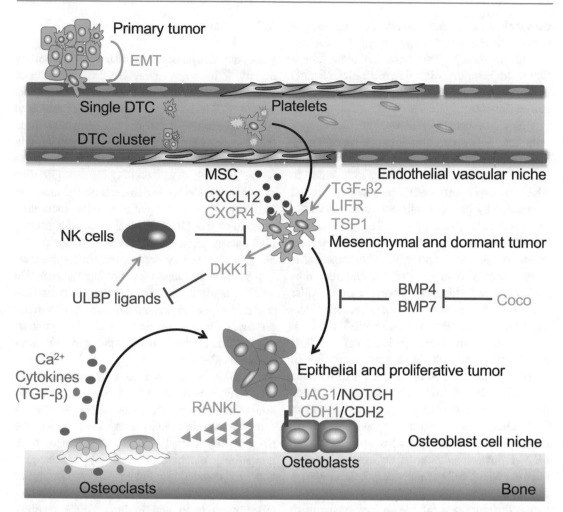

Fig. 2 Metastatic Niches in the Bone Marrow. Metastases are formed through a series of inter-connected steps that lead to successful tumor outgrowth at distant organs. Tumor cells escaping from primary site intravasate into the circulation through single cell movement or collective cell movement. In blood circulation, platelets coagulate around CTCs to provide physical protection and prevent immune cell attack. Homing of tumor cells at distant bone appears to be regulated by CXCL12 secreted from perivascular niche cells. Studies suggest that cancer cells within the bone can maintain a mesenchymal and indolent state called "tumor dormancy", regulated by high TSP1, TGF-β signaling, and LIFR among other niche factors. Dormant tumor cells express WNT inhibitor DKK1 to decrease ULBP ligands expression to suppress functions of NK cells. Upon interactions with osteoblasts through JAG1–NOTCH and N-E-cadherin, tumor cells utilize osteoblastic niche to survive and proliferate. BMP4 and BMP7 could induce dormancy to inhibit this progress, while COCO can wake up the dormant tumor cells by inhibiting BMP4. After tumor cell expansion, tumor cells induce RANKL secretion by osteoblasts. This stimulates the osteoclast maturation and promotes the bone matrix degradation to release calcium and growth factors such as TGF-β, which in turn feeds back to tumor cells to promote the survival and proliferation of metastatic tumor cells, generating an eventual irreversible "vicious cycle" of late stage bone metastasis

pathways for their low proliferation rate with that of HSCs. For example, TGF-β family ligand BMP4 (bone morphogenetic protein-4) keep metastatic breast cancer quiescent while BMP ligands antagonist, Coco wakes up tumor cells to form metastasis [86, 87]. Treatment of BMP7 in a murine prostate bone metastasis model significantly increased the survival time, accompanied with increased p38 and reduced ERK signaling in tumor cells [88, 89]. TGF-β2, endothelial-derived thrombospondin 1 (TSP1) and leukemia inhibitory factor receptor (LIFR) are other niche factors to keep breast cancer cell dormant in perivascular niches [90–92].

Being quiescent at the distant site, these tumor cells are under constant immune-surveillance. Speculations exist that some dormant colonies are in "immune mediated dormant" state, in which some tumor cells are recognized and eliminated by immune cells, while others within the same colony are proliferating slowly. The overall outcome is the relative stable cell population within the same colony [93]. Although less is experimentally demonstrated for immune dormancy, recent studies suggest nature killer (NK) cells might be critical in monitoring dormant tumor cells. By expressing WNT inhibitor DKK1, dormant tumor cells enter a slow proliferating state with decreased expression of ULBP ligands for NK cells, leading to immune evasion of NK cells. These tumor cells thus keep their quiescence and remain dormant [94]. Disseminated pancreatic cancer cells in mouse liver and latent DTCs from patients utilize MHC-I down-regulation as an evasion mechanism to avoid T cell surveillance [95, 96]. Because a long latency exists for dormant tumor cells before developing a full-blown metastasis, there is an apparent therapeutic window for intervention. In theory, two possible but drastically different strategies could be considered. First, "quiescent" niche factors can be disrupted to wake up DTCs. Active proliferating DTCs can be eliminated with additional therapies including chemotherapies and targeted therapies. Alternatively, it might be better to keep these cells quiescent. For example, through normalizing the immune system to prevent outgrowth of the dormant DTCs [97]. Immune dormant tumor cells could be waked up when T cells were depleted or major tumor killing cytokine was blocked, suggesting keeping the adaptive immune system functional could be critical to maintain the tumor dormancy [98].

3.2   Targeting Metastasis

Once the metastases established in secondary organs, it becomes extremely difficult to treat. Besides aforementioned approved therapies, another palliative but not curable treatment is to utilize radioactive substrates that emit α-particles (since α-particles are high energy particles that kill tumor cells but spare the surrounding tissues because of their short travelling distance [99]. For example, clinical trials demonstrate the usage of Radium-223 benefits patients with metastatic prostate cancer [100]. To make sure the emitted high energy particles only destroy tumor tissues but no other healthy organs, targeted alpha therapy is also developed by linking radionuclide with an antibody against specific protein which is either highly expressed on tumor cells or on resident cells of target organ [101]. Similar strategy by linking chemotherapeutic agent with antibody was also developed.

In 2019, FDA approved a first-in-class drug for residual invasive HER2-positive patients-- ado- trastuzumab emtansine (also known as Trastuzumab-emtansine, prescribed under the trade name Kadcyla) [102]. It consists of a humanized monoclonal antibody trastuzumab and a covalently linked cytotoxic agent mertansine (DM1). In this case, trastuzumab specifically binds to and inhibits HER2 receptor tyrosine kinase signaling on breast cancer cells; meanwhile, its binding with HER2 receptor enriches the local concentration and facilitates the internalization of ado-trastuzumab emtansine into tumor cells for more effective cytotoxic tumor killing [103].

Another recent therapy development for metastatic disease is radiofrequency thermal ablation (RFA) [104]. This is a minimally invasive, live-imaging guided therapy for cancers or metastasis arising from liver and few other organs. During the operation, tumor region is revealed by ultrasound, computed tomography or other techniques, a fine needle electrode front is inserted into tumor tissue. Physical ionic vibration at the tip generates a refined heating area and kills tumor cells within. This technique has been widely adopted in treating patients with local and metastatic diseases [105]. For example, a clinical study revealed benefits of RFA procedure in treating liver metastases of colorectal cancer. Patients with combined RFA treatment and chemotherapy had significant better progression-free survival than those with chemotherapy alone [104].

4 Systemic Changes in Metastasis: Metabolic Changes

Metabolism is a biochemical reaction network that converts nutrients into small molecule metabolites [106]. About a century ago, metabolic reprogramming has become a hallmark of cancer, and aerobic glycolysis (known as Warburg effect) is probably the most typical example [107–109]. Recently, the advances in new analytical technologies have propelled a renaissance of research on metabolic phenotypes and their underlying mechanisms in cancer [110, 111]. Indeed, the past decade has witnessed significant progress in our understanding of reprogrammed cancer metabolism, which is distinct from the adjacent normal tissues. Cancer cells reprogram the core cellular metabolism to provide support for the three needs of dividing cells: the rapid generation of bioenergetics (ATP), increased macromolecule biosynthesis, and precise maintenance of redox homeostasis [112]. Specific dysregulated pathways such as aerobic glycolysis, glutaminolysis, biosynthesis of macromolecules, and redox balance have been shown to support the tumorigenesis, growth, and proliferation in cancer [113, 114]. These pathways are regulated by oncogenic signaling networks and can often be exploited as targets to treat cancer [106].

Metastatic cancer cells usually undergo cell death as a result of metabolic stresses including reduced glucose uptake, ATP depletion and increased oxidative stress [115]. On the other hand, rapid proliferating cells consume nutrient and deplete fuels such as glucose and glutamine for immune cells, which often lead to compromised immune surveillance on cancer cells [116]. It has become increasingly clear that metabolic reprogramming is also crucial to the metastasis progression [117]. Cancer cells may acquire metabolic dysregulations that either make themselves more invasive or alter the microenvironment that could facilitate the metastasis [106]. The idea of selectively targeting the metabolic vulnerabilities during metastasis has provided increasing promise in new therapy development [118]. Some inhibitors targeting these metabolic vulnerabilities or dependencies have been proved or are in clinical trials to treat various cancers [119]. Previous studies show that all three major functions of metabolic network could be reprogrammed during the metastasis process. In this section, we review the progress of cancer metabolism-related to metastasis in recent years.

4.1 Bioenergetics

In particular, this occurs in brain metastases because brain itself is the most energy-consuming organ. Brain metastatic cells may gain the metabolic adaptation to survive the glucose deficiency environment by utilizing other sources of bioenergy such as acetate [120], glutamine [121], ketone bodies [122], and short-chain fatty acids [123]. For example, Mashimo et al. showed that brain metastases could convert acetate to acetyl-CoA to fuel the tricarboxylic acid (TCA) cycle by up-regulating acetyl-CoA synthetase enzyme 2. Moreover, by upregulating the fructose-1,6- bisphosphatase 2, a key enzyme in the gluconeogenesis pathway, metastatic cells can even gain the capability to produce glucose for their own use [123]. During lung metastasis of breast cancer, proline cycle was used to produce ATP at the expense of NADPH [124]. Breast cancer cells also increased PGC-1α activity to enhance their bioenergetics flexibility to metastasize to lungs [125]. These metabolic changes rely on the mitochondria to fuel energy needs. However, colon cancers metastasizing to the liver depend on the extracellular metabolic energetics [126]. It may be that extracellular ATP provides a short time window for tumor cells to ramping-up the cell-intrinsic ATP production [117].

4.2   Biosynthesis in Metastasis

Metabolic adaption may contribute to epithelial-mesenchymal transition (EMT), which plays a critical role in cell migration, invasion and metastasis [106, 127]. Glucose is the most important metabolic resource for primary cancer cells. The influence of glycolysis on metastases formation has been extensively studied and comprehensively reviewed [117]. Glycogen [128], amino acids such as proline [124] and asparagine [129] also promote cancer metastasis. In lung cancer, a recent study showed that the up-regulation of uridine 5′-diphosphate (UDP)–glucose 6- dehydrogenase (UGDH) depletes UDP-glucose. This led to the enhanced expression of the EMT inducing transcription factor SNAIL and the eventual metastasis in mice [130]. Based on the discovered specific metabolic dependencies, it is possible that lung cancer metastasis can be treated by phosphoglycerate mutase 1 (PGAM1) inhibitors [131], breast cancer metastasis can be inhibited by aldo-keto reductase AKR1B10 [132], while colorectal cancer metastasis may be treated with aldolase B (ALDOB) inhibitor to suppress fructose usage [133].

Recent studies also indicate that some metastatic cancer cells prefer fatty acids over other fuels [106]. TGF-β2 induced by acidosis could promote EMT and facilitate the uptake of fatty acid to support the metastasis [134]. Triple-negative breast cancer depends on fatty acid oxidation to maintain aberrant Src activity, which promotes metastasis [135]. Dietary lipids can drive metastasis by supporting the growth of metastasis initiating cells expressing high levels of the fatty acid receptor CD36 in human melanoma and breast cancer [136]. More recently, it was shown in mouse tumor models that cancer cells undergone a metabolic shift toward fatty acid oxidation to survive during lymph node metastasis. This was because yes-associated protein (YAP) was selectively activated in lymph node metastases and thereby led to the induction

of fatty acid oxidation. Pharmacological inhibition of fatty acid oxidation using etomoxir or deletion of YAP from the cancer cells inhibited lymph node metastasis [137].

4.3   Redox Hemostasis in Metastasis

During invasion and blood circulation, reduced cell-matrix interaction would result in ROS-accumulation and anoikis [138]. Thus, metastatic cells usually produce more reduced nicotinamide adenine dinucleotide phosphate (NADPH) to regenerate ROS-detoxifying metabolites, such as glutathione. Cultured cell spheroids can transfer the NADPH from cytosol into mitochondria to enhance cell growth [139]. In vivo, melanoma cells adapted reversible metabolic changes in order to survive the hostile environment during metastasis, such as exposure to ROS [140]. By upregulating ALDH1L2, a key NADPH generating enzyme in the folate pathway, metastatic cells generated more NADPH to cope with the increased ROS during metastasis. The increased dependency upon the metabolic pathway that confer oxidative stress resistance in metastatic cancer cells may provide a target for treating metastasis. In fact, the FDA approved agent methotrexate, an inhibitor to the rate-limiting enzyme dihydrofolate reductase (DHFR) in the folate pathway, can suppress metastasis progression [140]. Others also provide evidence that suppressing oxidative stress promotes metastasis in genetically modified mouse models [141–144]. These studies indicate that the inhibition of ROS-detoxification by utilizing the clinically approved agents may prevent the formation of the metastasis and therefore possibly render more cancer cases curable. It is also worth noting that the metabolic heterogeneity within primary tumor can regulate both organotropism and the overall metastatic efficiency [106]. For example, in a melanoma mouse model, inhibiting OXPHOS reduces metastasis to brain, but not to lung, possibly due to the enriched expression of OXPHOS in these tumor cells during brain metastasis [145]. Very recently, a study shows that melanoma cells expressing monocarboxylate transporter–1 (MCT1) defines a subpopulation of cells with high metastatic efficiency because of their hyper-activated oxidative pentose phosphate pathway and suppressed production of reactive oxygen species [146].

4.4 Systemic Metabolism Influences on Tumor Progression

Distant organs for metastasis are enriched of different nutrients. Metastatic cancer cells might rewire their metabolic program to generate efficient metastatic outgrowth in the new environment. For instance, in contrast to primary breast cancer, lung metastases increase pyruvate carboxylase activity to fully benefit from pyruvate-rich lung environment [147]. In a lack of serine and glycine brain environment, brain metastases are characterized by increased serine biosynthesis [148]. Recently, a study showed that exosomes containing miR-122 released by primary breast tumors could contribute to a permissive premetastatic niche through suppressing glucose metabolism of resident cells [149]. The relationship between metastasis cells and metabolic changes in distant organs need to further investigated.

Beyond metabolic changes in cancer cells or organs, overall metabolic state may affect tumor growth and metastasis. In recent years, modulation of amino acid composition through dietary interventions have been showed to dictate tumor progression or metastasis [150]. In some tumor models, the serine/glycine restricted diet has been found to increase serine biosynthesis, OXPHOS and ROS, which lead to decreased tumor growth [151]. However, CD8$^+$ T cell functions depend on these nutrients, which complicated the net-effect of these nutrition-restricted diet on tumor progression [152]. Similarly, methionine-restricted diet affects CD4$^+$ T cells proliferation and function [153], and the tumor growth will be inhibited [154]. Additional examples include decreased breast cancer metastasis caused by asparagine restriction and enhanced colorectal and breast cancer metastasis caused by creatine [129]

[155]. Finally, global caloric restriction could remodel PDAC lipid metabolism to constrain PDAC progression [156]. Together, these findings build a strong case that systemic metabolism through dietary interventions influence tumor progression and metastasis and provide possible pharmacological intervention windows for treating cancer.

5 Conclusions and Future Perspectives

Metastasis is still the major lethal threat to cancer patients. With a few examples of therapies against metastasis in hand, we still remind us that most therapies for metastatic patients are palpable but not curable. We hope that with the continued study for detailed molecular mechanism from different aspects, including tumor-stromal interactions, immune regulations, and metabolic reprograming, novel molecule targets and metastasis specific therapies will be developed to extend the patient survival and possible cure in the future.

Acknowledgements We apologize to researchers whose studies are unable to be covered in this chapter due to space limitation. We thank Zheng lab members for reading the draft and providing advises. The research in the lab is in part supported by the National Key Research and Development Program of China (2020YFA0509400), the National Science Foundation of China (81772981 and 81972462), and Tsinghua-Peking Center for Life Sciences.

References

1. Wild CP, Weiderpass E, Stewart BW (2020). World Cancer Report: Cancer Research for Cancer Prevention.
2. Fidler, I.J.: Metastasis: quantitative analysis of distribution and fate of tumor emboli labeled with 125 I-5-iodo-2'-deoxyuridine. J Natl Cancer Inst. **45**, 773–782 (1970)
3. Chambers, A.F., Groom, A.C., MacDonald, I.C.: Dissemination and growth of cancer cells in metastatic sites. Nat Rev Cancer. **2**, 563–572 (2002)
4. Steeg, P.S.: Targeting metastasis. Nature Reviews Cancer. **16**, 201 (2016)

5. Mundy, G.R.: Metastasis to bone: causes, consequences and therapeutic opportunities. Nature reviews Cancer. **2**, 584–593 (2002)

6. Weilbaecher, K.N., Guise, T.A., McCauley, L.K.: Cancer to bone: a fatal attraction. Nature reviews Cancer. **11**, 411–425 (2011)

7. Ren, G., Esposito, M., Kang, Y.: Bone metastasis and the metastatic niche. Journal of molecular medicine. **93**, 1203–1212 (2015)

8. Paget, S.: The distribution of secondary growths in cancer of the breast. Lancet. **133**, 3 (1889)

9. Ell, B., Kang, Y.: SnapShot: Bone Metastasis. Cell. **151**(690-690), e691 (2012)

10. Kang, Y.: Dissecting Tumor-Stromal Interactions in Breast Cancer Bone Metastasis. Endocrinol Metab (Seoul). **31**, 206–212 (2016)

11. Waning, D.L., Guise, T.A.: Molecular mechanisms of bone metastasis and associated muscle weakness. Clin Cancer Res. **20**, 3071–3077 (2014)

12. Juppner, H., Abou-Samra, A.B., Freeman, M., Kong, X.F., Schipani, E., Richards, J., Kolakowski Jr., L.F., Hock, J., Potts Jr., J.T., Kronenberg, H.M., et al.: A G protein- linked receptor for parathyroid hormone and parathyroid hormone-related peptide. Science. **254**, 1024–1026 (1991)

13. Sethi, N., Dai, X., Winter, C.G., Kang, Y.: Tumor-derived JAGGED1 promotes osteolytic bone metastasis of breast cancer by engaging notch signaling in bone cells. Cancer Cell. **19**, 192–205 (2011)

14. Kohno, N., Aogi, K., Minami, H., Nakamura, S., Asaga, T., Iino, Y., Watanabe, T., Goessl, C., Ohashi, Y., Takashima, S.: Zoledronic acid significantly reduces skeletal complications compared with placebo in Japanese women with bone metastases from breast cancer: a randomized, placebo-controlled trial. J Clin Oncol. **23**, 3314–3321 (2005)

15. Baron, R., Ferrari, S., Russell, R.G.: Denosumab and bisphosphonates: different mechanisms of action and effects. Bone. **48**, 677–692 (2011)

16. Fizazi, K., Carducci, M., Smith, M., Damiao, R., Brown, J., Karsh, L., Milecki, P., Shore, N., Rader, M., Wang, H., et al.: Denosumab versus zoledronic acid for treatment of bone metastases in men with castration-resistant prostate cancer: a randomised, double-blind study. Lancet. **377**, 813–822 (2011)

17. Stopeck, A.T., Lipton, A., Body, J.J., Steger, G.G., Tonkin, K., de Boer, R.H., Lichinitser, M., Fujiwara, Y., Yardley, D.A., Viniegra, M., et al.: Denosumab compared with zoledronic acid for the treatment of bone metastases in patients with advanced breast cancer: a randomized, double-blind study. J Clin Oncol. **28**, 5132–5139 (2010)

18. Durgeau, A., Virk, Y., Corgnac, S., Mami-Chouaib, F.: Recent Advances in Targeting CD8 T-Cell Immunity for More Effective Cancer Immunotherapy. Front Immunol. **9**, 14 (2018)

19. van der Leun, A.M., Thommen, D.S., Schumacher, T.N.: CD8(+) T cell states in human cancer: insights from single-cell analysis. Nat Rev Cancer. **20**, 218–232 (2020)

20. Vinay, D.S., Ryan, E.P., Pawelec, G., Talib, W.H., Stagg, J., Elkord, E., Lichtor, T., Decker, W.K., Whelan, R.L., Kumara, H., et al.: Immune evasion in cancer: Mechanistic basis and therapeutic strategies. Semin Cancer Biol. **35**(Suppl), S185–S198 (2015)

21. Chamoto, K., Hatae, R., Honjo, T.: Current issues and perspectives in PD-1 blockade cancer immunotherapy. Int J Clin Oncol. (2020)

22. Chen, L., Flies, D.B.: Molecular mechanisms of T cell co-stimulation and co- inhibition. Nat Rev Immunol. **13**, 227–242 (2013)

23. Sharma, P., Allison, J.P.: Dissecting the mechanisms of immune checkpoint therapy. Nat Rev Immunol. **20**, 75–76 (2020)

24. Akinleye, A., Rasool, Z.: Immune checkpoint inhibitors of PD-L1 as cancer therapeutics. J Hematol Oncol. **12**, 92 (2019)

25. Hodi, F.S., O'Day, S.J., McDermott, D.F., Weber, R. W., Sosman, J.A., Haanen, J.B., Gonzalez, R., Robert, C., Schadendorf, D., Hassel, J.C., et al.: Improved survival with ipilimumab in patients with metastatic melanoma. N Engl J Med. **363**, 711–723 (2010)

26. McDermott, D., Lebbe, C., Hodi, F.S., Maio, M., Weber, J.S., Wolchok, J.D., Thompson, J.A., Balch, C.M.: Durable benefit and the potential for long-term survival with immunotherapy in advanced melanoma. Cancer Treat Rev. **40**, 1056–1064 (2014)

27. Wolchok, J.D., Chiarion-Sileni, V., Gonzalez, R., Rutkowski, P., Grob, J.J., Cowey, C.L., Lao, C.D., Wagstaff, J., Schadendorf, D., Ferrucci, P.F., et al.: Overall Survival with Combined Nivolumab and Ipilimumab in Advanced Melanoma. N Engl J Med. **377**, 1345–1356 (2017)

28. Hara, H., Fukuoka, S., Takahashi, N., Kojima, T., Kawazoe, A., Asayama, M., Yoshii, T., Kotani, D., Tamura, H., Mikamoto, Y., et al.: Regorafenib plus nivolumab in patients with advanced colorectal or gastric cancer: an open-label, dose-finding, and dose-expansion phase 1b trial (REGONIVO, EPOC1603). Ann Oncol. **30 Suppl 4**, iv124 (2019)

29. Brown, C.E., Mackall, C.L.: CAR T cell therapy: inroads to response and resistance. Nat Rev Immunol. **19**, 73–74 (2019)

30. Fesnak, A.D., June, C.H., Levine, B.L.: Engineered T cells: the promise and challenges of cancer immunotherapy. Nat Rev Cancer. **16**, 566–581 (2016)

31. Finn, O.J.: Cancer vaccines: between the idea and the reality. Nat Rev Immunol. **3**, 630–641 (2003)

32. Sahin, U., Tureci, O.: Personalized vaccines for cancer immunotherapy. Science. **359**, 1355–1360 (2018)

33. Ott, P.A., Hu, Z., Keskin, D.B., Shukla, S.A., Sun, J., Bozym, D.J., Zhang, W., Luoma, A., Giobbie-Hurder, A., Peter, L., et al.: An immunogenic personal neoantigen vaccine for patients with melanoma. Nature. **547**, 217–221 (2017)

34. Sahin, U., Derhovanessian, E., Miller, M., Kloke, B. P., Simon, P., Lower, M., Bukur, V., Tadmor, A.D., Luxemburger, U., Schrors, B., et al.: Personalized RNA mutanome vaccines mobilize poly-specific therapeutic immunity against cancer. Nature. **547**, 222–226 (2017)

35. Schumacher, T.N., Schreiber, R.D.: Neoantigens in cancer immunotherapy, vol. 348, pp. 69–74. Science (2015)

36. Hong, M.K., Macintyre, G., Wedge, D.C., Van Loo, P., Patel, K., Lunke, S., Alexandrov, L.B., Sloggett, C., Cmero, M., Marass, F., et al.: Tracking the origins and drivers of subclonal metastatic expansion in prostate cancer. Nat Commun. **6**, 6605 (2015)

37. Vermaat, J.S., Nijman, I.J., Koudijs, M.J., Gerritse, F. L., Scherer, S.J., Mokry, M., Roessingh, W.M., Lansu, N., de Bruijn, E., van Hillegersberg, R., et al.: Primary colorectal cancers and their subsequent hepatic metastases are genetically different: implications for selection of patients for targeted treatment. Clin Cancer Res. **18**, 688–699 (2012)

38. Robinson, D.R., Wu, Y.M., Vats, P., Su, F., Lonigro, R.J., Cao, X., Kalyana-Sundaram, S., Wang, R., Ning, Y., Hodges, L., et al.: Activating ESR1 mutations in hormone-resistant metastatic breast cancer. Nat Genet. **45**, 1446–1451 (2013)

39. Toy, W., Shen, Y., Won, H., Green, B., Sakr, R.A., Will, M., Li, Z., Gala, K., Fanning, S., King, T.A., et al.: ESR1 ligand-binding domain mutations in hormone-resistant breast cancer. Nat Genet. **45**, 1439–1445 (2013)

40. Gonzalez, T.L., Hancock, M., Sun, S., Gersch, C.L., Larios, J.M., David, W., Hu, J., Hayes, D.F., Wang, S., Rae, J.M.: Targeted degradation of activating estrogen receptor alpha ligand-binding domain mutations in human breast cancer. Breast Cancer Res Treat. **180**, 611–622 (2020)

41. Priestley, P., Baber, J., Lolkema, M.P., Steeghs, N., de Bruijn, E., Shale, C., Duyvesteyn, K., Haidari, S., van Hoeck, A., Onstenk, W., et al.: Pan-cancer whole-genome analyses of metastatic solid tumours. Nature. **575**, 210–216 (2019)

42. Ferrara, N., Hillan, K.J., Gerber, H.P., Novotny, W.: Discovery and development of bevacizumab, an anti-VEGF antibody for treating cancer. Nat Rev Drug Discov. **3**, 391–400 (2004)

43. McCormack, P.L., Keam, S.J.: Bevacizumab: a review of its use in metastatic colorectal cancer. Drugs. **68**, 487–506 (2008)

44. Ebos, J.M., Lee, C.R., Cruz-Munoz, W., Bjarnason, G.A., Christensen, J.G., Kerbel, R.S.: Accelerated metastasis after short-term treatment with a potent inhibitor of tumor angiogenesis. Cancer Cell. **15**, 232–239 (2009)

45. Loges, S., Mazzone, M., Hohensinner, P., Carmeliet, P.: Silencing or fueling metastasis with VEGF inhibitors: antiangiogenesis revisited. Cancer Cell. **15**, 167–170 (2009)

46. Nguyen, D.X., Bos, P.D., Massague, J.: Metastasis: from dissemination to organ- specific colonization. Nat Rev Cancer. **9**, 274–284 (2009)

47. Obenauf, A.C., Massague, J.: Surviving at a Distance: Organ-Specific Metastasis. Trends Cancer. **1**, 76–91 (2015)

48. Lambert, A.W., Pattabiraman, D.R., Weinberg, R.A.: Emerging Biological Principles of Metastasis. Cell. **168**, 670–691 (2017)

49. Massague, J., Obenauf, A.C.: Metastatic colonization by circulating tumour cells. Nature. **529**, 298–306 (2016)

50. Miller, K.D., Nogueira, L., Mariotto, A.B., Rowland, J.H., Yabroff, K.R., Alfano, C.M., Jemal, A., Kramer, J.L., Siegel, R.L.: Cancer treatment and survivorship statistics, 2019. CA Cancer J Clin. **69**, 363–385 (2019)

51. Hosseini, H., Obradovic, M.M.S., Hoffmann, M., Harper, K.L., Sosa, M.S., Werner-Klein, M., Nanduri, L.K., Werno, C., Ehrl, C., Maneck, M., et al.: Early dissemination seeds metastasis in breast cancer. Nature. **540**, 552–558 (2016)

52. Hu, Z., Ding, J., Ma, Z., Sun, R., Seoane, J.A., Scott Shaffer, J., Suarez, C.J., Berghoff, A.S., Cremolini, C., Falcone, A., et al.: Quantitative evidence for early metastatic seeding in colorectal cancer. Nat Genet. **51**, 1113–1122 (2019)

53. Pantel, K., Alix-Panabieres, C.: The potential of circulating tumor cells as a liquid biopsy to guide therapy in prostate cancer. Cancer Discov. **2**, 974–975 (2012)

54. Pantel, K., Deneve, E., Nocca, D., Coffy, A., Vendrell, J.P., Maudelonde, T., Riethdorf, S., Alix-Panabieres, C.: Circulating epithelial cells in patients with benign colon diseases. Clin Chem. **58**, 936–940 (2012)

55. Rhim, A.D., Mirek, E.T., Aiello, N.M., Maitra, A., Bailey, J.M., McAllister, F., Reichert, M., Beatty, G. L., Rustgi, A.K., Vonderheide, R.H., et al.: EMT and dissemination precede pancreatic tumor formation. Cell. **148**, 349–361 (2012)

56. Dongre, A., Weinberg, R.A.: New insights into the mechanisms of epithelial- mesenchymal transition and implications for cancer. Nat Rev Mol Cell Biol. **20**, 69–84 (2019)

57. Friedl, P., Locker, J., Sahai, E., Segall, J.E.: Classifying collective cancer cell invasion. Nat Cell Biol. **14**, 777–783 (2012)

58. Bonadonna, G., Rossi, A., Valagussa, P.: Adjuvant CMF chemotherapy in operable breast cancer: ten years later. Lancet. **1**, 976–977 (1985)

59. Foss, A., Munoz-Sagredo, L., Sleeman, J., Thiele, W.: The contribution of platelets to intravascular arrest, extravasation, and outgrowth of disseminated tumor cells. Clin Exp Metastasis. **37**, 47–67 (2020)

60. Padua, D., Zhang, X.H., Wang, Q., Nadal, C., Gerald, W.L., Gomis, R.R., Massague, J.: TGFbeta primes breast tumors for lung metastasis seeding through angiopoietin-like 4. Cell. **133**, 66–77 (2008)

61. Goddard, E.T., Bozic, I., Riddell, S.R., Ghajar, C.M.: Dormant tumour cells, their niches and the influence of immunity. Nat Cell Biol. **20**, 1240–1249 (2018)

62. Hoshino, A., Costa-Silva, B., Shen, T.L., Rodrigues, G., Hashimoto, A., Tesic Mark, M., Molina, H., Kohsaka, S., Di Giannatale, A., Ceder, S., et al.: Tumour exosome integrins determine organotropic metastasis. Nature. **527**, 329–335 (2015)

63. Murgai, M., Ju, W., Eason, M., Kline, J., Beury, D. W., Kaczanowska, S., Miettinen, M.M., Kruhlak, M., Lei, H., Shern, J.F., et al.: KLF4-dependent perivascular cell plasticity mediates pre-metastatic niche formation and metastasis. Nat Med. **23**, 1176–1190 (2017)

64. Peinado, H., Aleckovic, M., Lavotshkin, S., Matei, I., Costa-Silva, B., Moreno-Bueno, G., Hergueta-Redondo, M., Williams, C., Garcia-Santos, G., Ghajar, C., et al.: Melanoma exosomes educate bone marrow progenitor cells toward a pro-metastatic phenotype through MET. Nat Med. **18**, 883–891 (2012)

65. Rodrigues, G., Hoshino, A., Kenific, C.M., Matei, I. R., Steiner, L., Freitas, D., Kim, H.S., Oxley, P.R., Scandariato, I., Casanova-Salas, I., et al.: Tumour exosomal CEMIP protein promotes cancer cell colonization in brain metastasis. Nat Cell Biol. **21**, 1403–1412 (2019)

66. Ruppender, N.S., Morrissey, C., Lange, P.H., Vessella, R.L.: Dormancy in solid tumors: implications for prostate cancer. Cancer Metastasis Rev. **32**, 501–509 (2013)

67. Zhang, X.H., Giuliano, M., Trivedi, M.V., Schiff, R., Osborne, C.K.: Metastasis dormancy in estrogen receptor-positive breast cancer. Clin Cancer Res. **19**, 6389–6397 (2013)

68. Aslakson, C.J., Miller, F.R.: Selective events in the metastatic process defined by analysis of the sequential dissemination of subpopulations of a mouse mammary tumor. Cancer Res. **52**, 1399–1405 (1992)

69. Kang, Y., Siegel, P.M., Shu, W., Drobnjak, M., Kakonen, S.M., Cordon-Cardo, C., Guise, T.A., Massague, J.: A multigenic program mediating breast cancer metastasis to bone. Cancer Cell. **3**, 537–549 (2003)

70. Lu, X., Mu, E., Wei, Y., Riethdorf, S., Yang, Q., Yuan, M., Yan, J., Hua, Y., Tiede, B.J., Lu, X., et al.: VCAM-1 promotes osteolytic expansion of indolent bone micrometastasis of breast cancer by engaging alpha4beta1-positive osteoclast progenitors. Cancer Cell. **20**, 701–714 (2011)

71. Linde, N., Fluegen, G., Aguirre-Ghiso, J.A.: The Relationship Between Dormant Cancer Cells and Their Microenvironment. Adv Cancer Res. **132**, 45–71 (2016)

72. Aguirre-Ghiso, J.A.: Models, mechanisms and clinical evidence for cancer dormancy. Nat Rev Cancer. **7**, 834–846 (2007)

73. Harper, K.L., Sosa, M.S., Entenberg, D., Hosseini, H., Cheung, J.F., Nobre, R., Avivar-Valderas, A., Nagi, C., Girnius, N., Davis, R.J., et al.: Mechanism of early dissemination and metastasis in Her2(+) mammary cancer. Nature. **540**, 588–592 (2016)

74. Schlimok, G., Funke, I., Holzmann, B., Gottlinger, G., Schmidt, G., Hauser, H., Swierkot, S., Warnecke, H.H., Schneider, B., Koprowski, H., et al.: Micrometastatic cancer cells in bone marrow: in vitro detection with anti-cytokeratin and in vivo labeling with anti-17-1A monoclonal antibodies. Proc Natl Acad Sci U S A. **84**, 8672–8676 (1987)

75. Celia-Terrassa, T., Kang, Y.: Metastatic niche functions and therapeutic opportunities. Nat Cell Biol. **20**, 868–877 (2018)

76. Ghajar, C.M.: Metastasis prevention by targeting the dormant niche. Nat Rev Cancer. **15**, 238–247 (2015)

77. Shiozawa, Y., Pedersen, E.A., Havens, A.M., Jung, Y., Mishra, A., Joseph, J., Kim, J.K., Patel, L.R., Ying, C., Ziegler, A.M., et al.: Human prostate cancer metastases target the hematopoietic stem cell niche to establish footholds in mouse bone marrow. J Clin Invest. **121**, 1298–1312 (2011)

78. Lawson, M.A., McDonald, M.M., Kovacic, N., Hua Khoo, W., Terry, R.L., Down, J., Kaplan, W., Paton-Hough, J., Fellows, C., Pettitt, J.A., et al.: Osteoclasts control reactivation of dormant myeloma cells by remodelling the endosteal niche. Nat Commun. **6**, 8983 (2015)

79. Wang, H., Tian, L., Liu, J., Goldstein, A., Bado, I., Zhang, W., Arenkiel, B.R., Li, Z., Yang, M., Du, S., et al.: The Osteogenic Niche Is a Calcium Reservoir of Bone Micrometastases and Confers Unexpected Therapeutic Vulnerability. Cancer Cell. **34**(823-839), e827 (2018)

80. Wang, H., Yu, C., Gao, X., Welte, T., Muscarella, A. M., Tian, L., Zhao, H., Zhao, Z., Du, S., Tao, J., et al.: The osteogenic niche promotes early-stage bone colonization of disseminated breast cancer cells. Cancer Cell. **27**, 193–210 (2015)

81. Zheng, H., Bae, Y., Kasimir-Bauer, S., Tang, R., Chen, J., Ren, G., Yuan, M., Esposito, M., Li, W., Wei, Y., et al.: Therapeutic Antibody Targeting Tumor- and Osteoblastic Niche-Derived Jagged1 Sensitizes Bone Metastasis to Chemotherapy. Cancer Cell. **32**(731-747), e736 (2017)

82. Ding, L., Morrison, S.J.: Haematopoietic stem cells and early lymphoid progenitors occupy distinct bone marrow niches. Nature. **495**, 231–235 (2013)

83. Morrison, S.J., Scadden, D.T.: The bone marrow niche for haematopoietic stem cells. Nature. **505**, 327–334 (2014)

84. Muller, A., Homey, B., Soto, H., Ge, N., Catron, D., Buchanan, M.E., McClanahan, T., Murphy, E., Yuan, W., Wagner, S.N., et al.: Involvement of chemokine receptors in breast cancer metastasis. Nature. **410**, 50–56 (2001)

85. Zlotnik, A., Burkhardt, A.M., Homey, B.: Homeostatic chemokine receptors and organ-specific metastasis. Nat Rev Immunol. **11**, 597–606 (2011)

86. Gao, H., Chakraborty, G., Lee-Lim, A.P., Mo, Q., Decker, M., Vonica, A., Shen, R., Brogi, E., Brivanlou, A.H., Giancotti, F.G.: The BMP inhibitor Coco reactivates breast cancer cells at lung metastatic sites. Cell. **150**, 764–779 (2012)

87. Wei, Q., Frenette, P.S.: Niches for Hematopoietic Stem Cells and Their Progeny. Immunity. **48**, 632–648 (2018)

88. Aguirre-Ghiso, J.A., Liu, D., Mignatti, A., Kovalski, K., Ossowski, L.: Urokinase receptor and fibronectin regulate the ERK (MAPK) to p38(MAPK) activity ratios that determine carcinoma cell proliferation or dormancy in vivo. Mol Biol Cell. **12**, 863–879 (2001)

89. Kobayashi, A., Okuda, H., Xing, F., Pandey, P.R., Watabe, M., Hirota, S., Pai, S.K., Liu, W., Fukuda, K., Chambers, C., et al.: Bone morphogenetic protein 7 in dormancy and metastasis of prostate cancer stem-like cells in bone. J Exp Med. **208**, 2641–2655 (2011)

90. Bragado, P., Estrada, Y., Parikh, F., Krause, S., Capobianco, C., Farina, H.G., Schewe, D.M., Aguirre-Ghiso, J.A.: TGF-beta2 dictates disseminated tumour cell fate in target organs through TGF-beta-RIII and p38alpha/beta signalling. Nat Cell Biol. **15**, 1351–1361 (2013)

91. Catena, R., Bhattacharya, N., El Rayes, T., Wang, S., Choi, H., Gao, D., Ryu, S., Joshi, N., Bielenberg, D., Lee, S.B., et al.: Bone marrow-derived Gr1+ cells can generate a metastasis-resistant microenvironment via induced secretion of thrombospondin-1. Cancer Discov. **3**, 578–589 (2013)

92. Johnson, R.W., Finger, E.C., Olcina, M.M., Vilalta, M., Aguilera, T., Miao, Y., Merkel, A.R., Johnson, J. R., Sterling, J.A., Wu, J.Y., et al.: Induction of LIFR confers a dormancy phenotype in breast cancer cells disseminated to the bone marrow. Nat Cell Biol. **18**, 1078–1089 (2016)

93. Romero, I., Garrido, F., Garcia-Lora, A.M.: Metastases in immune-mediated dormancy: a new opportunity for targeting cancer. Cancer Res. **74**, 6750–6757 (2014)

94. Malladi, S., Macalinao, D.G., Jin, X., He, L., Basnet, H., Zou, Y., de Stanchina, E., Massague, J.: Metastatic Latency and Immune Evasion through Autocrine Inhibition of WNT. Cell. **165**, 45–60 (2016)

95. Pantel, K., Schlimok, G., Kutter, D., Schaller, G., Genz, T., Wiebecke, B., Backmann, R., Funke, I., Riethmuller, G.: Frequent down-regulation of major histocompatibility class I antigen expression on individual micrometastatic carcinoma cells. Cancer Res. **51**, 4712–4715 (1991)

96. Pommier, A., Anaparthy, N., Memos, N., Kelley, Z. L., Gouronnec, A., Yan, R., Auffray, C., Albrengues, J., Egeblad, M., Iacobuzio-Donahue, C.A., et al.: Unresolved endoplasmic reticulum stress engenders immune-resistant, latent pancreatic cancer metastases. Science. **360** (2018)

97. Sanmamed, M.F., Chen, L.: A Paradigm Shift in Cancer Immunotherapy: From Enhancement to Normalization. Cell. **175**, 313–326 (2018)

98. Koebel, C.M., Vermi, W., Swann, J.B., Zerafa, N., Rodig, S.J., Old, L.J., Smyth, M.J., Schreiber, R.D.: Adaptive immunity maintains occult cancer in an equilibrium state. Nature. **450**, 903–907 (2007)

99. Buroni, F.E., Persico, M.G., Pasi, F., Lodola, L., Nano, R., Aprile, C.: Radium-223: Insight and Perspectives in Bone-metastatic Castration-resistant Prostate Cancer. Anticancer Res. **36**, 5719–5730 (2016)

100. Parker, C., Nilsson, S., Heinrich, D., Helle, S.I., O'Sullivan, J.M., Fossa, S.D., Chodacki, A., Wiechno, P., Logue, J., Seke, M., et al.: Alpha emitter radium-223 and survival in metastatic prostate cancer. N Engl J Med. **369**, 213–223 (2013)

101. Suominen, M.I., Wilson, T., Kakonen, S.M., Scholz, A.: The Mode-of-Action of Targeted Alpha Therapy Radium-223 as an Enabler for Novel Combinations to Treat Patients with Bone Metastasis. Int J Mol Sci. **20** (2019)

102. von Minckwitz, G., Huang, C.S., Mano, M.S., Loibl, S., Mamounas, E.P., Untch, M., Wolmark, N., Rastogi, P., Schneeweiss, A., Redondo, A., et al.: Trastuzumab Emtansine for Residual Invasive HER2-Positive Breast Cancer. N Engl J Med. **380**, 617–628 (2019)

103. Barok, M., Joensuu, H., Isola, J.: Trastuzumab emtansine: mechanisms of action and drug resistance. Breast Cancer Res. **16**, 209 (2014)

104. Yang, P.C., Lin, B.R., Chen, Y.C., Lin, Y.L., Lai, H. S., Huang, K.W., Liang, J.T.: Local Control by Radiofrequency Thermal Ablation Increased Overall Survival in Patients With Refractory Liver Metastases of Colorectal Cancer. Medicine (Baltimore). **95**, e3338 (2016)

105. Camacho, J.C., Petre, E.N., Sofocleous, C.T.: Thermal Ablation of Metastatic Colon Cancer to the Liver. Semin Intervent Radiol. **36**, 310–318 (2019)

106. Faubert, B., Solmonson, A., DeBerardinis, R.J.: Metabolic reprogramming and cancer progression, p. 368. Science (New York, NY (2020)

107. Hanahan, D., Weinberg, R.A.: Hallmarks of cancer: the next generation. Cell. **144**, 646–674 (2011)

108. Vander Heiden, M.G., Cantley, L.C., Thompson, C. B.: Understanding the Warburg effect: the metabolic requirements of cell proliferation. Science (New York, NY). **324**, 1029–1033 (2009)

109. Warburg, O., Wind, F., Negelein, E.: The Metabolism of Tumors in the Body. J Gen Physiol. **8**, 519–530 (1927)

110. Jang, C., Chen, L., Rabinowitz, J.D.: Metabolomics and Isotope Tracing. Cell. **173**, 822–837 (2018)

111. Kaushik, A.K., DeBerardinis, R.J.: Applications of metabolomics to study cancer metabolism. Biochim Biophys Acta Rev Cancer. **1870**, 2–14 (2018)

112. Cairns, R.A., Harris, I.S., Mak, T.W.: Regulation of cancer cell metabolism. Nat Rev Cancer. **11**, 85–95 (2011)

113. Shi, X., Tasdogan, A., Huang, F., Hu, Z., Morrison, S.J., DeBerardinis, R.J.: The abundance of

metabolites related to protein methylation correlates with the metastatic capacity of human melanoma xenografts. Sci Adv. **3**, eaao5268 (2017)

114. Vander Heiden, M.G., DeBerardinis, R.J.: Understanding the Intersections between Metabolism and Cancer Biology. Cell. **168**, 657–669 (2017)

115. Debnath, J., Brugge, J.S.: Modelling glandular epithelial cancers in three-dimensional cultures. Nat Rev Cancer. **5**, 675–688 (2005)

116. Kishton, R.J., Sukumar, M., Restifo, N.P.: Metabolic Regulation of T Cell Longevity and Function in Tumor Immunotherapy. Cell Metab. **26**, 94–109 (2017)

117. Elia, I., Doglioni, G., Fendt, S.M.: Metabolic Hallmarks of Metastasis Formation. Trends in cell biology. **28**, 673–684 (2018a)

118. Ubellacker, J.M., Morrison, S.J.: Metabolic Adaptation Fuels Lymph Node Metastasis. Cell Metab. **29**, 785–786 (2019)

119. DiNardo, C.D., Stein, E.M., de Botton, S., Roboz, G. J., Altman, J.K., Mims, A.S., Swords, R., Collins, R. H., Mannis, G.N., Pollyea, D.A., et al.: Durable Remissions with Ivosidenib in IDH1-Mutated Relapsed or Refractory AML. The New England journal of medicine. **378**, 2386–2398 (2018)

120. Mashimo, T., Pichumani, K., Vemireddy, V., Hatanpaa, K.J., Singh, D.K., Sirasanagandla, S., Nannepaga, S., Piccirillo, S.G., Kovacs, Z., Foong, C., et al.: Acetate is a bioenergetic substrate for human glioblastoma and brain metastases. Cell. **159**, 1603–1614 (2014)

121. Palmieri, E.M., Menga, A., Martin-Perez, R., Quinto, A., Riera-Domingo, C., De Tullio, G., Hooper, D.C., Lamers, W.H., Ghesquiere, B., McVicar, D.W., et al.: Pharmacologic or Genetic Targeting of Glutamine Synthetase Skews Macrophages toward an M1-like Phenotype and Inhibits Tumor Metastasis. Cell Rep. **20**, 1654–1666 (2017)

122. Wang, Y.H., Liu, C.L., Chiu, W.C., Twu, Y.C., Liao, Y.J.: HMGCS2 Mediates Ketone Production and Regulates the Proliferation and Metastasis of Hepatocellular Carcinoma. Cancers, 11 (2019b)

123. Chen, J., Lee, H.J., Wu, X., Huo, L., Kim, S.J., Xu, L., Wang, Y., He, J., Bollu, L.R., Gao, G., et al.: Gain of glucose-independent growth upon metastasis of breast cancer cells to the brain. Cancer Res. **75**, 554–565 (2015)

124. Elia, I., Broekaert, D., Christen, S., Boon, R., Radaelli, E., Orth, M.F., Verfaillie, C., Grunewald, T.G.P., Fendt, S.M.: Proline metabolism supports metastasis formation and could be inhibited to selectively target metastasizing cancer cells. Nat Commun. **8**, 15267 (2017a)

125. Andrzejewski, S., Klimcakova, E., Johnson, R.M., Tabaries, S., Annis, M.G., McGuirk, S., Northey, J. J., Chenard, V., Sriram, U., Papadopoli, D.J., et al.: PGC-1alpha Promotes Breast Cancer Metastasis and Confers Bioenergetic Flexibility against Metabolic Drugs. Cell Metab. **26**(778-787), e775 (2017)

126. Loo, J.M., Scherl, A., Nguyen, A., Man, F.Y., Weinberg, E., Zeng, Z.S., Saltz, L., Paty, P.B., Tavazoie, S.F.: Extracellular Metabolic Energetics Can Promote Cancer Progression. Cell. **160**, 393–406 (2015)

127. Jiang, L., Xiao, L., Sugiura, H., Huang, X., Ali, A., Kuro-o, M., Deberardinis, R.J., Boothman, D.A.: Metabolic reprogramming during TGFbeta1-induced epithelial-to- mesenchymal transition. Oncogene. **34**, 3908–3916 (2015)

128. Curtis, M., Kenny, H.A., Ashcroft, B., Mukherjee, A., Johnson, A., Zhang, Y., Helou, Y., Batlle, R., Liu, X., Gutierrez, N., et al.: Fibroblasts Mobilize Tumor Cell Glycogen to Promote Proliferation and Metastasis. Cell Metab. (2018)

129. Knott, S.R.V., Wagenblast, E., Khan, S., Kim, S.Y., Soto, M., Wagner, M., Turgeon, M.O., Fish, L., Erard, N., Gable, A.L., et al.: Asparagine bioavailability governs metastasis in a model of breast cancer. Nature. **554**, 378–381 (2018a)

130. Wang, X., Liu, R., Zhu, W., Chu, H., Yu, H., Wei, P., Wu, X., Zhu, H., Gao, H., Liang, J., et al.: UDP-glucose accelerates SNAI1 mRNA decay and impairs lung cancer metastasis. Nature. **571**, 127–131 (2019a)

131. Huang, K., Liang, Q., Zhou, Y., Jiang, L.L., Gu, W. M., Luo, M.Y., Tang, Y.B., Wang, Y., Lu, W., Huang, M., et al.: A Novel Allosteric Inhibitor of Phosphoglycerate Mutase 1 Suppresses Growth and Metastasis of Non-Small-Cell Lung Cancer. Cell Metab. **30**(1107-1119), e1108 (2019)

132. van Weverwijk, A., Koundouros, N., Iravani, M., Ashenden, M., Gao, Q., Poulogiannis, G., Jungwirth, U., Isacke, C.M.: Metabolic adaptability in metastatic breast cancer by AKR1B10-dependent balancing of glycolysis and fatty acid oxidation. Nat Commun. **10**, 2698 (2019)

133. Bu, P., Chen, K.Y., Xiang, K., Johnson, C., Crown, S.B., Rakhilin, N., Ai, Y., Wang, L., Xi, R., Astapova, I., et al.: Aldolase B-Mediated Fructose Metabolism Drives Metabolic Reprogramming of Colon Cancer Liver Metastasis. Cell Metab. **27**(1249-1262), e1244 (2018)

134. Corbet, C., Bastien, E., Santiago de Jesus, J.P., Dierge, E., Martherus, R., Vander Linden, C., Doix, B., Degavre, C., Guilbaud, C., Petit, L., et al.: TGFbeta2-induced formation of lipid droplets supports acidosis-driven EMT and the metastatic spreading of cancer cells. Nat Commun. **11**, 454 (2020)

135. Park, J.H., Vithayathil, S., Kumar, S., Sung, P.L., Dobrolecki, L.E., Putluri, V., Bhat, V.B., Bhowmik, S.K., Gupta, V., Arora, K., et al.: Fatty Acid Oxidation-Driven Src Links Mitochondrial Energy Reprogramming and Oncogenic Properties in

Triple-Negative Breast Cancer. Cell Rep. **14**, 2154–2165 (2016)

136. Pascual, G., Avgustinova, A., Mejetta, S., Martin, M., Castellanos, A., Attolini, C.S., Berenguer, A., Prats, N., Toll, A., Hueto, J.A., et al.: Targeting metastasis-initiating cells through the fatty acid receptor CD36. Nature. **541**, 41–45 (2017)

137. Lee, C.K., Jeong, S.H., Jang, C., Bae, H., Kim, Y.H., Park, I., Kim, S.K., Koh, G.Y.: Tumor metastasis to lymph nodes requires YAP-dependent metabolic adaptation. Science (New York, NY). **363**, 644–649 (2019)

138. Hawk, M.A., Schafer, Z.T.: Mechanisms of redox metabolism and cancer cell survival during extracellular matrix detachment. J Biol Chem. **293**, 7531–7537 (2018)

139. Jiang, L., Shestov, A.A., Swain, P., Yang, C., Parker, S.J., Wang, Q.A., Terada, L.S., Adams, N.D., McCabe, M.T., Pietrak, B., et al.: Reductive carboxylation supports redox homeostasis during anchorage-independent growth. Nature. **532**, 255–258 (2016)

140. Piskounova, E., Agathocleous, M., Murphy, M.M., Hu, Z., Huddlestun, S.E., Zhao, Z., Leitch, A.M., Johnson, T.M., DeBerardinis, R.J., Morrison, S.J.: Oxidative stress inhibits distant metastasis by human melanoma cells. Nature. **527**, 186–191 (2015)

141. Le Gal, K., Ibrahim, M.X., Wiel, C., Sayin, V.I., Akula, M.K., Karlsson, C., Dalin, M.G., Akyurek, L.M., Lindahl, P., Nilsson, J., et al.: Antioxidants can increase melanoma metastasis in mice. Science translational medicine. **7**, 308re308 (2015)

142. LeBleu, V.S., O'Connell, J.T., Gonzalez Herrera, K.N., Wikman, H., Pantel, K., Haigis, M.C., de Carvalho, F.M., Damascena, A., Domingos Chinen, L.T., Rocha, R.M., et al.: PGC-1alpha mediates mitochondrial biogenesis and oxidative phosphorylation in cancer cells to promote metastasis. Nat Cell Biol. **16**(992-1003), 1001–1015 (2014)

143. Wang, H., Liu, X., Long, M., Huang, Y., Zhang, L., Zhang, R., Zheng, Y., Liao, X., Wang, Y., Liao, Q., et al.: NRF2 activation by antioxidant antidiabetic agents accelerates tumor metastasis. Science translational medicine. **8**, 334ra351 (2016)

144. Wiel, C., Le Gal, K., Ibrahim, M.X., Jahangir, C.A., Kashif, M., Yao, H., Ziegler, D.V., Xu, X., Ghosh, T., Mondal, T., et al.: BACH1 Stabilization by Antioxidants Stimulates Lung Cancer Metastasis. Cell. **178**(330-345), e322 (2019)

145. Fischer, G.M., Jalali, A., Kircher, D.A., Lee, W.C., McQuade, J.L., Haydu, L.E., Joon, A.Y., Reuben, A., de Macedo, M.P., Carapeto, F.C.L., et al.: Molecular Profiling Reveals Unique Immune and Metabolic Features of Melanoma Brain Metastases. Cancer discovery. **9**, 628–645 (2019)

146. Tasdogan, A., Faubert, B., Ramesh, V., Ubellacker, J.M., Shen, B., Solmonson, A., Murphy, M.M., Gu, Z.,

Gu, W., Martin, M., et al.: Metabolic heterogeneity confers differences in melanoma metastatic potential. Nature. **577**, 115–120 (2020)

Additional References

147. Christen, S., Lorendeau, D., Schmieder, R., Broekaert, D., Metzger, K., Veys, K., Elia, I., Buescher, J.M., Orth, M.F., Davidson, S.M., et al.: Breast Cancer-Derived Lung Metastases Show Increased Pyruvate Carboxylase-Dependent Anaplerosis. Cell Reports. **17**, 837–848 (2016)

148. Ngo, B., Kim, E., Osorio-Vasquez, V., Doll, S., Bustraan, S., Liang, R.J., Luengo, A., Davidson, S.M., Ali, A., Ferraro, G.B., et al.: Limited Environmental Serine and Glycine Confer Brain Metastasis Sensitivity to PHGDH Inhibition. Cancer Discov. **10**, 1352–1373 (2020)

149. Fong, M.Y., Zhou, W.Y., Liu, L., Alontaga, A.Y., Chandra, M., Ashby, J., Chow, A., O'Connor, S.T.F., Li, S.S., Chin, A.R., et al.: Breast-cancer-secreted miR-122 reprograms glucose metabolism in premetastatic niche to promote metastasis. Nature Cell Biology. **17**, 183-+ (2015)

150. Elia, I., Haigis, M.C.: Metabolites and the tumour microenvironment: from cellular mechanisms to systemic metabolism. Nat Metab. **3**, 21–32 (2021)

151. Maddocks, O.D.K., Athineos, D., Cheung, E.C., Lee, P., Zhang, T., van den Broek, N.J.F., Mackay, G.M., Labuschagne, C.F., Gay, D., Kruiswijk, F., et al.: Modulating the therapeutic response of tumours to dietary serine and glycine starvation. Nature. **544**, 372–376 (2017)

152. Ma, E.H., Bantug, G., Griss, T., Condotta, S., Johnson, R.M., Samborska, B., Mainolfi, N., Suri, V., Guak, H., Balmer, M.L., et al.: Serine Is an Essential Metabolite for Effector T Cell Expansion. Cell Metab. **25**, 482 (2017)

153. Roy, D.G., Chen, J., Mamane, V., Ma, E.H., Muhire, B.M., Sheldon, R.D., Shorstova, T., Koning, R., Johnson, R.M., Esaulova, E., et al.: Methionine Metabolism Shapes T Helper Cell Responses through Regulation of Epigenetic Reprogramming. Cell Metab. **31**(250-266), e259 (2020)

154. Gao, X., Sanderson, S.M., Dai, Z., Reid, M.A., Cooper, D.E., Lu, M., Richie Jr., J.P., Ciccarella, A., Calcagnotto, A., Mikhael, P.G., et al.: Dietary methionine influences therapy in mouse cancer models and alters human metabolism. Nature. **572**, 397–401 (2019)

155. Zhang, L., Zhu, Z., Yan, H., Wang, W., Wu, Z., Zhang, F., Zhang, Q., Shi, G., Du, J., Cai, H., et al.: Creatine promotes cancer metastasis through activation of Smad2/3. Cell Metab. **33**(1111-1123), e1114 (2021)

156. Lien, E.C., Westermark, A.M., Zhang, Y., Yuan, C., Li, Z., Lau, A.N., Sapp, K.M., Wolpin, B.M., Vander Heiden, M.G.: Low glycaemic diets alter lipid metabolism to influence tumour growth. Nature. **599**, 302–307 (2021)

157. Elia, I., Broekaert, D., Christen, S., Boon, R., Radaelli, E., Orth, M.F., Verfaillie, C., Grunewald, T.G.P., Fendt, S.M.: Proline metabolism supports metastasis formation and could be inhibited to selectively target metastasizing cancer cells. Nat Commun. **8**, 15267 (2017b)

158. Elia, I., Doglioni, G., Fendt, S.M.: Metabolic Hallmarks of Metastasis Formation. Trends Cell Biol. **28**, 673–684 (2018b)

159. Knott, S.R.V., Wagenblast, E., Khan, S., Kim, S.Y., Soto, M., Wagner, M., Turgeon, M.O., Fish, L., Erard, N., Gable, A.L., et al.: Asparagine bioavailability governs metastasis in a model of breast cancer. Nature. **554**, 378–381 (2018b)

Part II

Cachexia: A Debilitating Systemic Effect of Cancer

Signaling Pathways That Promote Muscle Catabolism in Cachexia

J. E. Gilda and S. Cohen

Abstract

Cachexia is a severe loss of muscle mass in chronic disease, often accompanied by metabolic dysfunction, fatigue, loss of appetite, general weight loss, and poor patient outcomes. This debilitating condition is highly prevalent in cancer patients, lowering their quality of life and likelihood of survival. Prevention of muscle loss is expected to prolong survival, however, currently there are no accepted effective treatments, and cachexia is usually only diagnosed in advanced stages when therapeutic intervention is particularly challenging. Development and administration of effective therapies require foundational understanding of the underlying mechanisms that promote protein breakdown and cause wasting. Recently, the sequence of key molecular and cellular events driving catabolism, and the regulatory signals stimulating overall proteolysis, have become clearer from studies on animal models for muscle atrophy. Different types of atrophy share common transcriptional changes and proteolytic pathways that activate degradation of soluble and contractile myofibrillar proteins; consequently, identified key regulators of these common mechanisms should be attractive drug targets for treatment of numerous wasting conditions. In this chapter, we discuss our current knowledge on these catabolic events and key regulators, which orchestrate the continuous progression of the atrophy process from its initial cues. We also highlight promising avenues of research for better understanding and treating cancer cachexia.

Learning Objectives

- Outline the catabolic pathways mediating cancer-induced cachexia:
  - Systemic inflammation
  - Transforming Growth Factor-β signaling
  - Glucocorticoids
  - Metabolic alterations: insulin resistance and hypermetabolism
  - Decreased food intake and nutrient availability
- Discuss the molecular mechanisms contributing to muscle loss:
  - Reduced growth-promoting pathways
  - Enhanced proteolysis via the proteasome and autophagy
  - Excessive breakdown of myofibrils in an ordered fashion

(continued)

J. E. Gilda · S. Cohen (✉)
Technion-Israel Institute of Technology, Haifa, Israel
e-mail: shenhavc@technion.ac.il

© Springer Nature Switzerland AG 2022
S. Acharyya (ed.), *The Systemic Effects of Advanced Cancer*,
https://doi.org/10.1007/978-3-031-09518-4_5

- Degradation of myofibrillar proteins in a specific order
- Loss of desmin intermediate filament precedes and promotes myofibril destruction
- Early perturbations of structural and signaling modules on the muscle membrane facilitate desmin filament disassembly

1   Introduction

Skeletal muscle is the most abundant tissue in the human body, making up 40–45% of body mass, and is required for mechanical actuation, respiration, heat production, and posture. This tissue is a central pillar of whole body metabolism: it is the main site for glucose disposal and plays a major role in the adaptive physiological response to fasting. In times of scarcity, the fall in insulin signaling enhances proteolysis in muscle and causes atrophy, and the amino acids produced are converted to glucose in the liver to nurture the brain [1]. Therefore, it is unsurprising that maintenance of muscle mass is critical for health and survival, and loss of muscle mass has deleterious effects, including disability, morbidity, and death [2].

Systemic muscle wasting is often associated with chronic disease (i.e., cachexia) and generally accompanied by overall weight loss and other symptoms such as fatigue and loss of appetite. An estimated 40–87% of cancer patients are afflicted with cachexia, depending on cancer type, which contributes to the decline in physical function and quality of life, as well as the decreased efficacy and increased toxicity of chemotherapy [3, 4]. Skeletal muscle mass is recognized as an independent prognostic factor in advanced cancer patients [5], and cachexia directly causes 20–30% of cancer deaths [4, 6, 7]. Blocking cachexia in several mouse models

for cancer dramatically increased muscle mass and strength and even prolonged survival [7, 8]. Notably, in colon 26 (C26) tumor-bearing and Lewis lung carcinoma mice, survival was greatly improved even without any effect on rates of tumor growth, indicating that independently of cancer, prevention of cachexia should be highly beneficial to the patient [7, 9].

Despite the fact that cachexia is recognized as the major immediate cause of death in many patients with cancer, cancer-induced cachexia goes largely undiagnosed until advanced stages when wasting is severe, and is usually not treated. There is currently no effective treatment for preventing this systemic muscle wasting. Novel strategies propose combined nutritional supplementation and exercise, and some drugs show promise. Recent substantial progress in our understanding of the molecular mechanisms that mediate this loss of muscle mass in disease suggests several new key players as potential drug targets. In this chapter, we focus on these novel mechanisms, and highlight promising avenues of research for better understanding and treating cachexia.

2   Catabolic Pathways Mediating Cancer-Induced Cachexia

Cancer cachexia is a complex syndrome attributable to multiple factors. Increased tumor-derived signaling molecules, elevated pro-inflammatory cytokines, depressed appetite and food intake, and systemic metabolic changes such as glucose intolerance and accelerated whole body protein turnover all contribute to the systemic loss of muscle mass, making it a complicated condition to understand and treat [10, 11]. Chemotherapy can also accelerate cachexia, displaying distinct metabolic derangements that further complicate the understanding and treatment of cachexia [12, 13]. Currently, the only validated treatment for cancer-induced cachexia is exercise, which represses catabolic pathways and activates mechanisms that promote muscle growth.

2.1 Systemic Inflammation

Chronic inflammation is an important mediator of carcinogenesis by supplying the tumor with important biomolecules, including growth factors, angiogenic factors, and enzymes that support angiogenesis and metastasis [14, 15]. Inflammatory cells also contribute to mutagenesis within cancer cells by producing reactive oxygen and nitrogen species, cytokines, metalloproteases, and prostaglandin E2 (PGE2). These amplify the inflammatory response and lead to further genetic instability in cancer cells by downregulating DNA repair pathways and inducing double strand breaks and defective mitotic checkpoints [14]. Inflammation occurs systemically and also within the tumor microenvironment, consistently promoting negative outcomes for cancer patients [16, 17].

Many pro-inflammatory cytokines, including tumor necrosis factor-α (TNF-α), C-reactive protein (CRP), interleukin 6 (IL-6), and interleukin 1 beta (IL-1beta), are elevated in blood of cancer patients [18–22]. This increase in pro-inflammatory cytokines is mainly attributable to the immune response of the host to the tumor (Fig. 1). Inflammatory immune cells play critical functions in cancer-related inflammation [23], and histopathological analyses indicate that innate and adaptive immune cells are present in the majority of human tumors [24]. Multiple types of immune cells within the tumor microenvironment (e.g., macrophages and T helper cells) secrete pro-inflammatory cytokines such as interleukins, TNF-α, and IFNγ [23, 24]. However, inflammation may also be due in part to inflammatory cytokines secreted by the tumor cells themselves; some breast cancer cell lines produce IL-6 [25], and breast and ovarian cancer cell lines treated with chemotherapeutic taxanes show increased production of TNF-α (Fig. 1) [26, 27].

Fig. 1 Signaling pathways mediating cancer cachexia. Cancer cachexia is a complex, multifactorial syndrome involving a systemic crosstalk between multiple catabolic signaling pathways and organs. Tumor-secreted factors, the effects of immune response to the tumor, and impaired metabolism all contribute to decreased availability of nutrients, inefficient energy usage, and activation of proteolytic pathways in muscle, ultimately leading to wasting

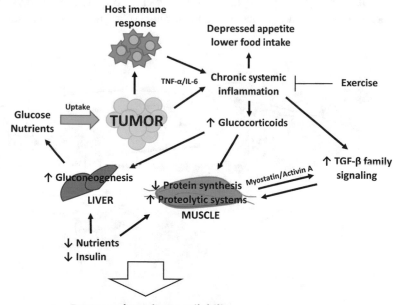

Pro-inflammatory cytokines can trigger the loss of muscle mass. Treatment of mice or cultured myotubes with TNF-α induces the expression of the atrophy-related genes (i.e., atrogenes), the ubiquitin ligases MAFbx/atrogin-1 and MuRF1 and reduces muscle fiber diameter [28–31]. Moreover, mice bearing tumors transfected with an expression vector for TNF-α lose more muscle mass than mice with control tumors (transfected with control plasmid) [32]. In terminally ill cancer patients, however, treatment with an antibody against TNF-α had no benefits [33], suggesting that TNF-α and other pro-inflammatory factors may together contribute to cachexia, and further studies in humans may be required to determine their specific roles. For example, IL-6 elevated levels correlate with weight loss and reduce survival in cancer patients [11, 34–37], and treatment with IL-6 receptor antibody ameliorates cachexia progression in mice by suppressing muscle proteolysis [38, 39]. Furthermore, treatment of lung cancer patients with an antibody to IL-6 was well-tolerated and reduced anemia and cachexia

[40]. Therefore, IL-6 shows promise as a target in treating cancer cachexia. TNF-related weak inducer of apoptosis (TWEAK) is another cytokine that increases atrogin-1 and MuRF1 levels and is a potent inducer of skeletal muscle wasting [41] and important mediator of denervation-induced muscle atrophy [42].

These pro-inflammatory factors, in particular IL-6, probably trigger muscle loss by activating signal transducer and activator of transcription 3 (STAT3). STAT3 is a transcription factor that plays key roles in the regulation of muscle mass and metabolism (Fig. 2) [43, 44], and is activated by IL-6 and other cytokines in the same family [44]. Upon activation of this signaling pathway, STAT3 is phosphorylated by Janus Kinases (JAKs), leading to its dimerization and translocation to the nucleus, where it activates transcription of target genes, including the transcription factor CCAAT/enhancer-binding protein δ (C/EBP δ) and acute phase response proteins [44–46]. It has been shown in vitro that C/EBP δ in turn drives myostatin expression, a negative regulator of muscle mass (discussed in the

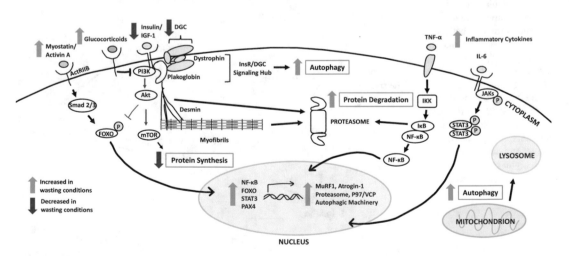

Fig. 2 Intracellular signaling pathways that promote muscle wasting. Input from multiple signaling pathways converges on key transcription factors, which in turn activate a transcriptional program that promotes muscle wasting. Insulin resistance, elevated glucocorticoids, and reduced DGC integrity contribute to the fall in PI3K/Akt/mTOR signaling, leading to inhibition of mTOR and protein synthesis. This reduction in PI3K/Akt/mTOR activity

and the increased myostatin/activin A signaling trigger proteolysis and muscle wasting by activating FoxO transcription factors, and inflammatory signals activate NF-κB transcription factor. FoxO, NF-κB, STAT3, and PAX4 are important in causing muscle wasting by stimulating the expression of genes that promote proteolysis, including ubiquitin ligases, proteasome subunits, the p97/VCP AAA-ATPase, and autophagy genes following section) [46, 47]. In mouse models, deletion of STAT3 or its inhibition in muscle wasting due to chronic kidney disease (CKD) [46], pancreatic cancer [48], Lewis lung carcinoma [49], or C26 tumors [49, 50] attenuates loss of body weight and muscle mass.

Pro-inflammatory factors likely also trigger muscle loss via activation of the transcription factor NF-κB (Figs. 1 and 2) [51, 52]. For instance, NF-κB can be activated in vitro and in vivo by the inflammatory cytokine TNF-α [52, 53]. NF-κB signaling is also activated in muscles of cachectic patients and mice [51, 54, 55] and as a side effect of cisplatin chemotherapy treatment in mouse models for cancer cachexia [12], and blocking binding of NF-κB to target genes attenuates muscle loss in tumor-bearing mice [56]. The roles of pro-inflammatory cytokines in promoting muscle wasting likely also involve increased levels of glucocorticoids and depressed appetite (Fig. 1) [57, 58].

Implementation of exercise regimens improves physical performance and reduces muscle loss in cancer patients [59–64], and animal models for cancer [65, 66]. Exercise likely inhibits muscle wasting by counteracting the systemic inflammation characteristic of most cancer cachexia patients. Physical activity leads to a transient inflammatory response in the short term, but has an anti-inflammatory effect in the long term, lowering levels of pro-inflammatory cytokines and C-reactive protein [67–69].

2.2   Transforming Growth Factor-β Signaling

The Transforming Growth Factor-β (TGF-β) family is comprised of numerous secreted factors that mediate cell proliferation, differentiation, and growth. TGFs were originally discovered as factors secreted by cancer cells that have the ability to promote transformation [70]. TGF-β signaling abnormalities are carcinogenic; in early-stage cancer, TGF-β has tumor suppressor function by inhibiting cell cycle progression. However, in late-stage cancer, TGF-β signaling promotes cancer cell motility and invasiveness

[71]. Activin A and myostatin (also called growth differentiation factor 8/GDF-8) are TGF-β family members that have been proposed to play a critical role in cancer cachexia (Fig. 1) [7, 72].

Activin A, myostatin, and other TGF-β family ligands (e.g., GDF-11) are negative regulators of muscle mass and growth via binding to their specific receptors, in particular, activin type II receptor B (ActRIIB). Consequently, the transcription factors SMAD2 and SMAD3 are activated and induce Forkhead box (FoxO)-mediated transcription of genes that promote proteolysis (Fig. 2) [73]. Deletion, loss-of-function mutations, or inhibition of myostatin in mice, cattle, sheep, and humans leads to widespread and dramatic increases in muscle mass [74–78]. By contrast, myostatin overexpression or its administration induces muscle loss in rodents [79, 80], and circulating levels of myostatin are increased in various atrophy conditions [81–85]. The physiological and pathophysiological mechanisms that regulate myostatin secretion may involve an extensive crosstalk among the multiple pathways mediating cancer cachexia [86–88]. For example, while chronic inflammation may upregulate ActRIIB signaling [89], there is also evidence that inhibiting myostatin-activin A signaling may in turn decrease systemic inflammation [90].

Circulating and intramuscular levels of myostatin and activin A in patients with various types of cancer are significantly higher than in healthy subjects [91–94], and myostatin expression is increased in muscles of tumor-bearing cachectic rodents [7, 95]. Inhibition of myostatin/activin A signaling by administration of an ActRIIB decoy receptor dramatically increases muscle mass and strength and prolongs survival of multiple mouse models for cancer cachexia [7–9, 96], and mitigates chemotherapy-induced muscle wasting [97, 98]. These improvements in survival of mice with Lewis lung carcinoma, C26 tumors, or pancreatic cancer treated with ActRIIB decoy receptor or a TGF-β inhibitor occurred without any effects on tumor size, indicating that prevention of cachexia may be highly beneficial for patient outcomes, independently of direct treatment of the tumor itself

[7, 9, 99]. In fact, several clinical trials utilizing inhibition of myostatin/activin A signaling to treat muscle wasting associated with cancer and other disorders are in progress, and appear to be well tolerated and beneficial in increasing muscle mass and strength [100]. In a phase I trial, treatment of advanced cancer patients with the myostatin antibody LY2495655 showed increased thigh muscle volume and hand grip strength, with no unusual safety concerns [101]. The completion of further clinical studies using myostatin antibody will be beneficial in determining safety and outcomes for survival and quality of life. Unwanted side effects can be avoided by using more selective antibodies, like the one generated by Novartis Pharmaceuticals against the ActRIIB receptor, the BYM338 antibody, also called bimagrumab. This antibody binds with high affinity to ActRIIB receptors, hence specifically preventing association of myostatin, activin A, and GDF11 with these receptors. A randomized study of bimagrumab conducted by Novartis for treatment of cachexia in patients with lung cancer or pancreatic adenocarcinoma showed promising results; patients treated with bimagrumab had increased lean body mass and thigh muscle volume, though they also had decreased total body weight [102]. Exercise is also effective at inhibiting myostatin and activin A in states of health [103, 104], disease [105–108], and aging [109, 110] by reducing myostatin and ActRIIB expression and increasing the levels of the myostatin/activin A inhibitor, follistatin. These likely account, at least in part, for the beneficial effects of exercise on muscle in cancer patients.

2.3 Glucocorticoids

Glucocorticoids are anti-inflammatory and immunosuppressive agents that are increased and promote muscle loss in conditions of systemic inflammation, including cancer (Fig. 1) [57, 111–113]. Glucocorticoid signaling is required for cachexia caused by cancer, chemotherapy, or treatment with the endotoxin

lipopolysaccharide, as muscle-specific deletion of the glucocorticoid receptor greatly attenuates muscle atrophy in both cachexia models [57, 114]. Their debilitating effects on muscle mass result from decreased rates of protein synthesis and increased proteolysis (Fig. 1) [115]. Elevated glucocorticoids also likely promote muscle wasting by increasing hepatic gluconeogenesis (Fig. 1) [116]. The suppression of protein synthesis appears to be due mainly to inhibitory effects of glucocorticoids on the phosphoinositide 3-kinase (PI3K)/Akt/mammalian target of rapamycin (mTOR) pathway (Fig. 2) [115], by lowering expression and inhibitory phosphorylation of the upstream signaling component insulin receptor substrate 1 (IRS-1) [117–119], and by decreasing PI3K activity and association with IRS-1. Consequently, Akt activity is reduced and FoxO-mediated expression of the atrogene program is enhanced [120–122]. Glucocorticoids also induce myostatin production in muscle [123, 124], which in turn promotes atrophy also by inhibiting PI3K/Akt signaling [125].

2.4 Metabolic Alterations: Insulin Resistance and Hypermetabolism

Abnormal glucose metabolism and the resulting insulin resistance have long been recognized in cancer patients, and are among the earliest metabolic changes to take place [126–130]. Mice bearing colon 26 tumors are insulin resistant, and display decreased phosphorylation of Akt and increased expression of atrophy-related genes, including atrogin-1, MuRF-1, and Bnip3 in muscle, leading to activation of proteolysis via both the proteasome and autophagy [131]. Notably, treatment with an insulin sensitizer blocked the early stages of cachexia in these mice [131].

Increased energy expenditure has also been proposed as a major cause of muscle wasting due to cancer [132]. Cancer cells show increased dependence on glycolysis for growth and survival, and the accelerated uptake of glucose by tumors, combined with inflammation and reduced nutrient intake, contribute to widespread metabolic disturbances. Consequently, utilization of glucose by muscle decreases, and the resulting energy deficits stimulate breakdown of muscle proteins to be utilized for gluconeogenesis by the liver (Fig. 1). If sustained, this catabolic state leads to the loss of vital muscle proteins and impaired muscle function. Patients and animal models of cachexia show hypermetabolism and higher resting energy expenditure, which contribute to energy loss [133]. Similarly, inefficient oxidative phosphorylation and reduced ATP consumption by sarcoplasmic reticulum Ca^{2+} pumps contribute to impaired energy production and reduced muscle mass and strength [132].

2.5 Decreased Food Intake and Nutrient Availability

Cancer patients have aberrant appetite control, which contributes to decreased food intake. This lack of appetite may be due in part to inflammatory signals, such as IL-1, which contributes to anorexia in cancer (Fig. 1) [134]. Appetite stimulants and nutritional intervention have been shown to increase appetite and weight; however, they cannot fully reverse the loss of muscle mass and the resulting functional impairment in cancer patients. Early nutritional intervention may be beneficial to slow loss of muscle mass and strength. For example, in a study on the effects of nutritional intervention, newly diagnosed cancer patients administered with a medical food high in protein and enriched with fish oil showed an increase in body weight, although the long-term effects on survival were not measured [135]. In addition, treatment with appetite stimulants (such as progestins) seems to have minimal benefits for cancer patients, leading to improved symptoms but not survival [136–140]. Consequently, it is now widely recognized that cancer cachexia can be best managed within a multimodal approach, combining specialized nutrition support with clinical intervention.

3 Molecular Mechanisms Contributing to Muscle Loss

The reduction in muscle mass and strength during atrophy is primarily due to the loss of the fundamental contractile machinery in muscle, the myofibrils. Myofibrils comprise the majority of muscle proteins, and their loss during atrophy ultimately leads to weakness, frailty, and disability. Prior investigations, primarily in animal models, have demonstrated a coordinated induction of atrogenes [141, 142] that activate proteolysis in various atrophy conditions, including cancer cachexia, inactivity, denervation, and fasting. Among these genes are components of the ubiquitin–proteasome system (UPS), which promotes degradation of myofibrillar and soluble proteins, and autophagy, which is largely responsible for loss of mitochondria and other organelles (Fig. 2). Because muscles atrophying in response to diverse catabolic stimuli all show similar transcriptional adaptations and activation of protein degradation by the UPS and autophagy, it is likely that common mechanisms promote muscle wasting in fasting, diabetes, denervation, and cancer. This section will focus on what is known about the cellular and biochemical events leading to myofibril loss in various types of atrophy, including the systemic wasting induced by cancer.

3.1 Reduced Growth-Promoting Pathways

Cachexia seems to be driven by a reduction in growth-promoting pathways, such as the PI3K/Akt/mTOR signaling. Proteolysis via the UPS and autophagy increases during atrophy largely through the fall in PI3K/Akt/mTOR signaling, and the resulting activation of FoxO transcription factors. FoxO activation at an early phase during this process promotes the expression of the atrogene program, and activation of FoxO alone is sufficient to trigger protolysis and to cause substantial atrophy (Fig. 2) [143–145]. Given its central role in promoting proteolysis, FoxO activity is tightly controlled by growth-promoting pathways, especially PI3K/Akt/mTOR signaling, and also the exercise-induced transcription coactivator peroxisome proliferator-activated receptor-γ coactivator 1α (PGC1α) [146, 147], the transcription factor JUNB [148], and the NAD-dependent protein deacetylase sirtuin 1 (SIRT1) [149]. The loss of these anabolic factors in various forms of atrophy contributes to atrophy by activating FoxO.

In muscle, activation of PI3K/Akt/mTOR signaling by IGF-1 or insulin promotes protein synthesis through mTOR kinase, and inhibits overall proteolysis by suppressing FoxO [150], although mTOR can also suppress autophagy [151] and the UPS [152, 153]. However, on fasting (when blood glucose and insulin levels are low), insulin resistance states (e.g., type-2 diabetes, obesity), aging, and many catabolic diseases, including cancer cachexia, sepsis, chronic renal disease, and heart failure, PI3K/Akt/mTOR signaling falls and consequently proteolysis increases largely via FoxO-mediated expression of atrogenes (Fig. 2) [131, 154]. Insulin resistance, elevated glucocorticoids, and increased levels of myostatin all appear to contribute to inhibition of PI3K/Akt signaling in cancer cachexia [126, 127, 130, 131]. In addition, several other transcription factors seem to cooperate with or act by FoxO-mediated transcription and are important in causing muscle wasting, including SMAD2 and SMAD3, glucocorticoid receptors, and NF-κB. Their inhibition can reduce or block different types of atrophy, although their precise roles in altering the expression of specific genes remain unclear.

SMAD2 and SMAD3 mediate protein loss downstream of myostatin, activin A, and other TGF-β family members (e.g., GDF-11), which are highly catabolic in muscle. Their activation reduces PI3K/Akt signaling in cancer cachexia, promotes insulin resistance in muscle fibers [106, 155] and enhances protein degradation [11]. Thus, the elevated circulating myostatin and activin A in cancer appear to lead to systemic muscle loss via effects on multiple signaling pathways that converge on FoxO transcription factors, leading to increased expression of atrogenes.

3.2 Enhanced Proteolysis Via the Proteasome and Autophagy

In atrophy, the UPS is activated by the fall in PI3K/Akt signaling and is primarily responsible for degradation of myofibrils, which constitute more than 70% of muscle proteins. Proteasome expression and activity increase, and the two muscle-specific atrogenes, MuRF1 and atrogin-1, are induced in atrophying muscles of tumor-bearing cachectic rodents [54, 141, 156–159], and almost all other types of atrophy [160, 161]. Since the UPS plays critical roles in virtually all cellular processes throughout the body, and proteasome inhibition has moderate to severe cardiotoxic effects in humans and animals [162–170], therapeutics aimed at inhibiting UPS function will likely be most successful if they target specific key components that promote muscle atrophy. Accordingly, using proteasome inhibitors to treat muscle loss in tumor-bearing mice yields mixed results [171, 172]. However, inhibiting or downregulating MuRF1 or atrogin-1 attenuates skeletal muscle wasting in mouse models of cardiac cachexia [173], fasting [174], and glucocorticoid-induced atrophy [175].

3.2.1 UPS Activation in Human Cancer Cachexia

Several studies have demonstrated that UPS components are induced and activated in muscles of human patients with cancer cachexia and other wasting conditions, including sepsis [176, 177], acquired immunodeficiency syndrome (AIDS) [178], diabetes [179], immobilization [180], and nerve damage [181]. Expression of proteasome subunits and UPS components are elevated in muscles of cancer patients, and this UPS induction correlates with disease stage and severity of weight loss [182–187]. Proteasome gene expression was not elevated in muscles of cancer patients until weight loss reached 10%, at which point it increased gradually in parallel to loss of up to 19% of muscle mass [185]. As weight loss became more extensive, expression of proteasome genes returned to baseline [185]. The findings in human patients are in very good agreement with a study from the same research group on adenocarcinoma mice, which showed a correlation between the expression of the α6 proteasome subunit and proteasome activity with rates of weight loss [157]. Expression and activity increased linearly with weight loss (from ~11–20% weight loss), after which point expression declined [157].

Expression of MuRF1 and atrogin-1 is elevated in muscles of patients afflicted by malignant disease, an effect that seems to occur in the early stages of muscle loss. For instance, MuRF1 and atrogin-1 are induced in muscles of early-stage lung cancer patients [188], though with only few accompanying changes in proteasome expression or activity and no significant change in ubiquitin levels [188–190]. Similarly, expression of MuRF1 and atrogin-1 is elevated in the muscle of patients afflicted by malignant disease, and the majority of patients had elevated expression of atrogin-1 and MuRF1 even before weight loss [159]. These findings suggest MuRF1 and atrogin-1 induction as an important early event in muscle loss in cachectic cancer patients. However, a study on late-stage non-small-cell lung cancer cachexia demonstrated no differences in MuRF1, atrogin-1, proteasome activity, or proteasome subunit protein expression [191]. The failure to detect an induction of proteasome subunits [188, 189], atrogin-1 or MuRF1 [191] is likely attributable to the stage of disease studied. These data from cancer cachexia patients suggest that expression of MuRF1 and atrogin-1 peaks in the early stage of disease and muscle loss, then returns toward baseline in later stages of wasting (Fig. 3). This expression pattern has been observed for MuRF1 and atrogin-1 in other types of muscle atrophy in mice, such as denervation, spinal isolation [142], and hindlimb immobilization [192]. Proteasome expression and activity seem to increase later than MuRF1 and atrogin-1 in cancer cachexia patients, and return toward baseline in a similar manner in late stages of disease

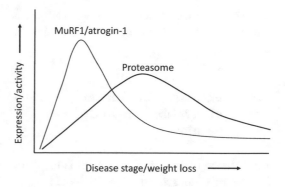

Fig. 3 Proposed correlation between disease stage and UPS activation in cancer cachexia. Clinical studies and findings in tumor-bearing mice and other types of atrophy indicate that the expression of the E3 ubiquitin ligases MuRF1 and atrogin-1 peaks early in the process of muscle loss, and returns to baseline in the later stages. Expression and activity of the proteasome seem to peak later in the process of muscle loss, and return to baseline in a manner similar to the E3 ligases

and muscle loss (Fig. 3) [185]. Together these data suggest that targeting the UPS in the early stages of muscle loss may be a promising therapeutic approach for maintaining muscle mass and function. However, the stage of disease, and potentially cancer type, will likely have an important impact on the effectiveness of UPS-targeted therapies. Targeting MuRF1 will prevent loss of myosin thick filaments, and may be most effective in combination with Trim32 inhibition, which is mostly responsible for the loss of actin thin filaments, Z-band components and desmin intermediate filaments [193–197].

While myofibrillar proteins are primarily degraded by the UPS, mitochondria and other organelles are degraded by autophagy, and their loss accounts for the decreased endurance capacity of cachectic patients (Fig. 2) [158, 198]. Expression of autophagy markers, including beclin-1 and LC3B-II, is increased in skeletal muscle of mouse models for cancer cachexia and cachectic cancer patients, and stimulating autophagy in mice by overexpressing the autophagy-regulating protein TP53INP2/ DOR exacerbates muscle loss [158, 188, 198, 199]. This increase in autophagy genes may be mediated by systemic inflammation, because IL-6 released from tumor cells [200] and TNF-α can stimulate autophagy in cultured muscle cells [198].

3.3 Excessive Breakdown of Myofibrils in an Ordered Fashion

Degradation of myofibrillar proteins is the defining feature of muscle atrophy. The basic contractile machinery in muscle, the myofibrils, is a highly organized structure whose primary constituents are myosin in thick and actin in thin filaments. These filaments are precisely aligned and organized into repeating units called sarcomeres [201]. Myosin makes up 50% of the myofibril by mass and is composed of two heavy chains (MyHC), two regulatory light chains (MyLC2), and two essential light chains (MyLC1). The globular myosin heavy chain head binds to actin in thin filaments to form cross-bridges and promote muscle contraction through ATP hydrolysis. Myosin light chains are important for maintaining thick filament stability and normal muscle contraction. Accordingly, depletion of cardiac myosin light chains in zebrafish leads to defects in cardiac contractility and disruption of sarcomere structure and alignment [202, 203], and removal of myosin light chains from chicken skeletal muscle reduces the velocity of sliding filaments [204]. Numerous other proteins are important for maintenance of myofibril stability, including myosin-binding protein C (MyBP-C), which binds the thick filament at the site of cross-bridges [205] and is important for thick filament stability and myofibril organization [206]. The thin filaments are stabilized at the boundaries of the sarcomere, the Z-bands, where desmin intermediate filaments (IF) are also aligned. Desmin IF represents the principle cytoskeletal network in muscle, which is critical for maintenance of muscle structural integrity and contractile capacity because it links the myofibrils laterally, and to organelles and the muscle membrane [207]. Due to their highly ordered and intricate structure, myofibrils must be disassembled and degraded during atrophy in a tightly regulated process, especially because muscles continue to contract even during rapid atrophy (e.g., fasting, cancer cachexia).

3.3.1 Degradation of Myofibrillar Proteins in a Specific Order

The mechanism for myofibril breakdown has recently become clearer primarily from work on animal models, demonstrating that specific ubiquitin ligases have a prominent role. These investigations demonstrated that proteins are lost from the myofibril in a specific order during atrophy. Initially, certain regulatory proteins that stabilize the thick filaments (MyLC1 and 2, MyBP-C) are ubiquitinated and selectively degraded by MuRF1, and subsequently myosin is lost by this enzyme [195, 208]. The differential loss of proteins that stabilize the myofibrils is a newly defined feature of atrophying muscles that seems to be of regulatory importance. However, MuRF1 is not the primary enzyme for degradation of actin and other thin filament components (e.g., tropomyosin) in vivo, and their destruction is linked to desmin IF loss by the ubiquitin ligase, Trim32 [196, 209]. Although MuRF1 can ubiquitinate actin in vitro and promote degradation of actin in cultured C2C12 myotubes [210], this enzyme seems to prefer MyHC as a substrate in vitro [211] and in vivo [195, 208]. By ubiquitinating myofibrillar components, MuRF1, Trim32 and probably other ubiquitin ligases facilitate destruction by the proteasome, although Trim32 can also activate autophagy [212, 213]. Therefore, during atrophy, the initial loss of structures that stabilize the myofibrils (e.g., desmin filaments, MyLC1 and 2, MyBP-C) probably loosens the myofibrils, rendering its constituents more accessible and susceptible to the catalytic activity of enzymes that promote proteolysis.

It is now eminently desirable to understand how the complex structure of myofibrils is disassembled during atrophy. Initial work in this direction aimed to identify the specific roles of UPS components in catalyzing myofibril destruction in different types of atrophy. The expression of atrogin-1 and other atrogenes is stimulated by FoxO transcription factors at an early phase in

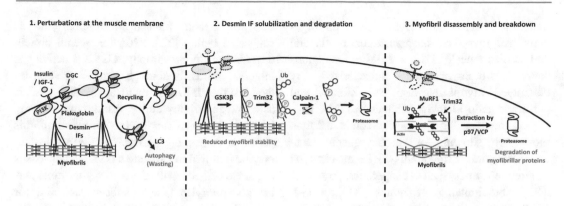

Fig. 4 Sequence of events leading to myofibril breakdown and atrophy. Myofibril disassembly and loss occur in a well-ordered and tightly regulated process. Initially, autophagy-mediated decrease in plakoglobin-DGC-insulin receptor clusters on the muscle membrane reduces tissue integrity by perturbing the linkage between desmin IF and the DGC. The simultaneous fall in insulin/PI3K/Akt signaling leads to activation of GSK3-β, and enhances FOXO-mediated proteolysis. Consequently, GSK3-β catalyzes desmin IF phosphorylation, and the resulting ubiquitination by Trim32, depolymerization by calpain-1 and degradation by the proteasome. This loss of desmin IF triggers myofibril destruction most likely by reducing myofibril integrity, and thus enhancing the susceptibility of actin, myosin, and other myofibrillar proteins to the catalytic activity of ubiquitin ligases (e.g., Trim32, MuRF1), p97/VCP AAA-ATPase, proteases, and the proteasome

various types of atrophy (Fig. 2) [214]. However, our recent studies demonstrate that the transcription factor PAX4 mediates a second phase of gene induction at a more delayed phase during atrophy [197], which seems to be required for the excessive destruction of myofibrils [197, 215]. For example, PAX4 promotes the induction of MuRF1 at 10 days after denervation, although the expression of this atrogene is stimulated by FoxO transcription factors already at an early phase (e.g., 3 days after denervation) in various types of atrophy. This second phase of gene induction by PAX4 at a more delayed phase occurs just as degradation of myofibrils is accelerated and is likely required to maintain high MuRF1 protein levels (and other components that promote proteolysis) in the cytosol to facilitate the destruction of myofibrillar components. These studies introduced a new concept into the muscle biology field, and presented atrophy as a two-phase process, involving early and delayed phases of gene induction by distinct transcription factors. In addition, early during atrophy induced by fasting or denervation, Trim32 catalyzes the initial loss of desmin filaments, and then at a later phase, it seems to act on the myofibrillar apparatus together with

MuRF1 and other enzymes induced by PAX4 (e.g., ubiquitin ligases, the AAA-ATPase p97/VCP complex) (Fig. 4) [197]. Because PAX4 is critical for the induction of genes that promote proteolysis and thus can affect muscle mass, it will be important to determine how PAX4 function is regulated, and whether there is also a delayed rapid loss of myofibrils along with a second phase of gene induction in more prolonged systemic wasting conditions (e.g., cancer cachexia, caloric restriction, type-2 diabetes, renal or cardiac failure).

3.3.2 Loss of Desmin IF Precedes and Promotes Myofibril Destruction

The recent progress in deciphering the mechanisms of muscle atrophy in rodents and human muscles suggests that this process is highly ordered, requiring the initial loss of desmin filaments, which ultimately leads to myofibril breakdown and atrophy (Fig. 4, panel 2) [194, 197, 216, 217]. In muscle, desmin forms IF that are localized between adjacent myofibrils, linking them laterally to the Z-lines, and between myofibrils and the sarcolemma, mitochondria, and nuclear membrane [218, 219]. Desmin null mutation in mice leads to sarcomere misalignment and myofibril disorganization in skeletal and cardiac muscle [220], and desmin mutations cause cardiomyopathies characterized by disarrayed myofibrils, displaced Z-lines, and perturbed actin filament architecture [221]. During atrophy, the desmin cytoskeleton is lost first, most likely allowing for the consequently loosened structure of myofibrils to be attacked by different enzymes, such as chaperones, proteases, and ubiquitinating enzymes [193, 194, 196]. Given the importance of desmin IF for the integrity of myofibrils, the reduced structural integrity of desmin filaments during atrophy is likely a key step in the destabilization of Z-band proteins and thin filaments, leading to the enhanced susceptibility to ubiquitination.

Dystrophin protein content is reduced in muscles of tumor-bearing mice, and this loss is selective to cachectic fibers (isolated by laser capture microdissection) [54]. A separate study corroborated these findings, showing by proteomic profiling that desmin is downregulated in gastrocnemius muscles of mice with cancer cachexia, which predominantly contain type II (fast-twitch) fibers, and is upregulated in soleus muscles, which contain mainly type I (slow-twitch) fibers [222]. Tumor-induced muscle loss is specific to type II fibers, indicating that dystrophin loss is specific to cachectic fibers [222]. Furthermore, dystrophin content is dramatically reduced in muscles of human cancer patients, and this loss is positively correlated with cachexia [54]. These findings indicate that loss of desmin (by reduced gene expression or protein degradation) is important for myofibril breakdown and muscle fiber atrophy in cancer cachexia [54]. Furthermore, it has been shown in rodents and humans that in atrophy induced by denervation, disassembly of desmin IF precedes and promotes myofibril destruction [197, 216, 223]. Early after denervation, desmin is phosphorylated by the protein kinase GSK3-β and is later cleaved by the Ca^{2+} specific protease calpain-1 (Fig. 4, panel 2) [194, 224]. Inhibiting GSK3-β or downregulating calpain-1 is sufficient to block atrophy on denervation or fasting. Therefore, activation of GSK3-β appears to be a critical early

step for desmin filament destruction and as suggested before [225, 226], for overall protein degradation during atrophy. GSK3-β activity is inhibited primarily through phosphorylation by the insulin/PI3K/Akt pathway [227], and is activated in atrophy during fasting and in many systemic human diseases by inhibition of this pathway. A priming phosphorylation of desmin may be required to generate a GSK3-β consensus sequence $SX_{3-4}S$, with the carboxy-terminal serine phosphorylated, as has been proposed for GSK3-β substrates [228, 229]. Interestingly, the desmin head domain contains a potential GSK3-β consensus sequence, $S_{28}PLSS_{32}$, where phosphorylation of desmin on serines 28, 32, and 68 occurs during fasting [196]. Therefore, phosphorylation by GSK3-β seems to be an early key step in the depolymerization of desmin filaments by calpain-1, and the resulting loss of myofibrils, and represents a potential therapeutic target in muscle wasting. The role of calpains in promoting cachexia in cancer is debated and yet to be determined [230, 231]. Therefore, further work will be required to clarify the mechanisms of desmin IF disassembly in cancer cachexia, and the specific roles of GSK3-β and calpain-1.

3.3.3 Early Perturbations of Structural and Signaling Modules on the Muscle Membrane Promote Desmin Filament Disassembly

The early phase of atrophy involves perturbations in signaling and structural modules at the muscle membrane (Fig. 4, panel 1) [232]. A key event in this step is a reduction in the membrane content of intact dystrophin-glycoprotein complex (DGC), a large protein complex whose integrity is critical for cardiac and skeletal muscle architecture and mechanotransduction [233]. It is composed of dystrophin, dystroglycans, sarcoglycans, sarcospan, dystrobrevins, and syntrophin, all of which are localized to membrane structures called costameres, which link the sarcomeric Z-bands to the sarcolemma [234]. Intracellular, transmembrane, and extracellular DGC subunits interact to transmit force generated by contraction of the myofibrils to the extracellular matrix. In the cytoplasm, the large, rod-shaped protein dystrophin binds via its N- and C- terminal domains to actin or desmin, and the transmembrane protein β-dystroglycan, respectively. In turn, β-dystroglycan is associated extracellularly with α-dystroglycan, and the linked laminin-2 in the extracellular matrix [54]. This structural link between the intracellular cytoskeleton and the extracellular space via intact DGC is vital for muscle architecture and function. Accordingly, mutations in various DGC components lead to different myopathies in human, including Duchenne muscular dystrophy (DMD), Becker muscular dystrophy (BMD), and X-linked dilated cardiomyopathy [233, 235–237].

In young rodents, forces generated in muscle are transmitted laterally to the muscle surface with little or no reduction in force [238]. However, in muscles of dystrophic *mdx* mice, which express truncated nonfunctional dystrophin, or old rats, which show greatly reduced dystrophin expression, lateral force transmission is severely diminished [238]. Intact DGC is not only important for force transmission, but it also seems to protect against contraction-induced injury [239]. For example, repeated muscle contractions lead to severe damage to muscle in *mdx* mice, and restoring dystrophin expression and DGC stability protects against this injury [240–242].

Recent findings indicate that in addition to its structural role, the DGC regulates signaling pathways that control cell growth and differentiation [54, 232, 243, 244]. This complex interacts with multiple signaling components, including growth factor receptor-bound protein 2 (GRB2), neuronal nitric oxide synthase (nNOS), Yap [243, 245], and calmodulin [243, 246–249]. For instance, inhibiting dystroglycan association with laminin in primary muscle cell cultures leads to Akt inactivation and apoptosis [250]. Our recent investigations offer a mechanistic explanation for these effects by identifying the DGC as a signaling hub for the insulin receptor in skeletal muscle, heart, and liver [232]. We showed that the desmosomal protein plakoglobin plays an important role in maintaining muscle size by linking the DGC and the insulin receptor structurally and functionally at costameres (Fig. 4, panel 1). Costameres link desmin IF and the bound

myofibrils to the ECM via laminin, hence contributing to muscle mechanical integrity [250]. Low insulin receptor activity reduces association with the DGC and plakoglobin and promotes loss of phosphorylated desmin filaments [232]. Likewise, compromised DGC integrity, as seen in DMD, cancer, type-2 diabetes, and aging, lowers insulin receptor signaling and leads to insulin resistance and muscle wasting [232]. However, overexpression of plakoglobin in muscles of mouse models for diabetes or DMD is sufficient to restore DGC-insulin receptor interaction, desmin stability, and insulin signaling, and prevent muscle wasting. Therefore, the DGC is a signaling hub whose interactions with the insulin receptor are mediated by plakoglobin, and perturbation of plakoglobin function (e.g., by ubiquitination by Trim32) [251] offers a likely mechanistic explanation for the insulin resistance seen in some DMD and Becker muscular dystrophy human patients [252–254], and metabolic disorders (e.g., untreated diabetes, obesity) [253, 254].

These studies also identified two LC3 interacting regions (LIR) motifs in β-dystroglycan, LQFIPV (714–719) and GEYTPL (846–851), that can potentially be recognized by the autophagy marker LC3 [255]. The co-assemblies of plakoglobin-DGC-insulin receptor at the muscle membrane appear to be dynamically regulated by recycling in autophagic vesicles. Under conditions of nutrient availability and normal insulin signaling, when growth and survival signals are transmitted, the active insulin/IGF-I/PI3K/Akt pathway inhibits proteolysis by autophagy [143, 145, 256–258]. Under these conditions, plakoglobin prevents the autophagic loss of the plakoglobin-DGC-insulin receptor co-assembly by masking β-dystroglycan's LIR domains. However, in catabolic states such as nutrient deprivation and disease, when IGF-I and insulin levels are low, LC3 levels increase, and autophagy is induced [143, 145, 214]. Then, LC3 competes with plakoglobin for binding to β-dystroglycan and targets the vesicles containing plakoglobin-DGC-insulin receptor co-assemblies to the lysosome [259] (Fig. 4). Thus, a bidirectional regulation of DGC and the insulin receptor appears to be necessary for DGC stability and insulin receptor activity.

Tumor-induced changes in DGC content on the muscle membrane (i.e., sarcolemma) are a key early event in cachexia (Fig. 4, panel 1). In tumor-bearing mice, dystrophin expression is reduced along the sarcolemma, and β-dystroglycan and β-sarcoglycan are hyperglycosylated prior to any reduction in fiber diameter [54]. In muscles from tumor-bearing mice, the binding between α-dystroglycan, laminin, and several other DGC components is reduced. These structural changes occur predominantly in type II muscle fibers, with very little effects seen in type I fibers. Because tumor-induced muscle loss occurs selectively in type II fibers, the reduced DGC integrity appears to be specific to cachectic fibers. Accordingly, loss of DGC enhances tumor-induced muscle loss [54]. This was shown using *mdx* mice that have a mutation for a premature stop codon in the dystrophin gene, leading to loss of full-length functional dystrophin and intact DGC. The *mdx* mice with tumor burden showed accelerated muscle loss compared to wild-type littermates [54]. Furthermore, in tumor-bearing dystrophin transgenic mice (expressing a dystrophin minigene), the muscle-specific expression of dystrophin attenuated MuRF1 and atrogin-1 induction and wasting, albeit with little to no change in grip strength, strongly supporting the notion that restoring DGC integrity and function can reduce tumor-induced muscle wasting [54, 232]. These beneficial effects on muscle are most likely due to enhanced DGC integrity and the resulting activation of the insulin receptor and PI3K/Akt signaling. DGC deregulation has also been reported in human cancer patients, who show a dramatic reduction in dystrophin expression and hyperglycosylated DGC proteins [54]. This study further demonstrated that the reduced integrity of DGC correlated positively with cachexia and negatively with survival in gastrointestinal cancer patients [54]. Since DGC deregulation is an early event in cancer cachexia and correlates with muscle loss, therapies aimed at restoring

DGC function in muscle may be a promising approach for combating cancer cachexia.

4 Conclusions and Future Perspectives

Cachexia is now understood to be a major cause of death in many patients with cancer, and studies in tumor-bearing mice show that treating muscle loss, even without slowing the rate of tumor growth, could be highly beneficial to prolonging survival. Therefore, understanding the mechanisms causing muscle loss is crucial for the development of rational therapies to prevent or slow cachexia and improve patient quality of life and survival. The molecular mechanisms of cancer cachexia are complex and involve a multitude of signaling molecules and catabolic pathways. Studies on human cancer patients and rodent models of cancer cachexia and other types of atrophy have provided valuable insights into the key catabolic mediators of cachexia (including inflammation, myostatin/activin A, and glucocorticoids), and advanced our understanding of the molecular mechanisms that mediate myofibril destruction and loss of muscle mass in disease. However, it is notable that while studies performed in animals have identified numerous potential therapeutic targets for treating cancer cachexia, around 100 clinical investigations have been conducted with no resulting approved therapy to date [260]. This disheartening outcome may be attributable to several factors. The subjects used in animal studies invariably constitute a homogeneous group, while genetic makeup, age, psychosocial factors, sex, fitness status, compounding chemotherapeutic treatments, and cancer type often vary widely within studies of human cancer patients. Rodent cancer models are also imperfect simulations of the complicated biology of human cancer. Implanted tumors do not metastasize, which poorly reflects the reality for many human patients [261]. They are also implanted into young animals, and grow quickly and aggressively, in contrast to the more gradual course of disease observed in humans [261]. Genetically altered mice give rise to spontaneous tumors, but the genetic alterations are present in all tissue, unlike what occurs in humans [261]. These differences highlight the need for more studies in humans to verify the molecular mechanisms elucidated in animals. Disease stage is also an important factor to consider when weighing treatment options, as the involvement of various mediators, such as the UPS, almost certainly varies depending on disease stage and extent of weight loss. The extent of cachexia is often measured as the loss of body weight, making it difficult to assess the stage of muscle loss, as well as assess and compare the efficacy of different treatments. More reliable methods of assessing cachexia should be implemented, such as computed tomography (CT) to measure body composition [260]. Many studies also use body weight and appetite as the endpoint measurements; assessing more impactful outcomes such as survival and physical fitness may be helpful in more clearly defining clinical outcomes [260].

Nonetheless, studies in animals and humans highlight several promising avenues of research. Further research on the mediators of cachexia and key players promoting myofibril breakdown will facilitate the development of effective therapies. For example, since phosphorylation by GSK3-β is an initial key step in the selective depolymerization of desmin IF in atrophy, its inhibition should be of major therapeutic promise for the treatment of numerous wasting conditions. Therefore, the sequence of events leading to loss of desmin IF in cancer cachexia, the nature of the subsequent steps, the identity of the priming kinases phosphorylating GSK3-β substrates (e.g., desmin IF), and the roles of calpains are important questions for future research. In addition, it will be of interest to explore whether similar to fasting or denervation, there is also a delayed rapid loss of myofibrils along with a second phase of gene induction in more prolonged systemic wasting conditions (e.g., cancer cachexia, caloric restriction, and renal or cardiac failure). Particularly, novel transcription factors (such as PAX4) that induce genes that promote proteolysis should be further investigated to clarify their roles in atrophy and how their activation is regulated. The outcome of such studies will likely have critical therapeutic implications because it may indicate that certain drugs against distinct targets should be applied at different stages during the atrophy process. Currently, targeting myostatin-activin A signaling is considered the most promising approach to combat systemic wasting. However, foundational understanding of the order of molecular and cellular events driving catabolism in cancer will facilitate finding of essential genes as new drug targets to block this debilitating loss of muscle.

References

1. Argilés, J.M., Campos, N., Lopez-Pedrosa, J.M., et al.: Skeletal muscle regulates metabolism via inter-organ crosstalk: roles in health and disease. J. Am. Med. Dir. Assoc. **17**(9), 789–796 (2016)
2. Bonaldo, P., Sandri, M.: Cellular and molecular mechanisms of muscle atrophy. DMM Disease Models and Mechanisms. (2013)
3. Loberg, R.D., Bradley, D.A., Tomlins, S.A., et al.: The Lethal Phenotype of Cancer: The Molecular Basis of Death Due to Malignancy. CA Cancer J. Clin. (2007). https://doi.org/10.3322/canjclin.57.4.225
4. Tomasin, R., Martin, A.C.B.M., Cominetti, M.R.: Metastasis and cachexia: alongside in clinics, but not so in animal models. J. Cachexia Sarcopenia Muscle (2019)
5. Parsons, H.A., Baracos, V.E., Dhillon, N., et al.: Body composition, symptoms, and survival in advanced cancer patients referred to a phase I service. PLoS One. (2012)
6. Tisdale, M.J.: Mechanisms of cancer cachexia. Physiol. Rev. (2009)
7. Zhou, X., Wang, J.L., Lu, J., et al.: Reversal of cancer cachexia and muscle wasting by ActRIIB antagonism leads to prolonged survival. Cell. **142**, 531–543 (2010). https://doi.org/10.1016/j.cell.2010.07.011
8. Busquets, S., Toledo, M., Orpí, M., et al.: Myostatin blockage using actRIIB antagonism in mice bearing the Lewis lung carcinoma results in the improvement of muscle wasting and physical performance. J. Cachexia. Sarcopenia Muscle. (2012). https://doi.org/10.1007/s13539-011-0049-z
9. Toledo, M., Busquets, S., Penna, F., et al.: Complete reversal of muscle wasting in experimental cancer cachexia: additive effects of activin type II receptor inhibition and β-2 agonist. Int. J. Cancer. (2016). https://doi.org/10.1002/ijc.29930

10. Argilés, J.M., Alvarez, B., López-Soriano, F.J.: The metabolic basis of cancer cachexia. Med. Res. Rev. (1997)

11. Fearon, K.C.H., Glass, D.J., Guttridge, D.C.: Cancer cachexia: mediators, signaling, and metabolic pathways. Cell Metab. (2012)

12. Damrauer, J.S., Stadler, M.E., Acharyya, S., et al.: Chemotherapy-induced muscle wasting: association with NF-κB and cancer cachexia. Eur. J. Transl. Myol. (2018). https://doi.org/10.4081/ejtm.2018.7590

13. Pin, F., Barreto, R., Couch, M.E., et al.: Cachexia induced by cancer and chemotherapy yield distinct perturbations to energy metabolism. J. Cachexia. Sarcopenia Muscle. (2019). https://doi.org/10.1002/jcsm.12360

14. Colotta, F., Allavena, P., Sica, A., et al.: Cancer-related inflammation, the seventh hallmark of cancer: links to genetic instability. Carcinogenesis. (2009)

15. Hanahan, D., Weinberg, R.A.: Hallmarks of cancer: the next generation. Cell. (2011)

16. Diakos, C.I., Charles, K.A., McMillan, D.C., Clarke, S.J.: Cancer-related inflammation and treatment effectiveness. Lancet Oncol. (2014)

17. Roxburgh, C.S.D., McMillan, D.C.: Cancer and systemic inflammation: treat the tumour and treat the host. Br. J. Cancer. (2014). https://doi.org/10.1038/bjc.2014.90

18. Deans, D.A.C., Tan, B.H., Wigmore, S.J., et al.: The influence of systemic inflammation, dietary intake and stage of disease on rate of weight loss in patients with gastro-oesophageal cancer. Br. J. Cancer. (2009). https://doi.org/10.1038/sj.bjc.6604828

19. Dülger, H., Alici, S., Şekeroğlu, M.R., et al.: Serum levels of leptin and proinflammatory cytokines in patients with gastrointestinal cancer. Int. J. Clin. Pract. (2004). https://doi.org/10.1111/j.1368-5031.2004.00149.x

20. Mantovani, G., Macciò, A., Madeddu, C., et al.: Serum values of proinflammatory cytokines are inversely correlated with serum leptin levels in patients with advanced stage cancer at different sites. J. Mol. Med. (2001). https://doi.org/10.1007/s001090100234

21. Staal-van Den Brekel, A.J., Dentener, M.A., Schols, A.M.W.J., et al.: Increased resting energy expenditure and weight loss are related to a systemic inflammatory response in lung cancer patients. J. Clin. Oncol. (1995). https://doi.org/10.1200/JCO.1995.13.10.2600

22. Tas, F., Duranyildiz, D., Argon, A., et al.: Serum levels of leptin and proinflammatory cytokines in advanced-stage non-small cell lung cancer. Med. Oncol. (2005). https://doi.org/10.1385/MO:22:4:353

23. Gonzalez, H., Hagerling, C., Werb, Z.: Roles of the immune system in cancer: From tumor initiation to metastatic progression. Genes Dev. (2018)

24. Fridman, W.H., Pagès, F., Saut̀s-Fridman, C., Galon, J.: The immune contexture in human tumours: impact on clinical outcome. Nat. Rev. Cancer. (2012)

25. Casneuf, T., Axel, A.E., King, P., et al.: Interleukin-6 is a potential therapeutic target in interleukin-6 dependent, estrogen receptor-α-positive breast cancer. Breast Cancer Targets Therapy. (2016). https://doi.org/10.2147/BCTT.S92414

26. Edwardson, D.W., Boudreau, J., Mapletoft, J., et al.: Inflammatory cytokine production in tumor cells upon chemotherapy drug exposure or upon selection for drug resistance. PLoS One. (2017). https://doi.org/10.1371/journal.pone.0183662

27. Sprowl, J.A., Reed, K., Armstrong, S.R., et al.: Alterations in tumor necrosis factor signaling pathways are associated with cytotoxicity and resistance to taxanes: a study in isogenic resistant tumor cells. Breast Cancer Res. (2012). https://doi.org/10.1186/bcr3083

28. Frost, R.A., Nystrom, G.J., Jefferson, L.S., Lang, C.H.: Hormone, cytokine, and nutritional regulation of sepsis-induced increases in atrogin-1 and MuRF1 in skeletal muscle. Am. J. Physiol. Endocrinol. Metab. (2007). https://doi.org/10.1152/ajpendo.00359.2006

29. Li, Y.P., Chen, Y., John, J., et al.: TNF-α acts via p38 MAPK to stimulate expression of the ubiquitin ligase atrogin1/MAFbx in skeletal muscle. FASEB J. (2005). https://doi.org/10.1096/fj.04-2364com

30. Moylan, J.S., Smith, J.D., Chambers, M.A., et al.: TNF induction of atrogin-1/MAFbx mRNA depends on Foxo4 expression but not AKT-Foxo1/3 signaling. Am. J. Physiol. Cell Physiol. (2008). https://doi.org/10.1152/ajpcell.00041.2008

31. Sishi, B.J.N., Engelbrecht, A.M.: Tumor necrosis factor alpha (TNF-α) inactivates the PI3-kinase/PKB pathway and induces atrophy and apoptosis in L6 myotubes. Cytokine. (2011). https://doi.org/10.1016/j.cyto.2011.01.009

32. Oliff, A., Defeo-Jones, D., Boyer, M., et al.: Tumors secreting human TNF/cachectin induce cachexia in mice. Cell. (1987). https://doi.org/10.1016/0092-8674(87)90028-6

33. Jatoi, A., Ritter, H.L., Dueck, A., et al.: A placebo-controlled, double-blind trial of infliximab for cancer-associated weight loss in elderly and/or poor performance non-small cell lung cancer patients (N01C9). Lung Cancer. (2010). https://doi.org/10.1016/j.lungcan.2009.06.020

34. Baltgalvis, K.A., Berger, F.G., Pena, M.M.O., et al.: Interleukin-6 and cachexia in ApcMin/+ mice. Am. J. Physiol. Regul. Integr. Comp. Physiol. (2008). https://doi.org/10.1152/ajpregu.00716.2007

35. Carson, J.A., Baltgalvis, K.A.: Interleukin 6 as a key regulator of muscle mass during cachexia. Exerc. Sport Sci. Rev. (2010). https://doi.org/10.1097/JES.0b013e3181f44f11

36. Moses, A.G.W., Maingay, J., Sangster, K., et al.: Pro-inflammatory cytokine release by peripheral blood mononuclear cells from patients with advanced

pancreatic cancer: relationship to acute phase response and survival. Oncol. Rep. (2009). https://doi.org/10.3892/or_00000328

37. Scott, H.R., McMillan, D.C., Crilly, A., et al.: The relationship between weight loss and interleukin 6 in non-small-cell lung cancer. Br. J. Cancer. (1996). https://doi.org/10.1038/bjc.1996.294

38. Fujlta, J., Tsujinaka, T., Yano, M., et al.: Anti-interleukin-6 receptor antibody prevents muscle atrophy in colon-26 adenocarcinoma-bearing mice with modulation of lysosomal and ATP-ubiquitin-dependent proteolytic pathways. Int. J. Cancer. (1996). 10.1002/(SICI)1097-0215(19961127)68:5<637::AID-IJC14>3.0.CO;2-Z

39. White, J.P., Baynes, J.W., Welle, S.L., et al.: The regulation of skeletal muscle protein turnover during the progression of cancer cachexia in the Apc Min/+ mouse. PLoS One. (2011). https://doi.org/10.1371/journal.pone.0024650

40. Bayliss, T.J., Smith, J.T., Schuster, M., et al.: A humanized anti-IL-6 antibody (ALD518) in non-small cell lung cancer. Expert. Opin. Biol. Ther. (2011). https://doi.org/10.1517/14712598.2011.627850

41. Dogra, C., Changotra, H., Wedhas, N., et al.: TNF-related weak inducer of apoptosis (TWEAK) is a potent skeletal muscle-wasting cytokine. FASEB J. (2007). https://doi.org/10.1096/fj.06-7537com

42. Mittal, A., Bhatnagar, S., Kumar, A., et al.: The TWEAK-Fn14 system is a critical regulator of denervation-induced skeletal muscle atrophy in mice. J. Cell Biol. (2010). https://doi.org/10.1083/jcb.200909117

43. Guadagnin, E., Mázala, D., Chen, Y.W.: STAT3 in skeletal muscle function and disorders. Int. J. Mol. Sci. (2018)

44. Sala, D., Sacco, A.: Signal transducer and activator of transcription 3 signaling as a potential target to treat muscle wasting diseases. Curr. Opin. Clin. Nutr. Mctab. Care. (2016)

45. Bonetto, A., Aydogdu, T., Kunzevitzky, N., et al.: STAT3 activation in skeletal muscle links muscle wasting and the acute phase response in cancer cachexia. PLoS One. (2011). https://doi.org/10.1371/journal.pone.0022538

46. Zhang, L., Pan, J., Dong, Y., et al.: Stat3 activation links a C/EBPδ to myostatin pathway to stimulate loss of muscle mass. Cell Metab. (2013a). https://doi.org/10.1016/j.cmet.2013.07.012

47. Allen, D.L., Cleary, A.S., Hanson, A.M., et al.: CCAAT/enhancer binding protein-δ expression is increased in fast skeletal muscle by food deprivation and regulates myostatin transcription in vitro. Am. J. Physiol. Regul. Integr. Comp. Physiol. (2010). https://doi.org/10.1152/ajpregu.00247.2010

48. Gilabert, M., Calvo, E., Airoldi, A., et al.: Pancreatic cancer-induced cachexia is Jak2-dependent in

mice. J. Cell. Physiol. (2014). https://doi.org/10.1002/jcp.24580

49. Silva, K.A.S., Dong, J., Dong, Y., et al.: Inhibition of Stat3 activation suppresses caspase-3 and the ubiquitin-proteasome system, leading to preservation of muscle mass in cancer cachexia. J. Biol. Chem. (2015). https://doi.org/10.1074/jbc.M115.641514

50. Chen, L., Xu, W., Yang, Q., et al.: Imperatorin alleviates cancer cachexia and prevents muscle wasting via directly inhibiting STAT3. Pharmacol. Res. (2020). https://doi.org/10.1016/j.phrs.2020.104871

51. Cai, D., Frantz Jr., J., Nicholas Jr., E.T., et al.: IKK [beta]/NF-[kappa] B activation causes severe muscle wasting in mice. Cell. (2004)

52. Thoma, A., Lightfoot, A.P.: Nf-kb and inflammatory cytokine signalling: role in skeletal muscle atrophy. In: Advances in experimental medicine and biology (2018)

53. Ashall, L., Horton, C.A., Nelson, D.E., et al.: Pulsatile stimulation determines timing and specificity of NF-κB-dependent transcription. Science. (2009). https://doi.org/10.1126/science.1164860

54. Acharyya, S., Butchbach, M.E.R., Sahenk, Z., et al.: Dystrophin glycoprotein complex dysfunction: a regulatory link between muscular dystrophy and cancer cachexia. Cancer Cell. (2005). https://doi.org/10.1016/j.ccr.2005.10.004

55. Rhoads, M.G., Kandarian, S.C., Pacelli, F., et al.: Expression of NF-κB and IκB proteins in skeletal muscle of gastric cancer patients. Eur. J. Cancer. (2010). https://doi.org/10.1016/j.ejca.2009.10.008

56. Kawamura, I., Morishita, R., Tomita, N., et al.: Intratumoral injection of oligonucleotides to the NFκB binding site inhibits cachexia in a mouse tumor model. Gene Ther. (1999). https://doi.org/10.1038/sj.gt.3300819

57. Braun, T.P., Grossberg, A.J., Krasnow, S.M., et al.: Cancer- and endotoxin-induced cachexia require intact glucocorticoid signaling in skeletal muscle. FASEB J. (2013). https://doi.org/10.1096/fj.13-230375

58. Gautron, L.: Neurobiology of inflammation-associated anorexia. Front. Neurosci. (2009). https://doi.org/10.3389/neuro.23.003.2009

59. Cole, C.L., Kleckner, I.R., Jatoi, A., et al.: The role of systemic inflammation in cancer-associated muscle wasting and rationale for exercise as a therapeutic intervention. JCSM Clin. Reports. (2018). https://doi.org/10.17987/jcsm-cr.v3i2.65

60. Fuller, J.T., Hartland, M.C., Maloney, L.T., Davison, K.: Therapeutic effects of aerobic and resistance exercises for cancer survivors: a systematic review of meta-analyses of clinical trials. Br. J. Sports Med. (2018)

61. Lønbro, S., Dalgas, U., Primdahl, H., et al.: Progressive resistance training rebuilds lean body mass in head and neck cancer patients after radiotherapy – results from the randomized DAHANCA 25B trial.

Radiother. Oncol. (2013). https://doi.org/10.1016/j.radonc.2013.07.002

62. Mishra, S.I., Scherer, R.W., Geigle, P.M., et al.: Exercise interventions on health-related quality of life for cancer survivors. Cochrane Database Syst. Rev. (2012)

63. Nadler, M.B., Desnoyers, A., Langelier, D.M., Amir, E.: The effect of exercise on quality of life, fatigue, physical function, and safety in advanced solid tumor cancers: a meta-analysis of randomized control trials. J. Pain Sympt. Manag. (2019)

64. Oldervoll, L.M., Loge, J.H., Lydersen, S., et al.: Physical exercise for cancer patients with advanced disease: a randomized controlled trial. Oncologist. (2011). https://doi.org/10.1634/theoncologist.2011-0133

65. Alves, C.R.R., Da Cunha, T.F., Da Paixão, N.A., Brum, P.C.: Aerobic exercise training as therapy for cardiac and cancer cachexia. Life Sci. (2015)

66. Tanaka, M., Sugimoto, K., Fujimoto, T., et al.: Preventive effects of low-intensity exercise on cancer cachexia-induced muscle atrophy. FASEB J. (2019). https://doi.org/10.1096/fj.201802430R

67. Gleeson, M., Bishop, N.C., Stensel, D.J., et al.: The anti-inflammatory effects of exercise: mechanisms and implications for the prevention and treatment of disease. Nat. Rev. Immunol. (2011)

68. Kasapis, C., Thompson, P.D.: The effects of physical activity on serum C-reactive protein and inflammatory markers: a systematic review. J. Am. Coll. Cardiol. (2005)

69. Petersen, A.M.W., Pedersen, B.K.: The anti-inflammatory effect of exercise. J. Appl. Physiol. (2005)

70. Moses, H.L.: The discovery of TGF-β: a historical perspective. In: The TGF-β Family (2008)

71. Syed, V.: TGF-β signaling in cancer. J. Cell. Biochem. (2016). https://doi.org/10.1002/jcb.25496

72. Han, H.Q., Mitch, W.E.: Targeting the myostatin signaling pathway to treat muscle wasting diseases. Curr. Opin. Support. Palliative Care. (2011)

73. Han, H.Q., Zhou, X., Mitch, W.E., Goldberg, A.L.: Myostatin/activin pathway antagonism: molecular basis and therapeutic potential. Int. J. Biochem. Cell Biol. (2013)

74. Cassar-Malek, I., Passelaigue, F., Bernard, C., et al.: Target genes of myostatin loss-of-function in muscles of late bovine fetuses. BMC Genomics. (2007). https://doi.org/10.1186/1471-2164-8-63

75. Clop, A., Marcq, F., Takeda, H., et al.: A mutation creating a potential illegitimate microRNA target site in the myostatin gene affects muscularity in sheep. Nat. Genet. (2006). https://doi.org/10.1038/ng1810

76. McPherron, A.C., Lawler, A.M., Lee, S.J.: Regulation of skeletal muscle mass in mice by a new TGF-β superfamily member. Nature. (1997). https://doi.org/10.1038/387083a0

77. Mosher, D.S., Quignon, P., Bustamante, C.D., et al.: A mutation in the myostatin gene increases muscle mass and enhances racing performance in heterozygote dogs. PLoS Genet. (2007). https://doi.org/10.1371/journal.pgen.0030079

78. Schuelke, M., Wagner, K.R., Stolz, L.E., et al.: Myostatin mutation associated with gross muscle hypertrophy in a child. N. Engl. J. Med. (2004). https://doi.org/10.1056/NEJMoa040933

79. Durieux, A.C., Amirouche, A., Banzet, S., et al.: Ectopic expression of myostatin induces atrophy of adult skeletal muscle by decreasing muscle gene expression. Endocrinology. (2007). https://doi.org/10.1210/en.2006-1500

80. Zimmers, T.A., Davies, M.V., Koniaris, L.G., et al.: Induction of cachexia in mice by systemically administered myostatin. Science. **296**, 1486–1488 (2002). https://doi.org/10.1126/science.1069525

81. Breitbart, A., Auger-Messier, M., Molkentin, J.D., Heineke, J.: Myostatin from the heart: local and systemic actions in cardiac failure and muscle wasting. Am. J. Physiol. Heart Circ. Physiol. (2011). https://doi.org/10.1152/ajpheart.00200.2011

82. Breitbart, A., Scharf, G.M., Duncker, D., et al.: Highly specific detection of myostatin prodomain by an immunoradiometric sandwich assay in serum of healthy individuals and patients. PLoS One. (2013). https://doi.org/10.1371/journal.pone.0080454

83. Gonzalez-Cadavid, N.F., Taylor, W.E., Yarasheski, K., et al.: Organization of the human myostatin gene and expression in healthy men and HIV-infected men with muscle wasting. Proc. Natl. Acad. Sci. USA. (1998). https://doi.org/10.1073/pnas.95.25.14938

84. Loumaye, A., De Barsy, M., Nachit, M., et al.: Role of activin A and myostatin in human cancer cachexia. J. Clin. Endocrinol. Metab. (2015). https://doi.org/10.1210/jc.2014-4318

85. Yarasheski, K.E., Bhasin, S., Sinha-Hikim, I., et al.: Serum myostatin-immunoreactive protein is increased in 60–92 year old women and men with muscle wasting. J. Nutr. Health Aging. (2002)

86. de Kretser, D.M., O'Hehir, R.E., Hardy, C.L., Hedger, M.P.: The roles of activin A and its binding protein, follistatin, in inflammation and tissue repair. Mol. Cell. Endocrinol. (2012)

87. Hardy, C.L., Rolland, J.M., O'Hehir, R.E.: The immunoregulatory and fibrotic roles of activin A in allergic asthma. Clin. Exp. Allergy. (2015)

88. Jones, K.L., Mansell, A., Patella, S., et al.: Activin A is a critical component of the inflammatory response, and its binding protein, follistatin, reduces mortality in endotoxemia. Proc. Natl. Acad. Sci. USA. (2007). https://doi.org/10.1073/pnas.0705971104

89. Dankbar, B., Fennen, M., Brunert, D., et al.: Myostatin is a direct regulator of osteoclast differentiation and its inhibition reduces inflammatory joint destruction in mice. Nat. Med. (2015). https://doi.org/10.1038/nm.3917

90. Zhang, L., Rajan, V., Lin, E., et al.: Pharmacological inhibition of myostatin suppresses systemic

inflammation and muscle atrophy in mice with chronic kidney disease. FASEB J. (2011). https://doi.org/10.1096/fj.10-176917

91. Aversa, Z., Bonetto, A., Penna, F., et al.: Changes in myostatin signaling in non-weight-losing cancer patients. Ann. Surg. Oncol. (2012). https://doi.org/10.1245/s10434-011-1720-5

92. Harada, K., Shintani, Y., Sakamoto, Y., et al.: Serum immunoreactive activin A levels in normal subjects and patients with various diseases. J. Clin. Endocrinol. Metab. (1996). https://doi.org/10.1210/jc.81.6.2125

93. Incorvaia, L., Badalamenti, G., Rini, G., et al.: MMP-2, MMP-9 and activin a blood levels in patients with breast cancer or prostate cancer metastatic to the bone. Anticancer Res. (2007)

94. Leto, G., Incorvaia, L., Badalamenti, G., et al.: Activin A circulating levels in patients with bone metastasis from breast or prostate cancer. Clin. Exp. Metastasis. (2006). https://doi.org/10.1007/s10585-006-9010-5

95. Costelli, P., Muscaritoli, M., Bonetto, A., et al.: Muscle myostatin signalling is enhanced in experimental cancer cachexia. Eur. J. Clin. Investig. (2008). https://doi.org/10.1111/j.1365-2362.2008.01970.x

96. Benny Klimek, M.E., Aydogdu, T., Link, M.J., et al.: Acute inhibition of myostatin-family proteins preserves skeletal muscle in mouse models of cancer cachexia. Biochem. Biophys. Res. Commun. (2010). https://doi.org/10.1016/j.bbrc.2009.12.123

97. Barreto, R., Kitase, Y., Matsumoto, T., et al.: ACVR2B/Fc counteracts chemotherapy-induced loss of muscle and bone mass. Sci. Rep. (2017). https://doi.org/10.1038/s41598-017-15040-1

98. O'Connell, T.M., Pin, F., Couch, M.E., Bonetto, A.: Treatment with soluble activin receptor type IIB alters metabolic response in chemotherapy-induced cachexia. Cancers. (2019). https://doi.org/10.3390/cancers11091222

99. Greco, S.H., Tomkötter, L., Vahle, A.K., et al.: TGF-β blockade reduces mortality and metabolic changes in a validated murine model of pancreatic cancer cachexia. PLoS One. (2015). https://doi.org/10.1371/journal.pone.0132786

100. Smith, R.C., Lin, B.K.: Myostatin inhibitors as therapies for muscle wasting associated with cancer and other disorders. Curr. Opin. Support. Palliative Care. (2013)

101. Jameson, G.S., Von Hoff, D.D., Weiss, G.J., et al.: Safety of the antimyostatin monoclonal antibody LY2495655 in healthy subjects and patients with advanced cancer. J. Clin. Oncol. (2012). https://doi.org/10.1200/jco.2012.30.15_suppl.2516

102. Yakovenko, A., Cameron, M., Trevino, J.G.: Molecular therapeutic strategies targeting pancreatic cancer induced cachexia. World J. Gastroint. Surg. (2018). https://doi.org/10.4240/wjgs.v10.i9.95

103. Dieli-Conwright, C.M., Spektor, T.M., Rice, J.C., et al.: Hormone therapy and maximal eccentric exercise alters myostatin-related gene expression in postmenopausal women. J. Strength Cond. Res. (2012). https://doi.org/10.1519/JSC.0b013e318251083f

104. Matsakas, A., Friedel, A., Hertrampf, T., Diel, P.: Short-term endurance training results in a muscle-specific decrease of myostatin mRNA content in the rat. Acta Physiol. Scand. (2005). https://doi.org/10.1111/j.1365-201X.2005.01406.x

105. Bassi, D., Bueno, P.d.G., Nonaka, K.O., et al.: Exercise alters myostatin protein expression in sedentary and exercised streptozotocin-diabetic rats. Arch. Endocrinol. Metab. (2015). https://doi.org/10.1590/2359-3997000000028

106. Hittel, D.S., Axelson, M., Sarna, N., et al.: Myostatin decreases with aerobic exercise and associates with insulin resistance. Med. Sci. Sports Exerc. (2010). https://doi.org/10.1249/MSS.0b013e3181e0b9a8

107. Kopple, J.D., Cohen, A.H., Wang, H., et al.: Effect of exercise on mRNA levels for growth factors in skeletal muscle of hemodialysis patients. J. Ren. Nutr. (2006). https://doi.org/10.1053/j.jrn.2006.04.028

108. Lenk, K., Schur, R., Linke, A., et al.: Impact of exercise training on myostatin expression in the myocardium and skeletal muscle in a chronic heart failure model. Eur. J. Heart Fail. (2009). https://doi.org/10.1093/eurjhf/hfp020

109. Konopka, A.R., Douglass, M.D., Kaminsky, L.A., et al.: Molecular adaptations to aerobic exercise training in skeletal muscle of older women. J. Gerontol. Ser A Biol. Sci. Med. Sci. (2010). https://doi.org/10.1093/gerona/glq109

110. Ryan, A.S., Li, G., Blumenthal, J.B., Ortmeyer, H.K.: Aerobic exercise + weight loss decreases skeletal muscle myostatin expression and improves insulin sensitivity in older adults. Obesity. (2013). https://doi.org/10.1002/oby.20216

111. Coutinho, A.E., Chapman, K.E.: The anti-inflammatory and immunosuppressive effects of glucocorticoids, recent developments and mechanistic insights. Mol. Cell. Endocrinol. 335, 2–13 (2011). https://doi.org/10.1016/j.mce.2010.04.005

112. Tanaka, Y., Eda, H., Tanaka, T., et al.: Experimental cancer cachexia induced by transplantable colon 26 adenocarcinoma in mice. Cancer Res. (1990)

113. Tiao, G., Fagan, J., Roegner, V., et al.: Energy-ubiquitin-dependent muscle proteolysis during sepsis in rats is regulated by glucocorticoids. J. Clin. Investig. (1996). https://doi.org/10.1172/JCI118421

114. Braun, T.P., Szumowski, M., Levasseur, P.R., et al.: Muscle atrophy in response to cytotoxic chemotherapy is dependent on intact glucocorticoid signaling in skeletal muscle. PLoS One. (2014). https://doi.org/10.1371/journal.pone.0106489

115. Schakman, O., Kalista, S., Barbé, C., et al.: Glucocorticoid-induced skeletal muscle atrophy. Int. J. Biochem. Cell Biol. (2013)

116. Exton, J.H.: Regulation of gluconeogenesis by glucocorticoids. Monogr. Endocrinol. (1979)

117. Morgan, S.A., Sherlock, M., Gathercole, L.L., et al.: 11β-hydroxysteroid dehydrogenase type 1 regulates glucocorticoid-induced insulin resistance in skeletal muscle. Diabetes. (2009). https://doi.org/10.2337/db09-0525

118. Nakao, R., Hirasaka, K., Goto, J., et al.: Ubiquitin ligase Cbl-b is a negative regulator for insulin-like growth factor 1 signaling during muscle atrophy caused by unloading. Mol. Cell. Biol. (2009). https://doi.org/10.1128/mcb.01347-08

119. Zheng, B., Ohkawa, S., Li, H., et al.: FOXO3a mediates signaling crosstalk that coordinates ubiquitin and atrogin-1/MAFbx expression during glucocorticoid-induced skeletal muscle atrophy. FASEB J. (2010). https://doi.org/10.1096/fj.09-151480

120. Giorgino, F., Pedrini, M.T., Matera, L., Smithi, R.J.: Specific increase in p85α expression in response to dexamethasone is associated with inhibition of insulin-like growth factor-I stimulated phosphatidylinositol 3-kinase activity in cultured muscle cells. J. Biol. Chem. (1997). https://doi.org/10.1074/jbc.272.11.7455

121. Hu, Z., Wang, H., In, H.L., et al.: Endogenous glucocorticoids and impaired insulin signaling are both required to stimulate muscle wasting under pathophysiological conditions in mice. J. Clin. Investig. (2009). https://doi.org/10.1172/JCI38770

122. Kuo, T., Lew, M.J., Mayba, O., et al.: Genome-wide analysis of glucocorticoid receptor-binding sites in myotubes identifies gene networks modulating insulin signaling. Proc. Natl. Acad. Sci. USA. (2012). https://doi.org/10.1073/pnas.1111334109

123. Artaza, J.N., Bhasin, S., Mallidis, C., et al.: Endogenous expression and localization of myostatin and its relation to myosin heavy chain distribution in C2C12 skeletal muscle cells. J. Cell. Physiol. (2002). https://doi.org/10.1002/jcp.10044

124. Ma, K., Mallidis, C., Bhasin, S., et al.: Glucocorticoid-induced skeletal muscle atrophy is associated with upregulation of myostatin gene expression. Am. J. Physiol. Endocrinol. Metab. (2003). https://doi.org/10.1152/ajpendo.00487.2002

125. McFarlane, C., Plummer, E., Thomas, M., et al.: Myostatin induces cachexia by activating the ubiquitin proteolytic system through an NF-κB-independent, FoxO1-dependent mechanism. J. Cell. Physiol. (2006). https://doi.org/10.1002/jcp.20757

126. Cersosimo, E., Pisters, P.W.T., Pesola, G., et al.: The effect of graded doses of insulin on peripheral glucose uptake and lactate release in cancer cachexia. Surgery. (1991)

127. Dodesini, A.R., Benedini, S., Terruzzi, I., et al.: Protein, glucose and lipid metabolism in the cancer cachexia: a preliminary report [1]. Acta Oncol. (2007)

128. Glicksman, A.S., Rawson, R.W.: Diabetes and altered carbohydrate metabolism in patients with cancer. Cancer. (1956) 10.1002/1097-0142(195611/12)9:6<1127::AID-CNCR2820090610>3.0.CO;2-4

129. Rohdenburg, G.L., Bernhard, A., Krehbiel, O.: Sugar tolerance in cancer. J. Am. Med. Assoc. (1919). https://doi.org/10.1001/jama.1919.02610210024007

130. Tayek, J.A.: A review of cancer cachexia and abnormal glucose metabolism in humans with cancer. J. Am. Coll. Nutr. (1992). https://doi.org/10.1080/07315724.1992.10718249

131. Asp, M.L., Tian, M., Wendel, A.A., Belury, M.A.: Evidence for the contribution of insulin resistance to the development of cachexia in tumor-bearing mice. Int. J. Cancer. (2010). https://doi.org/10.1002/ijc.24784

132. Argilés, J.M., Fontes-Oliveira, C.C., Toledo, M., et al.: Cachexia: a problem of energetic inefficiency. J. Cachexia Sarcopenia Muscle. (2014)

133. Hyltander, A., Drott, C., Körner, U., et al.: Elevated energy expenditure in cancer patients with solid tumours. Eur. J. Cancer Clin. Oncol. (1991). https://doi.org/10.1016/0277-5379(91)90050-N

134. Aoyagi, T., Terracina, K.P., Raza, A., et al.: Cancer cachexia, mechanism and treatment. World J. Gastrointest. Oncol. (2015). https://doi.org/10.4251/wjgo.v7.i4.17

135. Faber, J., Uitdehaag, M.J., Spaander, M., et al.: Improved body weight and performance status and reduced serum PGE2 levels after nutritional intervention with a specific medical food in newly diagnosed patients with esophageal cancer or adenocarcinoma of the gastro-esophageal junction. J. Cachexia. Sarcopenia Muscle. (2015). https://doi.org/10.1002/jcsm.12009

136. Leśniak, W., Bała, M., Jaeschke, R., Krzakowski, M.: Effects of megestrol acetate in patients with cancer anorexia-cachexia syndrome – a systematic review and meta-analysis. Polskie Archiwum Medycyny Wewnetrznej. (2008). https://doi.org/10.20452/pamw.510

137. López AP, Roqué I Figuls M, Cuchi GU, et al (2004) Systematic review of megestrol acetate in the treatment of anorexia-cachexia syndrome. J. Pain Sympt. Manag.

138. Ruiz Garcia, V., López-Briz, E., Carbonell Sanchis, R., et al.: Megestrol acetate for treatment of anorexia-cachexia syndrome. Cochrane Database Syst. Rev. (2013)

139. Tugba, Y., Davis, M.P., Declan, W., et al.: Systematic review of the treatment of cancer-associated anorexia and weight loss. J. Clin. Oncol. **23**(33), 8500–8511 (2005)

140. Taylor, J.K., Pendleton, N.: Progesterone therapy for the treatment of non-cancer cachexia: a systematic review. BMJ Support. Palliative Care. (2016)

141. Lecker, S.H., Jagoe, R.T., Gilbert, A., et al.: Multiple types of skeletal muscle atrophy involve a common program of changes in gene expression. FASEB J. **18**, 39–51 (2004). https://doi.org/10.1096/fj.03-0610com

142. Sacheck, J.M., Hyatt, J.P., Raffaello, A., et al.: Rapid disuse and denervation atrophy involve transcriptional changes similar to those of muscle wasting during systemic diseases. FASEB J. **21**, 140–155 (2007). https://doi.org/10.1096/fj.06-6604com

143. Mammucari, C., Milan, G., Romanello, V., et al.: FoxO3 controls autophagy in skeletal muscle in vivo. Cell Metab. **6**, 458–471 (2007)

144. Sandri, M.: Regulation and involvement of the ubiquitin ligases in muscle atrophy. Free Radic. Biol. Med. (2014). https://doi.org/10.1016/j.freeradbiomed.2014.10.833

145. Zhao, J., Brault, J.J., Schild, A., et al.: FoxO3 coordinately activates protein degradation by the autophagic/lysosomal and proteasomal pathways in atrophying muscle cells. Cell Metab. (2007). https://doi.org/10.1016/j.cmet.2007.11.004

146. Brault, J.J., Jespersen, J.G., Goldberg, A.L.: Peroxisome proliferator-activated receptor gamma coactivator 1alpha or 1beta overexpression inhibits muscle protein degradation, induction of ubiquitin ligases, and disuse atrophy. J. Biol. Chem. **285**, 19460–19471 (2010). https://doi.org/10.1074/jbc.M110.113092

147. Sandri, M., Lin, J., Handschin, C., et al.: PGC-1α protects skeletal muscle from atrophy by suppressing FoxO3 action and atrophy-specific gene transcription. Proc. Natl. Acad. Sci. USA. (2006). https://doi.org/10.1073/pnas.0607795103

148. Raffaello, A., Milan, G., Masiero, E., et al.: JunB transcription factor maintains skeletal muscle mass and promotes hypertrophy. J. Cell Biol. **191**, 101–113 (2010). https://doi.org/10.1083/jcb.201001136

149. Lee, D., Goldberg, A.L.: SIRT1 protein, by blocking the activities of transcription factors FoxO1 and FoxO3, inhibits muscle atrophy and promotes muscle growth. J. Biol. Chem. (2013). https://doi.org/10.1074/jbc.M113.489716

150. Glass, D.J.: PI3 kinase regulation of skeletal muscle hypertrophy and atrophy. Curr. Top. Microbiol. Immunol. (2010). https://doi.org/10.1007/82-2010-78

151. Scott, R.C., Schuldiner, O., Neufeld, T.P.: Role and regulation of starvation-induced autophagy in the Drosophila fat body. Dev. Cell. (2004). https://doi.org/10.1016/j.devcel.2004.07.009

152. Latres, E., Amini, A.R., Amini, A.A., et al.: Insulin-like growth factor-1 (IGF-1) inversely regulates atrophy-induced genes via the phosphatidylinositol 3-kinase/Akt/mammalian target of rapamycin (PI3K/Akt/mTOR) pathway. J. Biol. Chem. (2005). https://doi.org/10.1074/jbc.M407517200

153. Shimizu, N., Yoshikawa, N., Ito, N., et al.: Crosstalk between glucocorticoid receptor and nutritional sensor mTOR in skeletal muscle. Cell Metab. (2011). https://doi.org/10.1016/j.cmet.2011.01.001

154. Wang, X., Hu, Z., Hu, J., et al.: Insulin resistance accelerates muscle protein degradation: activation of the ubiquitin-proteasome pathway by defects in muscle cell signaling. Endocrinology. (2006). https://doi.org/10.1210/en.2006-0251

155. Watts, R., McAinch, A.J., Dixon, J.B., et al.: Increased Smad signaling and reduced MRF expression in skeletal muscle from obese subjects. Obesity. (2013). https://doi.org/10.1002/oby.20070

156. Baracos, V.E., DeVivo, C., Hoyle, D.H.R., Goldberg, A.L.: Activation of the ATP-ubiquitin-proteasome pathway in skeletal muscle of cachectic rats bearing a hepatoma. Am. J. Physiol. Endocrinol. Metab. (1995). https://doi.org/10.1152/ajpendo.1995.268.5.e996

157. Khal, J., Wyke, S.M., Russell, S.T., et al.: Expression of the ubiquitin-proteasome pathway and muscle loss in experimental cancer cachexia. Br. J. Cancer. (2005b). https://doi.org/10.1038/sj.bjc.6602780

158. Penna, F., Ballarò, R., Martinez-Cristobal, P., et al.: Autophagy exacerbates muscle wasting in cancer cachexia and impairs mitochondrial function. J. Mol. Biol. (2019). https://doi.org/10.1016/j.jmb.2019.05.032

159. Yuan, L., Han, J., Meng, Q., et al.: Muscle-specific E3 ubiquitin ligases are involved in muscle atrophy of cancer cachexia: an in vitro and in vivo study. Oncol. Rep. (2015). https://doi.org/10.3892/or.2015.3845

160. Bodine, S.C., Latres, E., Baumhueter, S., et al.: Identification of ubiquitin ligases required for skeletal Muscle Atrophy. Science. (2001). https://doi.org/10.1126/science.1065874

161. Cohen, S., Nathan, J.A., Goldberg, A.L.: Muscle wasting in disease: molecular mechanisms and promising therapies. Nat. Rev. Drug Discov. **14**, 58–74 (2015). https://doi.org/10.1038/nrd4467

162. Chakraborty, R., Mukkamalla, S.K.R., Calderon, N.: Bortezomib induced reversible left ventricular systolic dysfunction: a case report and review of literature. Br. J. Med. Pract. (2013)

163. Grandin, E.W., Ky, B., Cornell, R.F., et al.: Patterns of cardiac toxicity associated with irreversible proteasome inhibition in the treatment of multiple myeloma. J. Card. Fail. (2015). https://doi.org/10.1016/j.cardfail.2014.11.008

164. Hacihanefioglu, A., Tarkun, P., Gonullu, E.: Acute severe cardiac failure in a myeloma patient due to proteasome inhibitor bortezomib. Int. J. Hematol. (2008). https://doi.org/10.1007/s12185-008-0139-7

165. Herrmann, J., Wohlert, C., Saguner, A.M., et al.: Primary proteasome inhibition results in cardiac dysfunction. Eur. J. Heart Fail. (2013). https://doi.org/10.1093/eurjhf/hft034

166. Jorge, R., Patricia, C., Asaf, R., et al.: Left ventricular dysfunction development after bortezomib treatment in patients with oncologic diagnoses. Circulation. (2013)

167. Jerkins, J.: Bortezomib-induced severe congestive heart failure. Cardiol. Res. (2010). https://doi.org/10.4021/cr105e

168. Orciuolo, E., Gabriele, B., Cecconi, N., et al.: Unexpected cardiotoxicity in haematological bortezomib treated patients [1]. Br. J. Haematol. (2007)

169. Takakuwa, T., Otomaru, I., Araki, T., et al.: The first autopsy case of fatal acute cardiac failure after administration of carfilzomib in a patient with multiple myeloma. Case Reports Hematol. (2019). https://doi.org/10.1155/2019/1816287

170. Voortman, J., Giaccone, G.: Severe reversible cardiac failure after bortezomib treatment combined with chemotherapy in a non-small cell lung cancer patient: a case report. BMC Cancer. (2006). https://doi.org/10.1186/1471-2407-6-129

171. Penna, F., Bonetto, A., Aversa, Z., et al.: Effect of the specific proteasome inhibitor bortezomib on cancer-related muscle wasting. J. Cachexia. Sarcopenia Muscle. (2016). https://doi.org/10.1002/jcsm.12050

172. Zhang, L., Tang, H., Kou, Y., et al.: MG132-mediated inhibition of the ubiquitin-proteasome pathway ameliorates cancer cachexia. J. Cancer Res. Clin. Oncol. (2013b). https://doi.org/10.1007/s00432-013-1412-6

173. Bowen, T.S., Adams, V., Werner, S., et al.: Small-molecule inhibition of MuRF1 attenuates skeletal muscle atrophy and dysfunction in cardiac cachexia. J. Cachexia. Sarcopenia Muscle. (2017). https://doi.org/10.1002/jcsm.12233

174. Cong, H., Sun, L., Liu, C., Tien, P.: Inhibition of atrogin-1/MAFbx expression by adenovirus-delivered small hairpin RNAs attenuates muscle atrophy in fasting mice. Hum. Gene Ther. (2011). https://doi.org/10.1089/hum.2010.057

175. Wada, S., Kato, Y., Okutsu, M., et al.: Translational suppression of atrophic regulators by Micro RNA-23a integrates resistance to skeletal muscle atrophy. J. Biol. Chem. (2011). https://doi.org/10.1074/jbc.M111.271270

176. Klaude, M., Fredriksson, K., Tjäder, I., et al.: Proteasome proteolytic activity in skeletal muscle is increased in patients with sepsis. Clin. Sci. (2007). https://doi.org/10.1042/CS20060265

177. Tiao, G., Hobler, S., Wang, J.J., et al.: Sepsis is associated with increased mRNAs of the ubiquitin-proteasome proteolytic pathway in human skeletal muscle. J. Clin. Investig. (1997). https://doi.org/10.1172/JCI119143

178. Llovera, M., Garcia-Martinez, C., Agell, N., et al.: Ubiquitin and proteasome gene expression is increased in skeletal muscle of slim AIDS patients. Int. J. Mol. Med. (1998). https://doi.org/10.3892/ijmm.2.1.69

179. Al-Khalili, L., de Castro, B.T., Östling, J., et al.: Proteasome inhibition in skeletal muscle cells unmasks metabolic derangements in type 2 diabetes. Am. J. Physiol. Cell Physiol. (2014). https://doi.org/10.1152/ajpcell.00110.2014

180. Suetta, C., Frandsen, U., Jensen, L., et al.: Aging affects the transcriptional regulation of human skeletal muscle disuse atrophy. PLoS One. (2012). https://doi.org/10.1371/journal.pone.0051238

181. Langer, H.T., Senden, J.M.G., Gijsen, A.P., et al.: Muscle atrophy due to nerve damage is accompanied by elevated myofibrillar protein synthesis rates. Front. Physiol. (2018). https://doi.org/10.3389/fphys.2018.01220

182. Bossola, M., Muscaritoli, M., Costelli, P., et al.: Increased muscle proteasome activity correlates with disease severity in gastric cancer patients. Ann. Surg. (2003). https://doi.org/10.1097/01.SLA.0000055225.96357.71

183. Bossola, M., Muscaritoli, M., Costelli, P., et al.: Increased muscle ubiquitin mrna levels in gastric cancer patients. Am. J. Physiol. Regul. Integr. Comp. Physiol. (2001). https://doi.org/10.1152/ajpregu.2001.280.5.r1518

184. Dejong, C.H.C., Busquets, S., Moses, A.G.W., et al.: Systemic inflammation correlates with increased expression of skeletal muscle ubiquitin but not uncoupling proteins in cancer cachexia. Oncol. Rep. (2005). https://doi.org/10.3892/or.14.1.257

185. Khal, J., Hine, A.V., Fearon, K.C.H., et al.: Increased expression of proteasome subunits in skeletal muscle of cancer patients with weight loss. Int. J. Biochem. Cell Biol. (2005a). https://doi.org/10.1016/j.biocel.2004.10.017

186. Sun, Y.S., Ye, Z.Y., Qian, Z.Y., et al.: Expression of TRAF6 and ubiquitin mRNA in skeletal muscle of gastric cancer patients. J. Exp. Clin. Cancer Res. (2012). https://doi.org/10.1186/1756-9966-31-81

187. Williams, A., Sun, X., Fischer, J.E., Hasselgren, P. O.: The expression of genes in the ubiquitin-proteasome proteolytic pathway is increased in skeletal muscle from patients with cancer. Surgery. (1999). https://doi.org/10.1016/S0039-6060(99)70131-5

188. Aniort, J., Stella, A., Philipponnet, C., et al.: Muscle wasting in patients with end-stage renal disease or early-stage lung cancer: common mechanisms at work. J. Cachexia. Sarcopenia Muscle. (2019). https://doi.org/10.1002/jcsm.12376

189. Jagoe, R.T., Redfern, C.P.F., Roberts, R.G., et al.: Skeletal muscle mRNA levels for cathepsin B, but not components of the ubiquitin-proteasome pathway, are increased in patients with lung cancer referred for thoracotomy. Clin. Sci. (2002). https://doi.org/10.1042/CS20010270

190. Op den Kamp, C.M., Langen, R.C., Minnaard, R., et al.: Pre-cachexia in patients with stages I-III non-small cell lung cancer: systemic inflammation and functional impairment without activation of skeletal muscle ubiquitin proteasome system. Lung Cancer. (2012). https://doi.org/10.1016/j.lungcan.2011.09.012

191. Murton, A.J., Maddocks, M., Stephens, F.B., et al.: Consequences of late-stage non-small-cell lung cancer cachexia on muscle metabolic processes. Clin.

Lung Cancer. (2017). https://doi.org/10.1016/j.cllc.2016.06.003

192. Okamoto, T., Torii, S., Machida, S.: Differential gene expression of muscle-specific ubiquitin ligase MAFbx/Atrogin-1 and MuRF1 in response to immobilization-induced atrophy of slow-twitch and fast-twitch muscles. J. Physiol. Sci. (2011)

193. Aweida, D., Cohen, S.: Breakdown of filamentous myofibrils by the UPS-step by step. Biomol. Ther. **11** (2021). https://doi.org/10.3390/biom11010110

194. Aweida, D., Rudesky, I., Volodin, A., et al.: GSK3-β promotes calpain-1-mediated desmin filament depolymerization and myofibril loss in atrophy. J. Cell Biol. **217**, 3698–3714 (2018). https://doi.org/10.1083/jcb.201802018

195. Cohen, S., Brault, J.J., Gygi, S.P., et al.: During muscle atrophy, thick, but not thin, filament components are degraded by MuRF1-dependent ubiquitylation. J. Cell Biol. (2009). https://doi.org/10.1083/jcb.200901052

196. Cohen, S., Zhai, B., Gygi, S.P., Goldberg, A.L.: Ubiquitylation by Trim32 causes coupled loss of desmin, Z-bands, and thin filaments in muscle atrophy. J. Cell Biol. **198**, 575–589 (2012). https://doi.org/10.1083/jcb.201110067

197. Volodin, A., Kosti, I., Goldberg, A.L., Cohen, S.: Myofibril breakdown during atrophy is a delayed response requiring the transcription factor PAX4 and desmin depolymerization. Proc. Natl. Acad. Sci. USA. (2017). https://doi.org/10.1073/pnas.1612988114

198. Penna, F., Costamagna, D., Pin, F., et al.: Autophagic degradation contributes to muscle wasting in cancer cachexia. Am. J. Pathol. (2013). https://doi.org/10.1016/j.ajpath.2012.12.023

199. Aversa, Z., Pin, F., Lucia, S., et al.: Autophagy is induced in the skeletal muscle of cachectic cancer patients. Sci. Rep. (2016). https://doi.org/10.1038/srep30340

200. Pettersen, K., Andersen, S., Degen, S., et al.: Cancer cachexia associates with a systemic autophagy-inducing activity mimicked by cancer cell-derived IL-6 trans-signaling. Sci. Rep. (2017). https://doi.org/10.1038/s41598-017-02088-2

201. Clark, K.A., McElhinny, A.S., Beckerle, M.C., Gregorio, C.C.: Striated muscle cytoarchitecture: an intricate web of form and function. Annu. Rev. Cell Dev. Biol. (2002). https://doi.org/10.1146/annurev.cellbio.18.012502.105840

202. Chen, Z., Huang, W., Dahme, T., et al.: Depletion of zebrafish essential and regulatory myosin light chains reduces cardiac function through distinct mechanisms. Cardiovasc. Res. (2008). https://doi.org/10.1093/cvr/cvn073

203. Rottbauer, W., Wessels, G., Dahme, T., et al.: Cardiac myosin light chain-2: a novel essential component of thick-myofilament assembly and contractility of the heart. Circ. Res. (2006). https://doi.org/10.1161/01.RES.0000234807.16034.fe

204. Lowey, S., Waller, G.S., Trybus, K.M.: Skeletal muscle myosin light chains are essential for physiological speeds of shortening. Nature. (1993). https://doi.org/10.1038/365454a0

205. Offer, G., Baker, H., Baker, L.: Interaction of monomeric and polymeric actin with myosin subfragment 1. J. Mol. Biol. (1972). https://doi.org/10.1016/0022-2836(72)90425-1

206. Yang, Q., Sanbe, A., Osinska, H., et al.: A mouse model of myosin binding protein C human familial hypertrophic cardiomyopathy. J. Clin. Investig. (1998). https://doi.org/10.1172/JCI3880

207. Paulin, D., Li, Z.: Desmin: a major intermediate filament protein essential for the structural integrity and function of muscle. Exp. Cell Res. (2004)

208. Clarke, B.A., Drujan, D., Willis, M.S., et al.: The E3 Ligase MuRF1 degrades myosin heavy chain protein in dexamethasone-treated skeletal muscle. Cell Metab. **6**, 376–385 (2007). https://doi.org/10.1016/j.cmet.2007.09.009

209. Panicucci, C., Traverso, M., Baratto, S., et al.: Novel TRIM32 mutation in sarcotubular myopathy. Acta Myologica. (2019)

210. Cécile, P., Anne-Elisabeth, H., Marianne, J., et al.: Muscle actin is polyubiquitinylated in vitro and in vivo and targeted for breakdown by the E3 ligase MuRF1. FASEB J. **25**, 3790–3802 (2011). https://doi.org/10.1096/FJ.11-180968

211. Dulce, P.-M., Mélodie, M., Agnès, C., et al.: UBE2L3, a partner of MuRF1/TRIM63, is involved in the degradation of myofibrillar actin and myosin. Cell. **10** (2021). https://doi.org/10.3390/CELLS10081974

212. Di Rienzo, M., Antonioli, M., Fusco, C., et al.: Autophagy induction in atrophic muscle cells requires ULK1 activation by TRIM32 through unanchored K63-linked polyubiquitin chains. Sci. Adv. (2019a). https://doi.org/10.1126/sciadv.aau8857

213. Di Rienzo, M., Piacentini, M., Fimia, G.M.: A TRIM32-AMBRA1-ULK1 complex initiates the autophagy response in atrophic muscle cells. Autophagy. (2019b)

214. Sandri, M., Sandri, C., Gilbert, A., et al.: Foxo transcription factors induce the atrophy-related ubiquitin ligase atrogin-1 and cause skeletal muscle atrophy. Cell. **117**, 399–412 (2004) https://doi.org/S0092867404004003

215. Piccirillo, R., Goldberg, A.L.: The p97/VCP ATPase is critical in muscle atrophy and the accelerated degradation of muscle proteins. EMBO J. (2012). https://doi.org/10.1038/emboj.2012.178

216. Boudriau, S., Côté, C.H., Vincent, M., et al.: Remodeling of the cytoskeletal lattice in denervated skeletal muscle. Muscle Nerve. (1996). https://doi.org/10.1002/(SICI)1097-4598(199611)19:11<1383::AID-MUS2>3.0.CO;2-8

217. Cohen, S.: Role of calpains in promoting desmin filaments depolymerization and muscle atrophy. Biochim. Biophys. Acta Mol. Cell Res. **1867**,

118788 (2020). https://doi.org/10.1016/j.bbamcr.2020.118788

218. Lazarides, E.: The distribution of desmin (100 Å) filaments in primary cultures of embryonic chick cardiac cells. Exp. Cell Res. (1978). https://doi.org/10.1016/0014-4827(78)90209-4

219. Lazarides, E., Hubbard, B.D.: Immunological characterization of the subunit of the 100 Å filaments from muscle cells. Proc. Natl. Acad. Sci. USA. (1976). https://doi.org/10.1073/pnas.73.12.4344

220. Milner, D.J., Weitzer, G., Tran, D., et al.: Disruption of muscle architecture and myocardial degeneration in mice lacking desmin. J. Cell Biol. (1996). https://doi.org/10.1083/jcb.134.5.1255

221. Conover, G.M., Henderson, S.N., Gregorio, C.C.: A myopathy-linked desmin mutation perturbs striated muscle actin filament architecture. Mol. Biol. Cell. (2009). https://doi.org/10.1091/mbc.E08-07-0753

222. Shum, A.M.Y., Poljak, A., Bentley, N.L., et al.: Proteomic profiling of skeletal and cardiac muscle in cancer cachexia: alterations in sarcomeric and mitochondrial protein expression. Oncotarget. (2018) https://doi.org/10.18632/oncotarget.25146

223. Helliwell, T.R., Gunhan, O., Edwards, R.H.T.: Lectin binding and desmin expression during necrosis, regeneration, and neurogenic atrophy of human skeletal muscle. J. Pathol. (1989). https://doi.org/10.1002/path.1711590111

224. Agnetti, G., Herrmann, H., Cohen, S.: New roles for desmin in maintenance of muscle homeostasis. FEBS J. **15864** (2021). https://doi.org/10.1111/febs.15864

225. Evenson, A.R., Fareed, M.U., Menconi, M.J., et al.: GSK-3β inhibitors reduce protein degradation in muscles from septic rats and in dexamethasone-treated myotubes. Int. J. Biochem. Cell Biol. (2005). https://doi.org/10.1016/j.biocel.2005.06.002

226. Verhees, K.J.P., Schols, A.M.W.J., Kelders, M.C.J. M., et al.: Glycogen synthase kinase-3β is required for the induction of skeletal muscle atrophy. Am. J. Physiol. Cell Physiol. (2011). https://doi.org/10.1152/ajpcell.00520.2010

227. Cross, D.A.E., Alessi, D.R., Cohen, P., et al.: Inhibition of glycogen synthase kinase-3 by insulin mediated by protein kinase B. Nature. (1995). https://doi.org/10.1038/378785a0

228. Cole, A.R., Sutherland, C.: Measuring GSK3 expression and activity in cells. Methods Mol. Biol. (2008). https://doi.org/10.1007/978-1-59745-249-6_4

229. Sutherland, C.: What are the bona fide GSK3 substrates? Int. J. Alzheimer's Dis. (2011)

230. Lin, X.Y., Chen, S.Z.: Calpain inhibitors ameliorate muscle wasting in a cachectic mouse model bearing CT26 colorectal adenocarcinoma. Oncol. Rep. (2017). https://doi.org/10.3892/or.2017.5396

231. Pin, F., Minero, V.G., Penna, F., et al.: Interference with Ca2+-dependent proteolysis does not alter the course of muscle wasting in experimental cancer cachexia. Front. Physiol. (2017). https://doi.org/10.3389/fphys.2017.00213

232. Eid Mutlak, Y., Aweida, D., Volodin, A., et al.: A signaling hub of insulin receptor, dystrophin glycoprotein complex and plakoglobin regulates muscle size. Nat. Commun. **11**, 1381 (2020). https://doi.org/10.1038/s41467-020-14895-9

233. Lapidos, K.A., Kakkar, R., McNally, E.M.: The dystrophin glycoprotein complex. Circ. Res. **94**, 1023–1031 (2004). https://doi.org/10.1161/01.RES.0000126574.61061.25

234. Peter, A.K., Cheng, H., Ross, R.S., et al.: The costamere bridges sarcomeres to the sarcolemma in striated muscle. Prog. Pediatr. Cardiol. (2011). https://doi.org/10.1016/j.ppedcard.2011.02.003

235. Ervasti, J.M., Campbell, K.P.: Membrane organization of the dystrophin-glycoprotein complex. Cell. (1991). https://doi.org/10.1016/0092-8674(91)90035-W

236. Madhavan, R., Massom, L.R., Jarrett, H.W.: Calmodulin specifically binds three proteins of the dystrophin-glycoprotein complex. Biochem. Biophys. Res. Commun. (1992). https://doi.org/10.1016/0006-291X(92)91690-R

237. Rando, T.A.: The dystrophin-glycoprotein complex, cellular signaling, and the regulation of cell survival in the muscular dystrophies. Muscle Nerve. **24**, 1575–1594 (2001). https://doi.org/10.1002/mus.1192

238. Ramaswamy, K.S., Palmer, M.L., van der Meulen, J. H., et al.: Lateral transmission of force is impaired in skeletal muscles of dystrophic mice and very old rats. J. Physiol. **589**, 1195–1208 (2011). https://doi.org/10.1113/jphysiol.2010.201921

239. Gumerson, J.D., Michele, D.E.: The dystrophin-glycoprotein complex in the prevention of muscle damage. J. Biomed. Biotechnol. (2011)

240. Claflin, D.R., Brooks, S.V.: Direct observation of failing fibers in muscles of dystrophic mice provides mechanistic insight into muscular dystrophy. Am. J. Physiol. Cell Physiol. (2008). https://doi.org/10.1152/ajpcell.00244.2007

241. Dellorusso, C., Crawford, R.W., Chamberlain, J.S., Brooks, S.V.: Tibialis anterior muscles in mdx mice are highly susceptible to contraction-induced injury. J. Muscle Res. Cell Motil. (2001). https://doi.org/10.1023/A:1014587918367

242. Li, S., Kimura, E., Ng, R., et al.: A highly functional mini-dystrophin/GFP fusion gene for cell and gene therapy studies of Duchenne muscular dystrophy. Hum. Mol. Genet. (2006). https://doi.org/10.1093/hmg/ddl082

243. Bassat, E., Mutlak, Y.E., Genzelinakh, A., et al.: The extracellular matrix protein agrin promotes heart regeneration in mice. Nature. **547**, 179–184 (2017). https://doi.org/10.1038/nature22978

244. Gawor, M., Prószyński, T.J.: The molecular cross talk of the dystrophin–glycoprotein complex. Ann. NY Acad. Sci. (2018)

245. Yatsenko, A.S., Kucherenko, M.M., Xie, Y., et al.: Profiling of the muscle-specific dystroglycan

interactome reveals the role of Hippo signaling in muscular dystrophy and age-dependent muscle atrophy. BMC Med. **18**, 8 (2020). https://doi.org/10.1186/s12916-019-1478-3

246. Anderson, J.T., Rogers, R.P., Jarrett, H.W.: Ca2+-−calmodulin binds to the carboxyl-terminal domain of dystrophin. J. Biol. Chem. (1996). https://doi.org/10.1074/jbc.271.12.6605

247. Lai, Y., Thomas, G.D., Yue, Y., et al.: Dystrophins carrying spectrin-like repeats 16 and 17 anchor nNOS to the sarcolemma and enhance exercise performance in a mouse model of muscular dystrophy. J. Clin. Investig. (2009). https://doi.org/10.1172/JCI36612

248. Oak, S.A., Russo, K., Petrucci, T.C., Jarrett, H.W.: Mouse α1-syntrophin binding to Grb2: further evidence of a role for syntrophin in cell signaling. Biochemistry. (2001). https://doi.org/10.1021/bi010490n

249. Yang, B., Jung, D., Motto, D., et al.: SH3 domain-mediated interaction of dystroglycan and Grb2. J. Biol. Chem. (1995). https://doi.org/10.1074/jbc.270.20.11711

250. Langenbach, K.J., Rando, T.A.: Inhibition of dystroglycan binding to laminin disrupts the PI3K/AKT pathway and survival signaling in muscle cells. Muscle Nerve. (2002). https://doi.org/10.1002/mus.10258

251. Cohen, S., Lee, D., Zhai, B., et al.: Trim32 reduces PI3K-Akt-FoxO signaling in muscle atrophy by promoting plakoglobin-PI3K dissociation. J. Cell Biol. **204**, 747–758 (2014). https://doi.org/10.1083/jcb.201304167

252. Cruz Guzmán, O.D.R., Chávez García, A.L., Rodríguez-Cruz, M.: Muscular dystrophies at different ages: metabolic and endocrine alterations. Int. J. Endocrinol. (2012)

253. Freidenberg, G.R., Olefsky, J.M.: Dissociation of insulin resistance and decreased insulin receptor binding in duchenne muscular dystrophy. J. Clin. Endocrinol. Metab. (1985). https://doi.org/10.1210/jcem-60-2-320

254. Rodriguez-Cruz, M., Sanchez, R., Escobar, R.E., et al.: Evidence of insulin resistance and other metabolic alterations in boys with Duchenne or becker muscular dystrophy. Int. J. Endocrinol. **2015**, 867273 (2015). https://doi.org/10.1155/2015/867273

255. Jacomin, A.C., Samavedam, S., Promponas, V., Nezis, I.P.: iLIR database: a web resource for LIR motif-containing proteins in eukaryotes. Autophagy. **12**, 1945–1953 (2016). https://doi.org/10.1080/15548627.2016.1207016

256. Glass, D.J.: Skeletal muscle hypertrophy and atrophy signaling pathways. Int. J. Biochem. Cell Biol. **37**, 1974–1984 (2005). https://doi.org/10.1016/j.biocel.2005.04.018

257. Sacheck, J.M., Ohtsuka, A., McLary, S.C., Goldberg, A.L.: IGF-I stimulates muscle growth by suppressing protein breakdown and expression of atrophy-related ubiquitin ligases, atrogin-1 and MuRF1. Am. J. Phys. Endocrinol. Metab. **287**, E591–E601 (2004). https://doi.org/10.1152/ajpendo.00073.2004

258. Stitt, T.N., Drujan, D., Clarke, B.A., et al.: The IGF-1/PI3K/Akt pathway prevents expression of muscle atrophy-induced ubiquitin ligases by inhibiting FOXO transcription factors. Mol. Cell. **14**, 395–403 (2004)

259. Birgisdottir, Å.B., Lamark, T., Johansen, T., et al.: The LIR motif – crucial for selective autophagy. J. Cell Sci. (2013). https://doi.org/10.1242/jcs.126128

260. Baracos, V.E.: Clinical trials of cancer cachexia therapy, now and hereafter. J. Clin. Oncol. (2013)

261. Mueller, T.C., Bachmann, J., Prokopchuk, O., et al.: Molecular pathways leading to loss of skeletal muscle mass in cancer cachexia – can findings from animal models be translated to humans? BMC Cancer. (2016)

The Role of Interleukin-6/GP130 Cytokines in Cancer Cachexia

Daenique H. A. Jengelley and Teresa A. Zimmers

Abstract

Cancer cachexia is characterized by the involuntary loss of skeletal muscle with or without adipose tissue loss in the presence of tumor burden. It is a prevalent yet clinically unmet need that results in devastating systemic wasting effects. Chronic inflammation triggers cancer progression and is observed in cachexia. This review will highlight the Interleukin-6 (IL-6) family of cytokines, including IL-6 itself, Leukemia Inhibitory Factor (LIF), Ciliary Neurotrophic Factor (CNTF), Cardiotrophin-1 (CT-1), Oncostatin M (OSM), Interleukin-11 (IL-11), Interleukin-27 (IL-27), and Cardiotrophin-like cytokine (CLC), which all share signaling through IL-6 Signal Transducer (IL6ST), also known as Glycoprotein 130 (GP130) to activate common downstream pathways including the JAK/STAT, MAPK, and AKT pathways. IL-6 has been long linked to cancer cachexia through both associative and functional studies; furthermore, anti-IL-6 therapies have been trialed in patients. Recently, LIF has emerged as a novel cachexia mediator in experimental systems. Far less is known about the other cytokines in cachexia, although they have suggestive properties on adipose and muscle tissues in other contexts. Future studies are required to determine the roles of these other factors in cancer cachexia and their potential for designing therapies.

D. H. A. Jengelley
Department of Biochemistry and Molecular Biology, Indianapolis, IN, USA

T. A. Zimmers (✉)
Department of Biochemistry and Molecular Biology, Indianapolis, IN, USA

Department of Surgery, Indianapolis, IN, USA

Department of Anatomy, Cell Biology, & Physiology, Indianapolis, IN, USA

Department of Otolaryngology—Head & Neck Surgery, Indiana University School of Medicine, Indianapolis, IN, USA

Research Service, Richard L. Roudebush Veterans Administration Medical Center, Indianapolis, IN, USA

Indiana Center for Musculoskeletal Health, IN, USA

Indiana University Simon Comprehensive Cancer Center, Indianapolis, IN, USA
e-mail: zimmerst@iu.edu

Learning Objectives

a. Define cancer cachexia
b. Summarize the Interleukin-6/GP130 cytokines and their common signaling receptor
c. Assess the involvement of Interleukin-6/GP130 cytokines in cancer cachexia

© Springer Nature Switzerland AG 2022
S. Acharyya (ed.), *The Systemic Effects of Advanced Cancer*,
https://doi.org/10.1007/978-3-031-09518-4_6

1   Introduction

Cancer cachexia is characterized by the involuntary loss of skeletal muscle with or without adipose tissue loss in the presence of tumor [1, 2]. The loss of skeletal muscle and fat is not always directly proportional to existing muscle and fat stores. Cancer patients with sarcopenic obesity have high fat mass but low muscle loss. As well, patients may have progressive muscle deterioration but exhibit fat gain during remission [3–5].

Cachexia is a prevalent yet clinically unmet need that results in devastating systemic wasting in approximately 50% of patients, particularly those with advanced-stage disease. Cachexia is progressive and patient status can be stratified into three states: pre-cachexia, cachexia, and refractory cachexia [2]. The clinical diagnostic criteria for cancer cachexia measures weight loss over time. The criteria defines greater than 5% weight loss over 6 months, or greater than 2% weight loss in individuals with visible depletion in their current body weight and height or skeletal muscle mass [2]. Cachexia is highly associated with colorectal, lung, head and neck, gastro-esophageal, and pancreatic cancers [1, 3, 6, 7]. The effects of cachexia are not only seen in skeletal muscle and fat, but in organ systems including the cardiovascular, gastrointestinal, hematopoietic, and central nervous systems. Indeed, it is a multi-organ disorder featuring cardiac dysfunction, malabsorption, inflammation, anorexia, altered thermogenesis, and dysmetabolism [7]. Cachexia cannot be fully reversed by nutritional support; thus, it is an active catabolic process. Its impact extends to treatment response, overall function, and ultimately survival [8]. It diminishes patient mobility and impairs chemotherapy, radiotherapy, and surgical therapy. Cachexia, itself, is estimated to cause up to 30% of all cancer deaths and currently, there are no effective FDA-approved therapies for cancer cachexia [1, 6, 7].

Due to decreased food intake and catabolism, body wasting in cachexia is driven by the host immuno-metabolic response to the presence of the tumor, which leads to negative protein and energy balance [2, 9]. Metabolic reprogramming in both the tumor microenvironment and the host macroenvironment conspires to promote cancer cell proliferation, survival, and metastasis while inducing wasting of muscle and fat tissues [10]. Muscle mass is regulated by the balance of protein synthesis against protein catabolism. Anabolic pathways promote muscle hypertrophy, whereas catabolic pathways result in muscle atrophy [1, 3, 11]. Similarly, adipose mass is regulated by the balance of adipogenesis, lipid uptake, and lipogenesis versus lipolysis [11]. Production of inflammatory cytokines and other mediators in the tumor and through the host response tips the balance to catabolism in both tissues. The tumor and host-immune system together release inflammatory mediators including IL-1, IL-6, TNF-α, and Activin which promote signaling of catabolic pathways including the Janus Kinase/Signal Transducer and Activator of Transcription (JAK/STAT), nuclear factor-κB (NF-κB), phosphoinositide 3-kinase (PI3K), and myostatin/activin-SMAD pathways, as well as miRNAs and HSP70/HSP90-carrying exosomes to induce Myd88 and p38 MAPK pathways [12–20]. In muscle, downstream activation of p38 MAPK, STAT3, NF-kB, and FOXO proteins leads to the unfolded protein response, oxidative stress, insulin resistance, translation inhibition, and impaired myogenesis along with activation of catabolism through the calpain, cathepsin, lysosomal, and proteasome-mediated degradation pathways [7, 13, 21]. Protein breakdown leads to loss of mass and function and to increased circulating branched chain amino acids. In adipose tissue, the tumor and host inflammatory response directly and indirectly activates lipolysis, which results in increased circulating fatty acids [7]. This cachexia-produced enrichment of serum amino acids and lipids likely drives tumor growth and progression, although definitive evidence for this is currently lacking.

2 The GP130 Cytokine Family

Chronic inflammation promotes cancer progression and is observed in cachexia. Typically, inflammation occurs in response to acute injury or infection and is self-limiting due to the balance of pro-inflammatory and anti-inflammatory cells and the healing of tissues. However, during chronic inflammation, there is a persistence of pro-inflammatory cells and cytokine production [22]. Inflammatory cells such as macrophages, dendritic cells, and lymphocytes release pro-inflammatory cytokines that can promote tumor cell growth, invasion, metastasis, as well as immune evasion, along with activating skeletal muscle and fat catabolism [22, 23]. Many of these functions are mediated or modulated by members of the IL-6/GP130 receptor family of cytokines.

The GP130 cytokine family consists of Interleukin-6 (IL-6), Leukemia Inhibitory Factor (LIF), Ciliary Neurotrophic Factor (CNTF), Cardiotrophin-1 (CT-1), Oncostatin M (OSM), Interleukin-11 (IL-11), Interleukin-27 (IL-27), and Cardiotrophin-like cytokine (CLC) [24]. The cytokines are structurally and functionally similar and are involved in disease states and biological processes including immune response, inflammation, hematopoiesis, cellular proliferation, metabolic regulation, and cardiovascular function [25, 26]. These cytokines share a common signal transmembrane receptor known as Glycoprotein 130 (GP130), encoded by the IL-6 signal transducer (IL6ST) gene, which accounts for their similar biological functions. All have overlapping or shared signal receptor complexes, including IL-6 with IL-6 receptor alpha (IL6Rα), IL-11 with IL-11 receptor alpha (IL11Rα), IL-27 with IL-27 receptor alpha (IL27Rα), OSM with OSM receptor beta (OSMRβ), LIF with LIF receptor beta (LIFRβ), and CNTF with CNTF receptor alpha (CNTFRα). LIF and CT-1 share the GP130/LIFRβ complex [27, 28], while OSM can bind to either GP130/LIFRβ or GP130/OSMRβ [29–32]. CNTF and CLC mediate signaling through CNTFRα [25, 33, 34] (Fig. 1).

Ligand binding to co-receptors and GP130 leads to the activation of signaling pathways and target genes. The receptors are classified into signaling and non-signaling receptors. The alpha receptors, which are non-signaling, interact with the signal-transducing receptors (gp130, LIFRβ, OSMRβ) to initiate JAK/STAT signaling which results in the activation of transcription factors STATs—STAT1, STAT3, STAT4, STAT5,

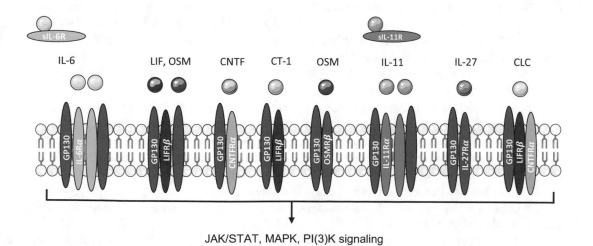

JAK/STAT, MAPK, PI(3)K signaling

Fig. 1 IL-6 family cytokines and signaling receptors. The members of the IL-6 family cytokines are IL-6, LIF, OSM, CNTF, CT-1, IL-11, IL-27, and CLC. All the cytokines have overlapping or shared signal receptor complexes. The cytokines share the common signaling transmembrane receptor GP130. The cytokines activate similar pathways and downstream activation of JAK/STAT, MAPK, PI(3)K signaling pathways and target genes

STAT6, as well as SHP-2 to activate ERK/MAPK and PI3K/AKT, and C/EBPβ. The multiple negative regulatory mechanisms include receptor internalization, deactivation of receptors and signaling intermediates, degradation of target mRNAs encoding cytokines or receptors, and STAT-driven induction of protein inhibitor of activated STAT (PIAS) protein, and suppressor of cytokine signaling (SOCS) proteins [12, 26, 35]. Activation of the signaling pathways alters cell proliferation, survival, differentiation, maturation, self-renewal, immune regulation, angiogenesis, oxidative stress, fate, and apoptosis [35].

3   Cytokines Documented in Cancer Cachexia

Only a few of the IL-6 family of cytokines have been studied in the context of cancer cachexia. We will focus on correlative evidence found in plasma levels of rodents, functional studies in cell and rodent models including overexpression, knockout, and cachexia model-based studies as well as evidence in patients and therapeutic options in cancer cachexia. Most existing evidence is related to IL-6 with emerging evidence for LIF. We could find no implicating evidence for CT-1, CLC, and IL-27 in cancer cachexia (Fig. 2 and Table 1).

Interleukin-6 (IL-6) was first identified as inducing differentiation in B lymphocytes and acute-phase protein synthesis in hepatocytes [36–38]. Now, it is widely understood that IL-6 functions in inflammation, immune response, and hematopoiesis [36, 39]. Circulating levels of IL-6 are low in normal conditions but become elevated in response to inflammation [26]. IL-6 can induce cell production, activation, and survival [35, 36, 39–42]. IL-6 has two modes of signaling: classical and trans-signaling [26, 36, 43]. Classical signaling is initiated through an IL-6, IL6R, and GP130 hetero-hexameric complex [36]. Trans-signaling occurs by IL-6 binding to the soluble IL6R complex with GP130. The soluble IL6R (sIL6R) is produced by protease-mediated shedding from the cell surface during conditions of inflammation, and less frequently by alternative splicing. GP130 is ubiquitously expressed, but IL6R is not; thus, elevated IL-6 and sIL6R trans-signaling endows cells with the ability to respond to IL-6 during conditions of inflammation [35, 36].

IL-6 has been well studied in cancer cachexia and various roles in cachexia progression have been determined. IL-6 is secreted primarily by tumor and immune cells; it activates inflammation, rapid progression of tumorigenesis, and metastatic spread [35, 41]. IL-6 induces the expression of STAT3 target genes and drives tumor cell proliferation and survival. IL-6 can induce activators of epithelial to mesenchymal transition (EMT) and act on cancer stem cells by promoting self-renewal and expansion [39]. In the microenvironment, tumor-associated macrophages, granulocytes, fibroblasts, and cancer cells secrete IL-6 [39].

Several cancer mouse models including lung, cervical, ovarian, thyroid, pancreatic, and colon have been used to study IL-6 in cancer cachexia [44–47]. Tamura et al. implanted a uterine cervical carcinoma derived cell line, Yumoto, in nude mice and observed cachexia along with expression of cytokines, including IL-6. Delivering a human anti-IL-6 antibody improved the effects on body weight, adipose, and skeletal muscle. The levels of IL-6 in the murine tumor were 3.13 ± 2.23 ng/g, and in the serum, IL-6 levels were 0.14 ± 0.05 ng/ml. From the Yumoto xenograft, the levels of human IL-6 were 15.2 ± 7.6 ng/g [45]. In a model of ovarian cancer, mice injected with ES-2 cells developed cachexia and had elevated levels of human IL-6 in plasma and ascites (26.3 and 279.6 pg/ml, respectively) [47]. Wang et al. implanted athymic nude mice with Thena-Nu, an anaplastic thyroid carcinoma xenograft. They reported fat wasting, decreased body weight, splenomegaly, and elevated circulating cytokines in the mice. The levels of IL-6 in the serum were 1156.4 ± 128.7 pg/ml and in the tumor, levels of IL-6 were 138.38 ± 23.76 ng/g tissue [42]. With clodronate treatment, body weight was sustained, and cytokine levels, tumor growth, and liver necrosis were all reduced. When clodronate treatment was discontinued tumor growth and weight loss were

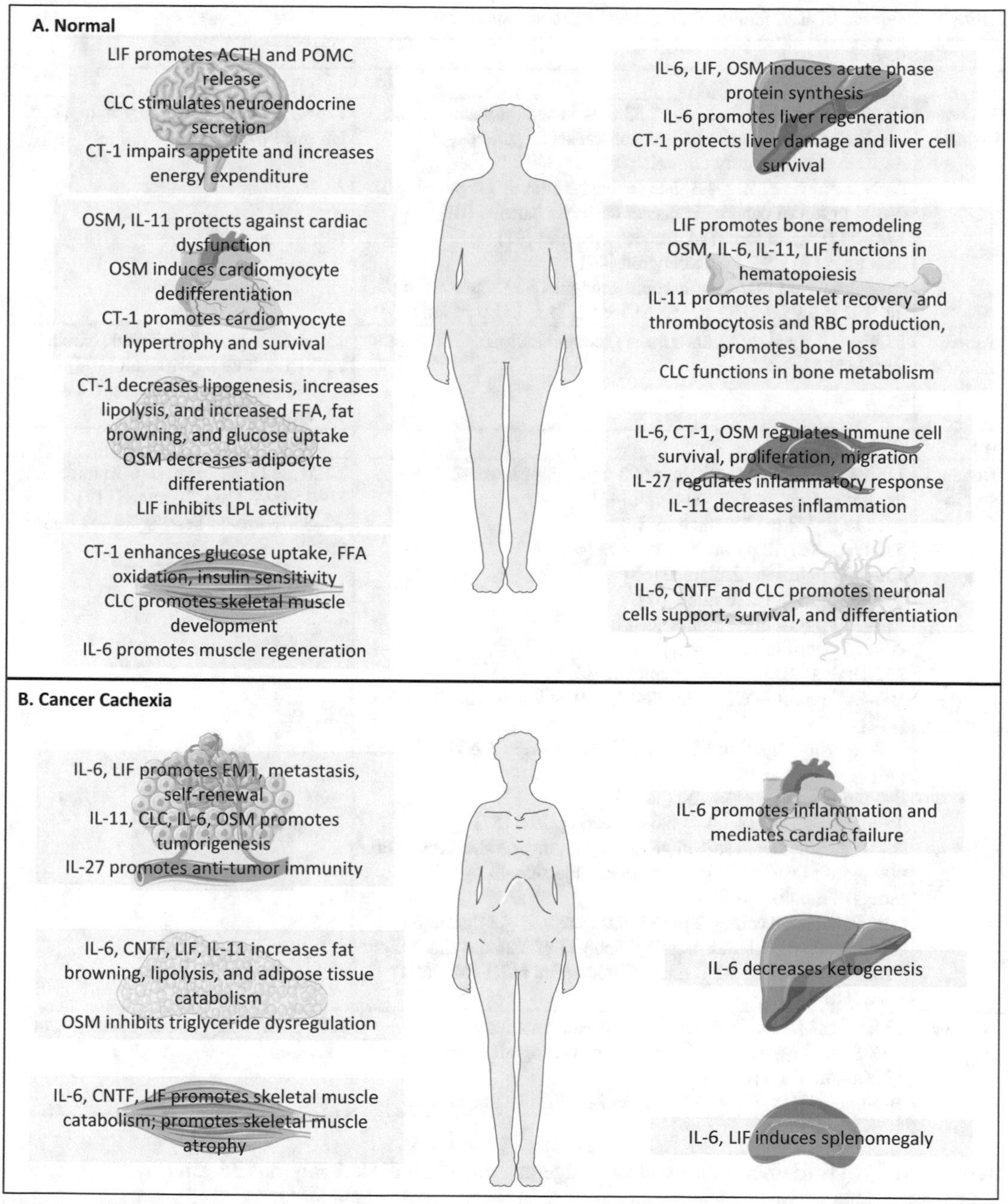

A. Normal

LIF promotes ACTH and POMC release
CLC stimulates neuroendocrine secretion
CT-1 impairs appetite and increases energy expenditure

IL-6, LIF, OSM induces acute phase protein synthesis
IL-6 promotes liver regeneration
CT-1 protects liver damage and liver cell survival

OSM, IL-11 protects against cardiac dysfunction
OSM induces cardiomyocyte dedifferentiation
CT-1 promotes cardiomyocyte hypertrophy and survival

LIF promotes bone remodeling
OSM, IL-6, IL-11, LIF functions in hematopoiesis
IL-11 promotes platelet recovery and thrombocytosis and RBC production, promotes bone loss
CLC functions in bone metabolism

CT-1 decreases lipogenesis, increases lipolysis, and increased FFA, fat browning, and glucose uptake
OSM decreases adipocyte differentiation
LIF inhibits LPL activity

IL-6, CT-1, OSM regulates immune cell survival, proliferation, migration
IL-27 regulates inflammatory response
IL-11 decreases inflammation

CT-1 enhances glucose uptake, FFA oxidation, insulin sensitivity
CLC promotes skeletal muscle development
IL-6 promotes muscle regeneration

IL-6, CNTF and CLC promotes neuronal cells support, survival, and differentiation

B. Cancer Cachexia

IL-6, LIF promotes EMT, metastasis, self-renewal
IL-11, CLC, IL-6, OSM promotes tumorigenesis
IL-27 promotes anti-tumor immunity

IL-6 promotes inflammation and mediates cardiac failure

IL-6, CNTF, LIF, IL-11 increases fat browning, lipolysis, and adipose tissue catabolism
OSM inhibits triglyceride dysregulation

IL-6 decreases ketogenesis

IL-6, CNTF, LIF promotes skeletal muscle catabolism; promotes skeletal muscle atrophy

IL-6, LIF induces splenomegaly

Fig. 2 Functions of IL-6 family of cytokines in organs in normal physiology and cancer cachexia. (**a**) Cytokines in physiological conditions. Summary of the documented roles of IL-6 family of cytokines in physiological conditions in the brain, heart, adipose tissue, skeletal muscle, liver, bone, immune, and neuronal cells. (**b**) Cytokines in cancer cachexia. Summary of the roles of IL-6 family of cytokines in cancer cachexia in the tumor, adipose tissue, skeletal muscle, heart, liver, and spleen. References are included in the text of this chapter

observed [42]. In a murine colon-26-adenocarcinoma (C26) mouse model, the serum levels of IL-6 were 600–800 pg/ml; and the mice demonstrated skeletal muscle wasting and exhibited signs of cachexia [48, 49]. In follow-up studies, treatment of Capecitabine, a

Table 1 Evidence of IL-6 family of cytokines in patients and rodents

	IL-6	LIF
(a)		
Patient serum	19.89 ± 8.25, 5.18 ± 2.81, 15.33 ± 9.54-fold in gastric, lung, and breast cancer cachexia and non-cachectic patients, respectively vs healthy controls [78] Early cachexia: 7.2 ± 4.3; late cachexia: 13.3 ± 7.4 pg/ml [80] 0–300 pg/ml of pancreatic cancer cachexia patients [81] 17.6–54.0 pg/ml serum of cachectic patients [77] 26.3 pg/ml in ES-2 host xenograft [47] Patient-derived PDAC orthotopic xenograft S002: >>700 pg/ml, S017: 1.95 ⊥ 0.95 pg/ml, S035: 513.9 ± 152.9 pg/ml [60]	1156.7 ± 114.7 serum pg/ml in Thena-Nu mice [42]
Patient tumor	15.2 ± 7.6 ng IL-6/g tumor mass (xenograft: clone 17 Yumoto line) [45]	132.5 ± 52 pg/10^6 pancreatic carcinoma cells in culture supernatant [103] 0.11 ± 0.04 ng/ml (xenograft: clone 17 Yumoto line) [45]
(b)		
Rodent serum	1156.4 ± 128.7 serum pg/ml; 600–800 pg/ml in anaplastic thyroid carcinoma cell xenograft [42] 80–110 pg/ml in C26 mice [63] 537.66 ± 417.18 pg/ml in C26 mice [64] 130–200 pg/ml in LLC mice [65] 80–100 pg/ml in C26 mice [67] 150–250 pg/ml in C26 mice [66] 20–100 pg/ml in LLC mice [69] 152.18 ± 15.01 ng/l in C26 mice [239] 100–400 pg/ml in C26 mice [240]; 600–800 pg/ml in C26 mice [112] 2574 ± 1664 pg/ml in C26 mice; 5–10 ng/ml in AAV-IL6 mice [74] 0–4 ng/ml in C26 clone 20 [50] 45 ± 9 pg/ml in HARA-B tumor-bearing mice [57] A431: subcutaneous injection 190 pg/ml; intracerebral injection 400 pg/ml; OVCAR3: intraperitoneal injection 43 pg/ml; intracerebral 40 pg/ml; GBLF: subcutaneous ≤2 pg/ml; intracerebral: ≤2 pg/ml [61] 0.14 ± 0.05 ng/ml mice bearing clone 17 of Yumoto line [45] ~100–700 pg/ml in KPC mice; 0–400 pg/ml in KPC-IL-6KO mice [60]	24.51 ± 11.26 pg/ml in C26 mice [64] 10–15 pg/ml in C26 mice [112] 25–55 pg/ml rat bearing 85As2 tumor [108]
Rodent tumor	134.38 ± 23.76 ng/g tumor tissue Thena-Nu mice [42] 3.13 ± 2.23 ng/g tumor tissue of mice (xenograft clone 17 Yumoto line) [45] 1–2-fold in KPC mice; 1–3-fold in KPC-IL-6KO mice [60]	2.72 ± 0.76 ng/g tissue Thena-Nu mice [42]

fluorinated pyrimidine anti-cancer agent, improved effects of wasting and decreased IL-6 levels [50]. In another C26 study, blocking STAT3 inhibited muscle atrophy by IL-6 or cancer [51]. However, not all C26 mouse models produce IL-6 or develop cachexia. In particular, clone 5 of the C26 mouse line secretes IL-6, but is reported to not induce cachexia in vivo [52]. Another C26 study observed cachexia, cardiac wasting, increased circulating levels of IL-6 [53]. In the colon cancer model, $Apc^{Min/+}$, mice

developed skeletal muscle atrophy, splenomegaly, and elevated IL-6 levels (13–20 pg/ml) [54]. In genetically engineered mouse models (GEMMs) of lung cancer and pancreatic cancer, IL-6 promotes fat browning, increased mitochondrial activity, and a systemic metabolic response [55]. In nude rats with LC-6-JCK, a human lung cancer xenograft, IL-6 levels were elevated compared to non-cancer controls [56]. IL-6 mRNA was observed in splenocytes in a Lewis lung carcinoma model in C57BL/6 J mice with cancer cachexia [44]. In a study using human lung cancer cells (HARA-B) implanted in BALB/c mice, the mice developed cachexia and had elevated circulating levels of IL-6 (45 ± 9 pg/ml) [57]. Zaki et al. determined that tumor cell-secreted IL-6 contributed to cachexia in a melanoma and prostate cancer xenograft model [58]. IL-6 impairs hepatic ketogenesis and increases a systemic metabolic response in an autochthonous colorectal cancer and pancreatic ductal adenocarcinoma models [59]. In a study using patient-derived PDAC orthotopic xenograft models, levels of avatar human IL-6 were measured and reported as (S002: >>700 pg/ml, S017:1.95 ± 0.95 pg/ml, S035: 513.9 ± 152.9 pg/ml). In the same study, patient weight loss grade correlated with weight loss in PDX mice [60]. Thus, multiple models of cancer have been used to study IL-6 and have determined that IL-6 induces cachexia.

Targeting or deleting IL-6 blocks cachexia in multiple models. When IL-6 producing human melanoma or prostate cancer cells were implanted, host mice developed cachexia; administration of a monoclonal human IL-6 antibody, CNTO 328, decreased tumor growth and cachexia [58]. Human epidermoid carcinoma, A431, and ovarian carcinoma, OVAR3, xenografts in mice induced wasting, and injection of murine IL-6 receptor ameliorated cachexia [61]. In $Apc^{Min/+}$ mice, 6-month-old mice showed severe cachexia and high serum IL-6 levels; the authors generated an $Apc^{Min/+}/IL-6^{-/-}$ knockout model, which did not develop cachexia or exhibit skeletal muscle atrophy. When IL-6 was added back systemically, skeletal muscle and adipose tissue wasting were observed [49, 62]. In a model of pancreatic cancer cachexia, Rupert et al. detected IL-6 in the KPC32908 cell line, further orthotopically implanting in wildtype mice and demonstrated that KPC IL-6^{KO} tumors were smaller; mice had attenuated muscle loss, decreased inflammation, lipid accumulation, and oxidative stress in skeletal muscle compared to KPC mice. However, adipose wasting and activation of STAT3 was still observed in KPC IL-6^{KO} tumor mice. Levels of plasma IL-6 were higher in KPC mice (~100–700 pg/ml) than KPC IL-6^{KO}

mice (0–400 pg/ml). Levels of plasma IL6rα was detected in KPC mice (0–60 ng/mL), but not detected in KPC IL-6^{KO} mice. Levels of $Il6$ mrna in the tumor of KPC IL-6^{KO} mice (1–3-fold) was greater compared to KPC mice (1–2-fold). Levels of $Il6r\alpha$ in tumor was decreased in KPC IL-6^{KO} mice (0.7–1.5-fold) compared to KPC mice (1-fold). Rupert et al. propose tumor derived IL-6 mediates muscle and adipose wasting and shedding of sIl6R from muscle to adipose and further release of IL-6 from adipose to muscle. This trans-signaling mechanism promotes local and systemic cachexia [60].

Treatments that indirectly reduce IL-6 also reduce cachexia. Lu et al. explored the therapeutic effects of Panax ginseng extract in the C26 cancer cachexia mouse model; they concluded that ginseng extract can reduce inflammation and improve symptoms of cancer cachexia. The levels of circulating IL-6 were 80–110 pg/ml, but in another cohort of mice, the measured levels of IL-6 ranged from 110–120 pg/ml and with different concentrations of ginseng extract the levels of circulating IL-6 ranged between 70–100 pg/ml, suggesting ginseng is a potential therapy for cancer cachexia [63]. In another C26 model of cancer cachexia, Liva et al. used a combined androgen receptor modulator (SARM) GTx-024 and histone deacetylase inhibitor (HDACi) AR-42, SARM/AR-42, as a potential anti-cachectic treatment. The levels of IL-6 in the experimental groups were vehicle: 537.66 ± 417.18 pg/ml, with GTx-024: 397.54 ± 341.43, AR-42: 256.59 ± 183.1, Combo: 448.16 ± 294.52, suggesting that combined therapy may improve anabolic response in patients [64]. In another study, Liu et al. aimed to explore if coix seed oil would have a therapeutic effect in the LLC cancer cachexia model. The measured circulating levels of IL-6 were 130–200 pg/ml, with coix treatment IL-6 levels decreased to 90–150 pg/ml, suggesting that coix seed oil can decrease circulating levels of IL-6 and other inflammatory markers as well as improve muscle and fat mass [65]. The administration of a ketogenic diet to C26 mice resulted in the reduction of circulating levels of IL-6 (100–150 pg/ml) compared to the C26 mice (150–250 pg/ml) [66]. Another study in

C26 mice reported that dietary intake of kimchi inhibits IL-6 mediated cancer cachexia [67]. Treatment with the IL-6 receptor antagonist (20S, 21-epoxy-resibufogenin-3-acetate) (ERBA) suppressed IL-6 activity and inhibited body weight loss in a C26 model [68]. In the LLC cancer cachexia model, the treatment of mice with selumetinib, a MEK inhibitor, decreased the IL-6 expression in the blood (−80.90%), and tumor (−35.04%), but not in the muscle compared to the tumor-bearing mice. These results suggest that selumetinib is effective for reducing tumor growth but not for cachexia symptoms [69]. The experimental therapeutics improved the effects of IL-6 mediated cachexia.

Overexpression of IL-6 induces cachexia. In a knock-in model of STAT3 overexpression, IL-6 levels were up-regulated in muscle and exacerbated levels of IL-6 promoted cachexia [70]. Prolonged activation of IL6/JAK/STAT pathways results in muscle wasting and activation of the acute-phase response specifically in the C26 model, $Apc^{Min/+}$, and in patients [10, 71–73]. In overexpression studies, IL-6 delivered by electroporation or adeno-associated viral vectors caused muscle and fat wasting in the absence of tumor [74, 75]. These data suggest IL-6 is sufficient to cause cachexia even in the absence of tumor.

In cancer patients, elevated IL-6 levels compared to healthy controls correlate with poor prognosis in patients [10, 72, 76, 77]. Specifically, gastric, lung, and breast cancer patients with cachexia show IL-6 levels, ranging from 19.89 ± 8.25, 5.18 ± 2.81, 15.33 ± 9.54-fold, respectively, compared to non-cancer controls [78]. Iwase et al. observed a correlation between the circulating levels of IL-6 and cachexia progression. IL-6 increases gradually during early cachexia and peaks near death; and levels of IL-6 were found in all cachectic patients [49, 79]. Han et al. measured circulating IL-6 levels in gastric and colorectal cancer patients and showed elevated levels of IL-6 in early cachexia (7.22 ± 4.3 pg/ml) or late cachexia (13.3 ± 7.4 pg/ml) [80]. In another study, pancreatic cancer patients with cachexia had elevated IL-6 levels ranging from 0–300 pg/ml [81]. In

patients with head and neck, colorectal, pancreatic, multiple myeloma, lung, and other type of cancers, the IL-6 median (interquartile range) levels were: pre-cachexia: 12.0 (6.6–15.3) pg/ml, cachexia: 31.8 (17.6–54.0) pg/ml, refractory cachexia: 45.8 (28.7–54.2) pg/ml [77]. In cancer patients with cachexia, levels of IL-6 ranged from 20–25 pg/ml and levels directly correlate with weight loss [82]. Utech et al. observed IL-6 levels ranging from 0.18–130.6 pg/ml in male cancer patients of different stages and tumor types. The authors reported an association between the IL-6 levels and mortality, but this trend was not observed when the patient's cancer stage, weight loss from baseline to follow-up, and serum albumin were considered [83]. Taken together, these clinical data provide evidence for IL-6 levels in cachexia progression/patients.

IL-6 inhibition has been trialed for cancer cachexia. Currently, the monoclonal antibodies tocilizumab and sarilumab, which block the interaction of IL-6 to IL-6R, have been shown to prolong overall survival in patients with lung cancer cachexia [35, 84]. In a case study, tocilizumab appeared to prolong the survival of a patient with advanced lung cancer cachexia by 9 months [84].

While some studies demonstrate a concordance between tumor burden and muscle loss, it remains unclear whether reducing tumor growth necessarily reduces cachexia severity or the converse—that reducing cachexia might reduce tumor growth. As yet, the protective effects observed in the experimental and clinical approaches targeting IL-6 cannot be specifically attributed to effects on tumor versus effects on muscle as these have not been definitively addressed, e.g., through genetic manipulation. Moreover, while anti-IL-6 interventions often lead to reduced tumor growth, the studies by Au et al. demonstrate reduced IL-6 expression and reduced tumor growth but more severe cachexia, indicating that tumor burden is not the sole determinant of muscle wasting. Tumor progression and growth could rely on the energetic stores liberated from muscle and fat tissues; thus, minimizing either tumor or cachexia could result in reduced disease burden. Further investigation is needed to determine which is primary or secondary; however, any therapeutic target should focus on ameliorating both tumor progression and cachexia.

Leukemia Inhibitory Factor (LIF) was originally identified as an inhibitor of differentiation and cell proliferation, an inducer of the acute-phase response, and an inhibitor of adipocyte lipoprotein lipase activity [85–89]. It is also widely understood that LIF functions in bone remodeling, cancer, hematopoiesis, and development [90]. LIF signals through LIFRβ forming a trimer with GP130 [91]. The receptor formation leads to the activation of JAK/STAT, MAPK, and PI(3)K pathways promoting differentiation, self-renewal, or survival [90]. The receptor is found on the surface of embryonic stem cells, the liver, and on monocytes/macrophages [90, 92–94]. LIF acts on liver, bone, uterus, kidney, myoblasts, and the central nervous system [94, 95]. In bone remodeling, LIF promotes differentiation of bone marrow stromal cells and inhibits the expression of sclerostin in osteocytes [96]. In the hypothalamus, LIF promotes the release of pro-opiomelanocortin and adrenocorticotrophic hormone from corticotropes [97]. In myoblasts, LIF stimulates proliferation but inhibits differentiation to myotubes [98]. LIF is pro-tumorigenic and promotes EMT cancer cell invasion, metastasis, self-renewal, and expansion of cancer stem cell populations [35, 99, 100]. Expression of LIF is elevated in breast, skin, colorectal, and nasopharyngeal cancer [90].

LIF expression is reported in a variety of cancer cell lines including melanoma, thyroid, human oral cavity, pancreas, and gastric cell lines [89, 101–103]. Expression has been reported to associate with cachexia. Tamura et al. implanted a uterine cervical carcinoma derived cell line, Yumoto, in nude mice and observed cachexia along with the expression of cytokines, including LIF. In Yumoto xenograft, human LIF was measured in the tumor (0.11 ± 0.04 ng/g) and in the serum (<0.05 ng/ml), but murine LIF levels were not measured [45]. Chang et al. reported high levels of LIF secreted and associated with cancer cachexia caused by an anaplastic thyroid cancer cell line,

Thena [104]. In another study, using Thena-Nu an anaplastic thyroid carcinoma xenograft, athymic nude mice showed fat wasting, decreased body weight, splenomegaly, and elevated circulating cytokines. LIF levels were measured in the serum (1156.7 ± 114.7 (pg/ml)) and tumor (2.72 ± 0.76 (ng/g)); however, with clodronate treatment, LIF levels were not reduced in the serum (1138.8 ± 264.6 (pg/ml)) or in the tumor (2.85 ± 0.42 (ng/g tissue)), but tumor cytostasis, attenuation of cachexia, improved liver necrosis, and prolonged survival were exhibited. When clodronate treatment was discontinued tumor growth and weight loss were observed [42]. In xenograft models using an oral cavity cancer cell line, OCC-1, a colonic cancer cell line, CO-3, a uterine cancer cell line, SW756, LIF was expressed in the tumor from non-cachectic and cachectic mice [105]. In a melanoma xenograft, SEKI and G361, and NAGAI, a neuroepithelioma cell line, LIF levels correlated with the development of cancer cachexia [102, 106]. Tanaka et al. identified severe cachexia in OCC-1 tumor-bearing nude mice; this human oral cavity tumor cell line expresses LIF and IL-6 [107]. In a study of rats bearing 85As2 tumor, LIF levels were measured from 25–55 pg/ml and higher levels were correlated with worsened cachexia symptoms [108]. Seto et al. identified serum levels of LIF (30–40 pg/ml) in a C26 model [109]. Thus, the evidence indicates that LIF is often present in cancer cachexia and may induce cachexia in cancer.

LIF is sufficient to cause cachexia. In an overexpression study, peripheral LIF administered to mice induced 50% loss of adipose and 10% loss of body weight [110]. In another study, recombinant human LIF administered to cynomolgus monkeys decreased body weight by 10%, induced thrombocytosis and splenomegaly, and caused a loss of subcutaneous fatty tissues [111].

Targeting LIF blocks cachexia in experimental models. LIF deletion in C26 cell lines reduced body weight, muscle loss, fat loss, and splenomegaly. The measured serum LIF levels of the C26 mice were 10–15 pg/ml. When LIF was knocked out of the C26 cells, levels of IL-6 dropped from 79-fold over normal in C26 tumor-bearing mice to five-fold over normal in the C26$^{Lif-/-}$ tumor-bearing mice [112]. Some have claimed anti-LIF effects of natural produce and small molecules. Rikkunshito, a Japanese medicine, decreased LIF levels in stomach cancer cells and ameliorated cachexia [113]. TNP470, a derivative of fumagillin, suppressed growth of hepatoma-bearing mice with cachexia, inhibited LIF production, and improved cachexia-like symptoms [114].

The link between LIF and human cancer cachexia is tenuous currently. Human pancreatic carcinoma cell lines express LIF and the amount of LIF protein was measured at 132.5 ± 52 pg/10^6 cells. In several of the pancreatic cancer cell lines, LIF, LIFR, and GP130 mRNA are also expressed [103]. Two patient-derived gastric cancer cell lines, 85As2 and MKN45cl85, express LIF and induced cachexia in nude mice [108, 113]. Another cell line, OCC-1C, an oral cavity carcinoma cell line, produces LIF and induced cachexia [115]. A melanoma cell line expressing LIF induced cancer cachexia in mice [89].

In vitro studies demonstrate a role for LIF in cancer cachexia. LIF enhanced myoblast proliferation *in vitro* and induced C2C12 satellite cell proliferation by activating JAK2 and STAT3 [116–118]. Seto et al. reported that LIF levels were elevated in myotubes treated with C26 conditioned medium compared to control. Treatment with a LIF blocking antibody abolished myotube atrophy and did not activate the phosphorylation of STAT3 [109]. Incubation of C2C12 myotubes with Lewis lung carcinoma derived media increased LIF expression and myotube atrophy, when supplemented with a recombinant LIF antibody there was decreased p-STAT3 and p-ERK levels [119]. Evidence of lipolysis was reported in colon cancer cells treated with recombinant LIF [110]. Marshall et al. reported that LIF treatment decreased LPL activity by 44% in cultured adipocytes, increased de novo synthesis of fatty acids, and increased lipolysis, suggesting LIF drives catabolic activity [120].

There is evidence of LIF in cancer cachexia, but more clinical-translational and functional studies are required to identify the specific contexts and mechanism by which LIF could be a major driver of the cachexia phenotype.

Ciliary Neurotrophic Factor (CNTF) was originally identified as a promoter of neural cell survival and differentiation. CNTF is classified as a neurotrophin; however, it is structurally and functionally similar to the IL-6 family of cytokines. CNTF binds to CNTFRα or LIFRβ and forms a heterodimer with GP130 resulting in the downstream activation of JAK/STAT signaling [121]. CNTF functions in neuronal cell support, survival, and differentiation on sympathetic, parasympathetic, sensory neurons, and immune cells [122–125]. CNTF is expressed in the brain, skeletal muscle, Schwann cells, and the embryo while the expression of the receptor, CNTFRα, is found primarily in neural precursor tissues [126–131]. CNTF has activity in metabolic disorders such as diabetes, obesity, and cancer [132].

CNTF is sufficient to cause cachexia. Recombinant human CNTF induced skeletal muscle atrophy in rats decreased body weight and reduced food intake. There was a sex-specific difference observed; male rats experienced greater skeletal muscle loss than female rats. Since levels of TNF, IL-6, and LIF were below detection in the serum, it was concluded that CNTF induced cachexia [133]. Moreover, mice implanted with C6 glioma cells over-expressing CNTF had increased adipose tissue and skeletal muscle catabolism, decreased levels of glucose and triglycerides, elevated red blood cell content, and exhibited cachexia. The wasting resulted in death over 7–10 days; in contrast to C6 glioma-bearing mice which did not exhibit these effects [134, 135]. Wild-type mice with subcutaneous injections of CNTF repeatedly administered over a 7-day period exhibited protein loss, an hepatic acute-phase response, and cachexia [136, 137].

We did not find any evidence linking CNTF to human cancer cachexia. Further studies are required to elucidate a role of CNTF in cancer cachexia.

Cardiotrophin-1 (CT-1) was originally discovered as a factor that can induce hypertrophy of cardiac myocytes. Now, it is known to function in the liver, blood, and nervous system [28, 33, 138]. CT-1 binds to GP130 and initiates the heterodimerization with LIFR to form the receptor complex and activate the JAK/STAT, MAPK signaling, and PI3K/AKT pathways. CT-1 is expressed in the heart, kidney, skeletal muscle, and liver [28, 33, 138]. Specifically, CT-1 promotes cell survival, reduces oxidative stress, induces cell migration and adhesion. It induces glucose uptake and increases energy expenditure [139–156]. In the heart, CT-1 promotes cardiomyocyte survival, activates heat shock proteins 70 and 90, induces the PI3K-Akt-BAD axis, p38MAPK, and reduces oxidative stress [139, 140]. CT-1 is elevated in the serum of patients with heart failure and hypertensive heart disease [157–161]. CT-1 protects against liver damage and improves liver cell survival [141, 142]. In inflammation, CT-1 prevents sepsis, activates IL-6 and TNF-α secretion, induces monocyte adhesion and migration, and promotes fibrosis [143–149]. CT-1 enhances fatty acid oxidation, AMPK, glucose uptake, and insulin sensitivity in skeletal muscle [150, 151]. In the liver, CT-1 decreases lipogenesis and increases fatty acid oxidation and AMPK [152, 153]. In white adipose tissue, CT-1 increases lipolysis, fatty acid oxidation, browning, and glucose uptake [154, 155]. Acting in the brain, CT-1 decreases food intake and increases energy expenditure [150, 156].

Currently, there is no evidence linking CT-1 to cancer cachexia.

Oncostatin M (OSM) was first characterized as a cell growth suppressor and an anti-cancer therapeutic in tumor cells; however, the understanding of OSM evolved to include its functions in regulating cell survival, proliferation, and tumorigenesis. OSM is produced primarily by immune cells, activated monocytes/macrophages, T lymphocytes, dendritic cells, and neutrophils [31, 162]. OSM mRNA expression is found in hematopoietic cells in the blood [31]. Intracellular signaling occurs as a consequence of extracellular binding of the ligand OSM to OSMR complex, formed from dimerization with receptor subunit GP130. GP130 can heterodimerize with either OSMR or LIFR in a species-specific manner. Ligand binding mediates downstream phosphorylation of JAK1/2 and activation of STAT3/5, MAPK, PI3K/AKT, ERK1/2, p38, or JNK. OSM functions in hematopoiesis, inflammation, cell differentiation, tissue remodeling, wound repair, and cancer [35, 163–170]. OSM is pro-tumorigenic and mediates EMT cancer cell invasion, metastasis, and resistance to therapy [35, 166]. In adipose tissue, OSM decreases adipocyte differentiation. OSM promotes fibrosis in skin. In the liver, OSM activates the acute-phase response proteins in inflammation. OSM induces inflammation by activating IL-6, AKT, PKC delta, and STAT5. OSM circulating levels are elevated in systemic sclerosis, liver diseases, systemic lupus erythematosus, and periodontitis [167–170]. OSMRβ is ubiquitously expressed, and its levels are elevated in healthy and cancer patients [171].

There is preliminary evidence of OSM in cancer cachexia in vivo and in vitro. In a murine lung model of cancer cachexia, levels of OSM mRNA was detected in the spleen of wild-type mice [44]. In an in vitro study, treatment of recombinant OSM suppressed C2C12 myotube formation, increased expression of atrophy genes (atrogin-1, MuRF-1, C/EBP theta) and OSMR, and decreased MyoD, myogenin, and STAT3 suggesting OSM might activate muscle wasting in vivo [172].

However, based on our search, there are currently no functional data for OSM in mouse cancer cachexia models and no data in patients with cancer cachexia.

Interleukin-11 (IL-11) was initially characterized as a hematopoietic growth factor. IL-11 promotes platelet recovery and decreases inflammation [173]. IL-11 signals by interacting with IL11Rα and GP130 to form the hexameric receptor complex [174–176]. Similar to IL6R, IL11R has a trans-signaling mediated pathway. The signaling activates JAK/STAT, MAPK/ERK, and PI3K/AKT pathways [177]. IL-11 functions in hematopoiesis and acts on the liver, GI, heart, central nervous system, bone, joint, and immune system [178–182]. In blood, IL-11 promotes thrombocytosis, T lymphocytes, macrophages, stimulates the red blood cell production, and megakaryocytes activation [179, 180]. IL11R is found in cardiomyocytes and plays a protective role in cardiac dysfunction [181–183]. IL-11 is an anti-inflammatory agent. In bone, IL-11 promotes bone loss [179]. In cancer, IL-11 promotes tumorigenesis, cell cycle progression, anti-apoptotic activities, and angiogenesis [35, 174, 177, 184–187]. IL-11 levels are elevated in the circulation and the tumor of cancer patients, in particular of patients with gastrointestinal cancers [174, 177, 186, 187].

Expression of IL-11 is present in models of cancer cachexia. In a murine lung model of cancer cachexia, IL-11 mRNA was detected in splenocytes [44]. In a human cell line, MMG-1, expressing IL-11, induces weight loss in tumor-bearing nude mice. In this model, IL-11 is linked to lipoprotein lipase inhibition and lipolysis [188]. In the oral cavity carcinoma cachexia-inducing cell line, OCC-1C, IL-11 is expressed [115].

Overexpression studies revealed that exogenous IL-11 induces cachexia and lipid catabolism in the absence of tumor [44]. In contrast to the overexpression studies, implantation of several human cell lines expressing IL-11 in mice did not show relevant signs of cachexia [188, 189]. IL-11 can block cancer cell proliferation and reduce effects of cachexia. Recombinant human IL-11 (rhIL-11) was used to treat chemotherapy-induced thrombocytopenia. rhIL-11 blocked cell proliferation of Lewis lung carcinoma cells. In mice implanted with Lewis lung carcinoma cells, treatment of rhIL-11 diminished signs of thrombocytopenia and cachexia [189].

Elevated IL-11 is found in the circulation of cancer patients with cachexia and in human cancer cell lines [190]. We could not find functional evidence of IL-11 in human studies or anti-11 therapeutics.

Interleukin-27 (IL-27) was first characterized based on its influence on the immune response. IL-27 forms a heterodimer with GP130 in the presence of IL27Rα and activates STAT, ERK, p38, MAPK, and AKT signaling [24, 191–193].

In physiological and pathological conditions, IL-27 controls the inflammatory response, response to pathogens, innate immune response, and immune cell differentiation. IL-27 is expressed in myeloid cell populations such as macrophages, inflammatory monocytes, microglia, dendritic cells, plasma cells, endothelial cells, and epithelial cells [193–200]. IL-27 levels are increased in patients with chronic immune thrombocytopenia [193, 201]. IL-27 induces expression of IL-21 and IL-10 [193, 202–204]. IL-27 has a role in Th-1 immunity and promotes growth for T-cells [195]. IL-27 upregulates T cell expression of lymphocyte function-associated antigen-1, intercellular adhesion molecule-1, and sphingosine-1-phosphate [205, 206]. In B cells, IL-27 promotes proliferation and interaction with T-cells [207, 208]. Also, IL-27 promotes neuroinflammation. In cancer, IL-27 promotes anti-tumor immunity and elevated IL-27 levels correlate with disease progression [196–200].

Currently, there is no evidence linking IL-27 to cancer cachexia.

Cardiotrophin-like Cytokine (CLC), also known as neutrophin-1/B cell-stimulating factor-3 (NNT1/BSF-3), was first characterized in promoting cell differentiation and B cell function. CLC signals through the LIFRα, CNTFRα, and GP130 complex [35, 209]. CLC is primarily expressed in lymph nodes, spleen, lymphocytes, bone marrow, and fetal liver. CLC promotes cell differentiation, organ development, and tumor progression. CLC contributes to tissue homeostasis, innate and adaptive immunity, hematopoiesis, regulating pain, autoimmunity, inflammation, and metabolism [35, 210–220]. CLC stimulates neuroendocrine hormone secretion [210–213]. In the nervous system, CLC supports motor and sympathetic neuron survival and astrocyte differentiation [210, 214–216]. In embryogenesis, CLC promotes skeletal muscle and kidney development [217–219]. In metabolism, CLC functions in bone and energy metabolism [35]. In cancer, CLC is involved in tumor progression [220].

Currently, there is no evidence for the implication of CLC in cancer cachexia.

Glycoprotein 130 (GP130) was identified as the transmembrane protein which is the founding member of the class of all cytokine receptors. It is often referred to as the common gp130 subunit and it is important for signal transduction following cytokine engagement. The common receptor subunit for this family has been interrogated in murine models of cancer cachexia, including Apc^{Min} and Lewis lung carcinoma models. In an overexpression study using the GP130F/F knock-in mouse model, there was hyper-activated transcription of STAT3, exacerbated weight loss, mortality, reduced skeletal muscle, and adipose tissue mass [70]. In mice with a skeletal muscle-specific GP130 deletion implanted with Lewis lung carcinoma cells, muscle loss was attenuated by 16%, as was activation of the atrophy-inducing signaling intermediates STAT3, p38, and FOXO3. There was no restoration in mTOR inhibition or AMPK expression in skeletal muscle [221].

4 Conclusions and Future Perspectives

Cancer cachexia is an inflammatory and metabolic condition mediated in part through members of the IL-6/GP130 family of cytokines. IL-6 itself has been firmly linked to cachexia through functional studies in mice and associative studies in patients and has been subjected to limited clinical trials for anti-cachexia therapy. Functional data demonstrate the capacity for LIF and CNTF to cause cachexia with more associative evidence for the former than the latter to date. While the other family members exhibit similar signaling and overlap in function, currently there are few data to link them to cancer cachexia to date. Future studies are needed to clarify the roles and therapeutic potential of this diverse family and its common receptor in cancer cachexia.

As of October 2020, there are 52 completed trials in cancer cachexia indexed in clinicaltrials.gov [222]. The current trials and therapeutic strategies involve targeting inflammation, nutrition, appetite, and metabolism [7, 223, 224]. Although there are promising candidates, still there is no approved therapy for cancer cachexia in the USA. The difficulties in identifying patient eligibility, study enrollment, study design, and endpoints of interest pose a challenge for anti-cachexia therapies. There is a lack of a universal diagnostic criterion for cancer cachexia patients, including a lack of consistency in reporting. For instance, in a clinical screening, body weight is measured, but cachexia or unintentional weight loss may not be defined nor evaluated [4, 9, 224–227]. In addition, patients may reach refractory cachexia and have dismal clinical outcomes, thus reducing eligibility for trials [3, 9]. Anti-cachexia trials have missing patient data often due to the low survivability in cancer cachexia, death, or lack of patient participation. Thus, increasing the difficulty in gaining a comprehensive evaluation of this disease and lowering the sample size and power of the study minimizing any statistical differences observed [4, 228]. There is variability in disease progression, the range of serum and mRNA expression levels of circulating cytokines and inflammatory markers, a lack of biomarkers, tumor heterogeneity, and the many activated signaling pathways [229]. The complexity of cancer itself compounded with cachexia increases the difficulty for a single targeted approach, thus highlighting the need for a multimodal approach. Currently, there are 5 multimodal interventions reported in cancer cachexia [230]. The path forward should focus on endpoints that reflect a clinical benefit, including stabilizing muscle and fat mass, improving nutrition, improving patient quality of life and physical function, while reducing tumor burden and toxicity, and ultimately improving survival [3, 9, 224, 229, 231, 232].

To date, the strategies targeting the IL-6 family of cytokines in cancer progression and cachexia include anti-inflammatory approaches as well as targeted therapies against either the ligand–receptor interaction or the downstream activation of the JAK/STAT pathway [223, 228]. General anti-inflammatory drugs trialed in cachexia including NSAIDs and thalidomide reduce IL-6 family cytokine expression [228]. In a trial of 124 patients with NSCLC, the IL-6 neutralizing antibody, clazakizumab, improved lung symptom score, reversed fatigue, and preserved lean body mass [224, 228, 233]. Anti-IL6R interventions have also been suggested for treatment of cancer cachexia [84, 234] with cautions [235]; however, there have been no prospective trials reported to date. PACTO, a Phase ½ trial (NCT02767557), randomizes patients with newly diagnosed metastatic pancreatic cancer to gemcitabine plus abraxane with or without Tocilizumab [236]. The primary endpoint of this study is 6-month survival; however, correlative studies will examine effects on body composition and other cachexia-relevant endpoints [236]. JAK and STAT inhibitors also have potential to target this family of cachexia-causing factors [9, 224, 228]. Sarilumab has been shown to inhibit the growth of tumor xenografts by blocking binding of IL-6 to IL-6R [237, 238]. The strengths of these anti-inflammation and anti-cytokine approaches are the potential to reduce tumor burden and cachexia simultaneously, thereby targeting the tumor and treating the patient and ultimately increasing physical strength and quality of life [223, 228]. Clearly, further investment and effort are needed to target IL-6, LIF, and other related cachexia-causing cytokines in the clinical setting.

Acknowledgements We would like to acknowledge Kellie N. Kaneshiro from the IU School of Medicine Ruth Lilly Medical Library for her contributions in the literature review search and library preparation. Art is adapted from Servier Medical Art (servier.com) used under a Creative Commons Attribution 3.0 Unported License. T.A.Z. is supported in part by the National Institute for Arthritis and Musculoskeletal and Skin Diseases (NIAMS grant R21AR074908), the National Cancer Institute (grant R01CA194593), and the Veterans Administration (grants I01BX004177 and I01CX002046). D.H.A.J. is funded in part by the Adam W. Herbert Ph. D. Fellowship from Indiana University and the Cancer Biology Training Program Fellowship from the IU Simon Cancer Center.

References

1. Baracos, V.E., Mazurak, V.C., Bhullar, A.S.: Cancer cachexia is defined by an ongoing loss of skeletal muscle mass. Ann Palliat Med. **8**(1), 3–12 (2019)
2. Fearon, K., et al.: Definition and classification of cancer cachexia: an international consensus. Lancet Oncol. **12**(5), 489–495 (2011)
3. Baracos, V.E., et al.: Cancer-associated cachexia. Nat. Rev. Dis. Primers. **4**, 17105 (2018)
4. Prado, C.M., et al.: Central tenet of cancer cachexia therapy: do patients with advanced cancer have exploitable anabolic potential? Am. J. Clin. Nutr. **98**(4), 1012–1019 (2013)
5. Baracos, V.E., Arribas, L.: Sarcopenic obesity: hidden muscle wasting and its impact for survival and complications of cancer therapy. Ann. Oncol. **29**, ii1–ii9 (2018)
6. Aoyagi, T., et al.: Cancer cachexia, mechanism and treatment. World J Gastrointest Oncol. **7**(4), 17–29 (2015)
7. Argiles, J.M., et al.: Cancer cachexia: understanding the molecular basis. Nat. Rev. Cancer. **14**(11), 754–762 (2014)
8. Fonseca, G., et al.: Cancer cachexia and related metabolic dysfunction. Int. J. Mol. Sci. **21**(7) (2020)
9. Fearon, K., Arends, J., Baracos, V.: Understanding the mechanisms and treatment options in cancer cachexia. Nat. Rev. Clin. Oncol. **10**(2), 90–99 (2013)
10. Porporato, P.E.: Understanding cachexia as a cancer metabolism syndrome. Oncogenesis. **5**, e200 (2016)
11. Tisdale, M.J.: Mechanisms of cancer cachexia. Physiol. Rev. **89**(2), 381–410 (2009)
12. Heinrich, P.C., et al.: Principles of interleukin (IL)-6-type cytokine signalling and its regulation. Biochem. J. **374**(Pt 1), 1–20 (2003)
13. Roy, A., Kumar, A.: ER stress and unfolded protein response in cancer cachexia. Cancers (Basel). **11**(12) (2019)
14. Langstein, H.N., Norton, J.A.: Mechanisms of cancer cachexia. Hematol. Oncol. Clin. North Am. **5**(1), 103–123 (1991)
15. Tisdale, M.J.: Catabolic mediators of cancer cachexia. Curr. Opin. Support. Palliat. Care. **2**(4), 256–261 (2008)
16. Zimmers, T.A., Fishel, M.L., Bonetto, A.: STAT3 in the systemic inflammation of cancer cachexia. Semin. Cell Dev. Biol. **54**, 28–41 (2016)
17. Benny Klimek, M.E., et al.: Acute inhibition of myostatin-family proteins preserves skeletal muscle in mouse models of cancer cachexia. Biochem. Biophys. Res. Commun. **391**(3), 1548–1554 (2010)
18. Acunzo, M., Croce, C.M.: MicroRNA in cancer and cachexia – a mini-review. J. Infect. Dis. **212**(Suppl 1), S74–S77 (2015)
19. Zhang, G., et al.: Tumor induces muscle wasting in mice through releasing extracellular Hsp70 and Hsp90. Nat. Commun. **8**(1), 589 (2017)
20. Michaelis, K.A., et al.: The TLR7/8 agonist R848 remodels tumor and host responses to promote survival in pancreatic cancer. Nat. Commun. **10**(1), 4682 (2019)
21. Argilés, J.M., et al.: Cachexia: a problem of energetic inefficiency. J. Cachexia. Sarcopenia Muscle. **5**(4), 279–286 (2014)
22. Coussens, L.M., Werb, Z.: Inflammation and cancer. Nature. **420**(6917), 860–867 (2002)

23. Singh, N., et al.: Inflammation and cancer. Ann. Afr. Med. **18**(3), 121–126 (2019)
24. West, N.R.: Coordination of immune-stroma crosstalk by IL-6 family cytokines. Front. Immunol. **10**, 1093 (2019)
25. White, U.A., Stephens, J.M.: The gp130 receptor cytokine family: regulators of adipocyte development and function. Curr. Pharm. Des. **17**(4), 340–346 (2011)
26. Rose-John, S.: Interleukin-6 family cytokines. Cold Spring Harb. Perspect. Biol. **10**(2) (2018)
27. Gearing, D.P., et al.: Leukemia inhibitory factor receptor is structurally related to the IL-6 signal transducer, gp130. EMBO J. **10**(10), 2839–2848 (1991)
28. Pennica, D., et al.: Cardiotrophin-1. Biological activities and binding to the leukemia inhibitory factor receptor/gp130 signaling complex. J. Biol. Chem. **270**(18), 10915–10922 (1995)
29. Gearing, D.P., et al.: The IL-6 signal transducer, gp130: an oncostatin M receptor and affinity converter for the LIF receptor. Science. **255**(5050), 1434–1437 (1992)
30. Ichihara, M., et al.: Oncostatin M and leukemia inhibitory factor do not use the same functional receptor in mice. Blood. **90**(1), 165–173 (1997)
31. Hermanns, H.M.: Oncostatin M and interleukin-31: Cytokines, receptors, signal transduction and physiology. Cytokine Growth Factor Rev. **26**(5), 545–558 (2015)
32. Mosley, B., et al.: Dual oncostatin M (OSM) receptors. Cloning and characterization of an alternative signaling subunit conferring OSM-specific receptor activation. J. Biol. Chem. **271**(51), 32635–32643 (1996)
33. Davis, S., et al.: LIFR beta and gp130 as heterodimerizing signal transducers of the tripartite CNTF receptor. Science. **260**(5115), 1805–1808 (1993)
34. Elson, G.C., et al.: CLF associates with CLC to form a functional heteromeric ligand for the CNTF receptor complex. Nat. Neurosci. **3**(9), 867–872 (2000)
35. Jones, S.A., Jenkins, B.J.: Recent insights into targeting the IL-6 cytokine family in inflammatory diseases and cancer. Nat. Rev. Immunol. **18**(12), 773–789 (2018)
36. Tanaka, T., Narazaki, M., Kishimoto, T.: IL-6 in inflammation, immunity, and disease. Cold Spring Harb. Perspect. Biol. **6**(10), a016295 (2014)
37. Kishimoto, T.: Factors affecting B-cell growth and differentiation. Annu. Rev. Immunol. **3**, 133–157 (1985)
38. Heinrich, P.C., Castell, J.V., Andus, T.: Interleukin-6 and the acute phase response. Biochem. J. **265**(3), 621–636 (1990)
39. Johnson, D.E., O'Keefe, R.A., Grandis, J.R.: Targeting the IL-6/JAK/STAT3 signalling axis in cancer. Nat. Rev. Clin. Oncol. **15**(4), 234–248 (2018)
40. Ishibashi, T., et al.: Interleukin-6 is a potent thrombopoietic factor in vivo in mice. Blood. **74**(4), 1241–1244 (1989)
41. Miyamoto, Y., et al.: Molecular pathways: cachexia signaling-A targeted approach to cancer treatment. Clin. Cancer Res. **22**(16), 3999–4004 (2016)
42. Wang, C.H., et al.: Clodronate alleviates cachexia and prolongs survival in nude mice xenografted with an anaplastic thyroid carcinoma cell line. J. Endocrinol. **190**(2), 415–423 (2006)
43. Rose-John, S.: IL-6 trans-signaling via the soluble IL-6 receptor: importance for the pro-inflammatory activities of IL-6. Int. J. Biol. Sci. **8**(9), 1237–1247 (2012)
44. Barton, B.E., Murphy, T.F.: Cancer cachexia is mediated in part by the induction of IL-6-like cytokines from the spleen. Cytokine. **16**(6), 251–257 (2001)
45. Tamura, S., et al.: Involvement of human interleukin 6 in experimental cachexia induced by a human uterine cervical carcinoma xenograft. Clin. Cancer Res. **1**(11), 1353–1358 (1995)
46. Bonetto, A., et al.: The colon-26 carcinoma tumor-bearing mouse as a model for the study of cancer cachexia. J. Vis. Exp. **117** (2016)
47. Pin, F., et al.: Growth of ovarian cancer xenografts causes loss of muscle and bone mass: a new model for the study of cancer cachexia. J. Cachexia. Sarcopenia Muscle. **9**(4), 685–700 (2018)
48. Strassmann, G., et al.: Evidence for the involvement of interleukin 6 in experimental cancer cachexia. J. Clin. Invest. **89**(5), 1681–1684 (1992)
49. Carson, J.A., Baltgalvis, K.A.: Interleukin 6 as a key regulator of muscle mass during cachexia. Exerc. Sport Sci. Rev. **38**(4), 168–176 (2010)
50. Fujimoto-Ouchi, K., et al.: Capecitabine improves cancer cachexia and normalizes IL-6 and PTHrP levels in mouse cancer cachexia models. Cancer Chemother. Pharmacol. **59**(6), 807–815 (2007)
51. Bonetto, A., et al.: JAK/STAT3 pathway inhibition blocks skeletal muscle wasting downstream of IL-6 and in experimental cancer cachexia. Am. J. Physiol. Endocrinol. Metab. **303**(3), E410–E421 (2012)
52. Fujimoto-Ouchi, K., et al.: Establishment and characterization of cachexia-inducing and -non-inducing clones of murine colon 26 carcinoma. Int. J. Cancer. **61**(4), 522–528 (1995)
53. Matsuyama, T., et al.: Tumor inoculation site affects the development of cancer cachexia and muscle wasting. Int. J. Cancer. **137**(11), 2558–2565 (2015)
54. Mehl, K.A., et al.: Myofiber degeneration/regeneration is induced in the cachectic ApcMin/+ mouse. J. Appl. Physiol. (1985), 2005. 99(6): 2379–2387
55. Petruzzelli, M., et al.: A switch from white to brown fat increases energy expenditure in cancer-associated cachexia. Cell Metab. **20**(3), 433–447 (2014)
56. Onuma, E., et al.: Parathyroid hormone-related protein (PTHrP) as a causative factor of cancer-associated wasting: possible involvement of PTHrP in the repression of locomotor activity in rats bearing human tumor xenografts. Int. J. Cancer. **116**(3), 471–478 (2005)

57. Iguchi, H., et al.: Involvement of parathyroid hormone-related protein in experimental cachexia induced by a human lung cancer-derived cell line established from a bone metastasis specimen. Int. J. Cancer. **94**(1), 24–27 (2001)

58. Zaki, M.H., Nemeth, J.A., Trikha, M.: CNTO 328, a monoclonal antibody to IL-6, inhibits human tumor-induced cachexia in nude mice. Int. J. Cancer. **111**(4), 592–595 (2004)

59. Flint, T.R., et al.: Tumor-induced IL-6 reprograms host metabolism to suppress anti-tumor immunity. Cell Metab. **24**(5), 672–684 (2016)

60. Rupert, J.E., et al.: Tumor-derived IL-6 and trans signaling among tumor, fat, and muscle mediate pancreatic cancer cachexia. J. Exp. Med. **218**(6), e20190450 (2021)

61. Negri, D.R., et al.: Role of cytokines in cancer cachexia in a murine model of intracerebral injection of human tumours. Cytokine. **15**(1), 27–38 (2001)

62. Baltgalvis, K.A., et al.: Interleukin-6 and cachexia in ApcMin/+ mice. Am. J. Physiol. Regul. Integr. Comp. Physiol. **294**(2), R393–R401 (2008)

63. Lu, S., et al.: Ginsenoside Rb1 can ameliorate the key inflammatory cytokines TNF-alpha and IL-6 in a cancer cachexia mouse model. BMC Complement Med Ther. **20**(1), 11 (2020)

64. Liva, S.G., et al.: Overcoming resistance to anabolic SARM therapy in experimental cancer cachexia with an HDAC inhibitor. EMBO Mol. Med. **12**(2), e9910 (2020)

65. Liu, H., et al.: Coix seed oil ameliorates cancer cachexia by counteracting muscle loss and fat lipolysis. BMC Complement. Altern. Med. **19**(1), 267 (2019)

66. Nakamura, K., et al.: A ketogenic formula prevents tumor progression and cancer cachexia by attenuating systemic inflammation in colon 26 tumor-bearing mice. Nutrients. **10**(2) (2018)

67. An, J.M., et al.: Dietary intake of probiotic kimchi ameliorated IL-6-driven cancer cachexia. J. Clin. Biochem. Nutr. **65**(2), 109–117 (2019)

68. Enomoto, A., et al.: Suppression of cancer cachexia by 20S,21-epoxy-resibufogenin-3-acetate-a novel nonpeptide IL-6 receptor antagonist. Biochem. Biophys. Res. Commun. **323**(3), 1096–1102 (2004)

69. Au, E.D., et al.: The MEK-inhibitor selumetinib attenuates tumor growth and reduces IL-6 expression but does not protect against muscle wasting in lewis lung cancer cachexia. Front. Physiol. **7**, 682 (2016)

70. Miller, A., et al.: Blockade of the IL-6 trans-signalling/STAT3 axis suppresses cachexia in Kras-induced lung adenocarcinoma. Oncogene. **36**(21), 3059–3066 (2017)

71. Bonetto, A., et al.: STAT3 activation in skeletal muscle links muscle wasting and the acute phase response in cancer cachexia. PLoS One. **6**(7), e22538 (2011)

72. Mantovani, G., et al.: Serum levels of leptin and proinflammatory cytokines in patients with advanced-stage cancer at different sites. J. Mol. Med. (Berl). **78**(10), 554–561 (2000)

73. White, J.P., et al.: The regulation of skeletal muscle protein turnover during the progression of cancer cachexia in the Apc(Min/+) mouse. PLoS One. **6**(9), e24650 (2011)

74. Chen, J.L., et al.: Differential effects of IL6 and activin A in the development of cancer-associated cachexia. Cancer Res. **76**(18), 5372–5382 (2016)

75. VanderVeen, B.N., et al.: The regulation of skeletal muscle fatigability and mitochondrial function by chronically elevated interleukin-6. Exp. Physiol. **104**(3), 385–397 (2019)

76. Kuroda, K., et al.: Interleukin 6 is associated with cachexia in patients with prostate cancer. Urology. **69**(1), 113–117 (2007)

77. Sato, H., et al.: Relationships between oxycodone pharmacokinetics, central symptoms, and serum interleukin-6 in cachectic cancer patients. Eur. J. Clin. Pharmacol. **72**(12), 1463–1470 (2016)

78. Eskiler, G.G., et al.: IL-6 mediated JAK/STAT3 signaling pathway in cancer patients with cachexia. Bratisl. Lek. Listy. **66**(11), 819–826 (2019)

79. Iwase, S., et al.: Steep elevation of blood interleukin-6 (IL-6) associated only with late stages of cachexia in cancer patients. Eur. Cytokine Netw. **15**(4), 312–316 (2004)

80. Han, J., et al.: Interleukin-6 induces fat loss in cancer cachexia by promoting white adipose tissue lipolysis and browning. Lipids Health Dis. **17**(1), 14 (2018)

81. Talbert, E.E., et al.: Circulating monocyte chemoattractant protein-1 (MCP-1) is associated with cachexia in treatment-naive pancreatic cancer patients. J. Cachexia. Sarcopenia Muscle. **9**(2), 358–368 (2018)

82. Garcia, J.M., et al.: Active ghrelin levels and active to total ghrelin ratio in cancer-induced cachexia. J. Clin. Endocrinol. Metab. **90**(5), 2920–2926 (2005)

83. Utech, A.E., et al.: Predicting survival in cancer patients: the role of cachexia and hormonal, nutritional and inflammatory markers. J. Cachexia. Sarcopenia Muscle. **3**(4), 245–251 (2012)

84. Ando, K., et al.: Possible role for tocilizumab, an anti-interleukin-6 receptor antibody, in treating cancer cachexia. J. Clin. Oncol. **31**(6), e69–e72 (2013)

85. Gearing, D.P., et al.: Molecular cloning and expression of cDNA encoding a murine myeloid leukaemia inhibitory factor (LIF). EMBO J. **6**(13), 3995–4002 (1987)

86. Smith, A.G., et al.: Inhibition of pluripotential embryonic stem cell differentiation by purified polypeptides. Nature. **336**(6200), 688–690 (1988)

87. Baumann, H., et al.: Distinct sets of acute phase plasma proteins are stimulated by separate human hepatocyte-stimulating factors and monokines in rat hepatoma cells. J. Biol. Chem. **262**(20), 9756–9768 (1987)

88. Patterson, P.H., Chun, L.L.: The induction of acetylcholine synthesis in primary cultures of dissociated

rat sympathetic neurons. II. Developmental aspects. Dev Biol. **60**(2), 473–481 (1977)

89. Mori, M., Yamaguchi, K., Abe, K.: Purification of a lipoprotein lipase-inhibiting protein produced by a melanoma cell line associated with cancer cachexia. Biochem. Biophys. Res. Commun. **160**(3), 1085–1092 (1989)

90. Nicola, N.A., Babon, J.J.: Leukemia inhibitory factor (LIF). Cytokine Growth Factor Rev. **26**(5), 533–544 (2015)

91. Robinson, R.C., et al.: The crystal structure and biological function of leukemia inhibitory factor: implications for receptor binding. Cell. **77**(7), 1101–1116 (1994)

92. Hilton, D.J., Nicola, N.A.: Kinetic analyses of the binding of leukemia inhibitory factor to receptor on cells and membranes and in detergent solution. J. Biol. Chem. **267**(15), 10238–10247 (1992)

93. Williams, R.L., et al.: Myeloid leukaemia inhibitory factor maintains the developmental potential of embryonic stem cells. Nature. **336**(6200), 684–687 (1988)

94. Hilton, D.J., Nicola, N.A., Metcalf, D.: Distribution and comparison of receptors for leukemia inhibitory factor on murine hemopoietic and hepatic cells. J. Cell. Physiol. **146**(2), 207–215 (1991)

95. Ni, H., et al.: Expression of leukemia inhibitory factor receptor and gp130 in mouse uterus during early pregnancy. Mol. Reprod. Dev. **63**(2), 143–150 (2002)

96. Walker, E.C., et al.: Oncostatin M promotes bone formation independently of resorption when signaling through leukemia inhibitory factor receptor in mice. J. Clin. Invest. **120**(2), 582–592 (2010)

97. Chesnokova, V., Auernhammer, C.J., Melmed, S.: Murine leukemia inhibitory factor gene disruption attenuates the hypothalamo-pituitary-adrenal axis stress response. Endocrinology. **139**(5), 2209–2216 (1998)

98. Jo, C., et al.: Leukemia inhibitory factor blocks early differentiation of skeletal muscle cells by activating ERK. Biochim. Biophys. Acta. **1743**(3), 187–197 (2005)

99. Li, X., et al.: LIF promotes tumorigenesis and metastasis of breast cancer through the AKT-mTOR pathway. Oncotarget. **5**(3), 788–801 (2014)

100. Yue, X., et al.: Leukemia inhibitory factor promotes EMT through STAT3-dependent miR-21 induction. Oncotarget. **7**(4), 3777–3790 (2016)

101. Metcalf, D., Gearing, D.P.: Fatal syndrome in mice engrafted with cells producing high levels of the leukemia inhibitory factor. Proc. Natl. Acad. Sci. USA. **86**(15), 5948–5952 (1989)

102. Mori, M., et al.: Cancer cachexia syndrome developed in nude mice bearing melanoma cells producing leukemia-inhibitory factor. Cancer Res. **51**(24), 6656–6659 (1991)

103. Kamohara, H., et al.: Leukemia inhibitory factor functions as a growth factor in pancreas carcinoma cells: involvement of regulation of LIF and its receptor expression. Int. J. Oncol. **30**(4), 977–983 (2007)

104. Chang, J.W., et al.: Production of multiple cytokines and induction of cachexia in athymic nude mice by a new anaplastic thyroid carcinoma cell line. J. Endocrinol. **179**(3), 387–394 (2003)

105. Kamoshida, S., et al.: Expression of cancer cachexia-related factors in human cancer xenografts: an immunohistochemical analysis. Biomed. Res. **27**(6), 275–281 (2006)

106. Iseki, H., et al.: Cytokine production in five tumor cell lines with activity to induce cancer cachexia syndrome in nude mice. Jpn. J. Cancer Res. **86**(6), 562–567 (1995)

107. Tanaka, R., et al.: Triple paraneoplastic syndrome of hypercalcemia, leukocytosis and cachexia in two human tumor xenografts in nude mice. Jpn. J. Clin. Oncol. **26**(2), 88–94 (1996)

108. Terawaki, K., et al.: Leukemia inhibitory factor via the Toll-like receptor 5 signaling pathway involves aggravation of cachexia induced by human gastric cancer-derived 85As2 cells in rats. Oncotarget. **9**(78), 34748–34764 (2018)

109. Seto, D.N., Kandarian, S.C., Jackman, R.W.: A key role for leukemia inhibitory factor in c26 cancer cachexia. J. Biol. Chem. **290**(32), 19976–19986 (2015)

110. Arora, G.K., et al.: Cachexia-associated adipose loss induced by tumor-secreted leukemia inhibitory factor is counterbalanced by decreased leptin. Jci Insight. **3**(14), 26 (2018)

111. Akiyama, Y., et al.: In vivo effect of recombinant human leukemia inhibitory factor in primates. Jpn. J. Cancer Res. **88**(6), 578–583 (1997)

112. Kandarian, S.C., et al.: Tumour-derived leukaemia inhibitory factor is a major driver of cancer cachexia and morbidity in C26 tumour-bearing mice. J. Cachexia. Sarcopenia Muscle. **9**(6), 1109–1120 (2018)

113. Terawaki, K., et al.: New cancer cachexia rat model generated by implantation of a peritoneal dissemination-derived human stomach cancer cell line. Am. J. Physiol. Endocrinol. Metab. **306**(4), E373–E387 (2014)

114. Billingsley, K.G., et al.: Macrophage-derived tumor necrosis factor and tumor-derived of leukemia inhibitory factor and interleukin-6: possible cellular mechanisms of cancer cachexia. Ann. Surg. Oncol. **3**(1), 29–35 (1996)

115. Kajimura, N., et al.: Toxohormones responsible for cancer cachexia syndrome in nude mice bearing human cancer cell lines. Cancer Chemother. Pharmacol. **38**(Suppl), S48–S52 (1996)

116. White, J.D., Davies, M., Grounds, M.D.: Leukaemia inhibitory factor increases myoblast replication and survival and affects extracellular matrix production: combined in vivo and in vitro studies in post-natal

skeletal muscle. Cell Tissue Res. **306**(1), 129–141 (2001)

117. Spangenburg, E.E., Booth, F.W.: Leukemia inhibitory factor restores the hypertrophic response to increased loading in the LIF(−/−) mouse. Cytokine. **34**(3–4), 125–130 (2006)

118. Lynch, G.S., Schertzer, J.D., Ryall, J.G.: Therapeutic approaches for muscle wasting disorders. Pharmacol. Ther. **113**(3), 461–487 (2007)

119. Gao, S., Carson, J.A.: Lewis lung carcinoma regulation of mechanical stretch-induced protein synthesis in cultured myotubes. Am. J. Physiol. Cell Physiol. **310**(1), C66–C79 (2016)

120. Marshall, M.K., et al.: Leukemia inhibitory factor induces changes in lipid metabolism in cultured adipocytes. Endocrinology. **135**(1), 141–147 (1994)

121. He, W., et al.: The N-terminal cytokine binding domain of LIFR is required for CNTF binding and signaling. FEBS Lett. **579**(20), 4317–4323 (2005)

122. Sendtner, M., et al.: Ciliary neurotrophic factor. J. Neurobiol. **25**(11), 1436–1453 (1994)

123. Lam, A., et al.: Sequence and structural organization of the human gene encoding ciliary neurotrophic factor. Gene. **102**(2), 271–276 (1991)

124. Barbin, G., Manthorpe, M., Varon, S.: Purification of the chick eye ciliary neuronotrophic factor. J. Neurochem. **43**(5), 1468–1478 (1984)

125. Hughes, S.M., et al.: Ciliary neurotrophic factor induces type-2 astrocyte differentiation in culture. Nature. **335**(6185), 70–73 (1988)

126. Davis, S., et al.: The receptor for ciliary neurotrophic factor. Science. **253**(5015), 59–63 (1991)

127. Ip, N.Y., et al.: The alpha component of the CNTF receptor is required for signaling and defines potential CNTF targets in the adult and during development. Neuron. **10**(1), 89–102 (1993)

128. Stockli, K.A., et al.: Regional distribution, developmental changes, and cellular localization of CNTF-mRNA and protein in the rat brain. J. Cell Biol. **115**(2), 447–459 (1991)

129. Rende, M., et al.: Immunolocalization of ciliary neuronotrophic factor in adult rat sciatic nerve. Glia. **5**(1), 25–32 (1992)

130. Friedman, B., et al.: Regulation of ciliary neurotrophic factor expression in myelin-related Schwann cells in vivo. Neuron. **9**(2), 295–305 (1992)

131. Sendtner, M., Stockli, K.A., Thoenen, H.: Synthesis and localization of ciliary neurotrophic factor in the sciatic nerve of the adult rat after lesion and during regeneration. J. Cell Biol. **118**(1), 139–148 (1992)

132. Pasquin, S., Sharma, M., Gauchat, J.F.: Cytokines of the LIF/CNTF family and metabolism. Cytokine. **82**, 122–124 (2016)

133. Martin, D., et al.: Cachectic effect of ciliary neurotrophic factor on innervated skeletal muscle. Am. J. Phys. **271**(5 Pt 2), R1422–R1428 (1996)

134. Henderson, J.T., Mullen, B.J., Roder, J.C.: Physiological effects of CNTF-induced wasting. Cytokine. **8**(10), 784–793 (1996)

135. Henderson, J.T., et al.: Systemic administration of ciliary neurotrophic factor induces cachexia in rodents. J. Clin. Investig. **93**(6), 2632–2638 (1994)

136. Espat, N.J., et al.: Ciliary neurotrophic factor is catabolic and shares with IL-6 the capacity to induce an acute phase response. Am. J. Phys. **271**(1 Pt 2), R185–R190 (1996)

137. Matthys, P., Billiau, A.: Cytokines and cachexia. Nutrition. **13**(9), 763–770 (1997)

138. Pennica, D., et al.: Expression cloning of cardiotrophin 1, a cytokine that induces cardiac myocyte hypertrophy. Proc. Natl. Acad. Sci. USA. **92**(4), 1142–1146 (1995)

139. Stephanou, A., et al.: Cardiotrophin-1 induces heat shock protein accumulation in cultured cardiac cells and protects them from stressful stimuli. J. Mol. Cell. Cardiol. **30**(4), 849–855 (1998)

140. Kuwahara, K., et al.: Cardiotrophin-1 phosphorylates akt and BAD, and prolongs cell survival via a PI3K-dependent pathway in cardiac myocytes. J. Mol. Cell. Cardiol. **32**(8), 1385–1394 (2000)

141. Bustos, M., et al.: Liver damage using suicide genes. A model for oval cell activation. Am. J. Pathol. **157**(2), 549–559 (2000)

142. Ho, D.W., et al.: Therapeutic potential of cardiotrophin 1 in fulminant hepatic failure: dual roles in antiapoptosis and cell repair. Arch. Surg. **141**(11), 1077–1084 (2006) discussion 1084

143. Fritzenwanger, M., et al.: Cardiotrophin-1 induces interleukin-6 synthesis in human umbilical vein endothelial cells. Cytokine. **36**(3–4), 101–106 (2006)

144. Lawrence, T., Fong, C.: The resolution of inflammation: anti-inflammatory roles for NF-kappaB. Int. J. Biochem. Cell Biol. **42**(4), 519–523 (2010)

145. Fritzenwanger, M., et al.: Cardiotrophin-1 induces tumor necrosis factor alpha synthesis in human peripheral blood mononuclear cells. Mediat. Inflamm. **2009**, 489802 (2009)

146. Ichiki, T., et al.: Cardiotrophin-1 stimulates intercellular adhesion molecule-1 and monocyte chemoattractant protein-1 in human aortic endothelial cells. Am. J. Physiol. Heart Circ. Physiol. **294**(2), H750–H763 (2008)

147. Fritzenwanger, M., et al.: Cardiotrophin-1 induces intercellular adhesion molecule-1 expression by nuclear factor kappaB activation in human umbilical vein endothelial cells. Chin. Med. J. **121**(24), 2592–2598 (2008)

148. Mauer, J., Denson, J.L., Bruning, J.C.: Versatile functions for IL-6 in metabolism and cancer. Trends Immunol. **36**(2), 92–101 (2015)

149. Lopez-Andres, N., et al.: Absence of cardiotrophin 1 is associated with decreased age-dependent arterial stiffness and increased longevity in mice. Hypertension. **61**(1), 120–129 (2013)

150. Moreno-Aliaga, M.J., et al.: Cardiotrophin-1 is a key regulator of glucose and lipid metabolism. Cell Metab. **14**(2), 242–253 (2011)

151. Limongelli, G., et al.: Cardiotrophin-1 and TNF-alpha circulating levels at rest and during cardiopulmonary exercise test in athletes and healthy individuals. Cytokine. **50**(3), 245–247 (2010)

152. Yang, Z.F., et al.: Cardiotrophin-1 enhances regeneration of cirrhotic liver remnant after hepatectomy through promotion of angiogenesis and cell proliferation. Liver Int. **28**(5), 622–631 (2008)

153. Castano, D., et al.: Cardiotrophin-1 eliminates hepatic steatosis in obese mice by mechanisms involving AMPK activation. J. Hepatol. **60**(5), 1017–1025 (2014)

154. Malavazos, A.E., et al.: Association of increased plasma cardiotrophin-1 with left ventricular mass indexes in normotensive morbid obesity. Hypertension. **51**(2), e8–e9 (2008) author reply e10

155. Rendo-Urteaga, T., et al.: Decreased cardiotrophin-1 levels are associated with a lower risk of developing the metabolic syndrome in overweight/obese children after a weight loss program. Metabolism. **62**(10), 1429–1436 (2013)

156. Barnabe-Heider, F., et al.: Evidence that embryonic neurons regulate the onset of cortical gliogenesis via cardiotrophin-1. Neuron. **48**(2), 253–265 (2005)

157. Freed, D.H., et al.: Emerging evidence for the role of cardiotrophin-1 in cardiac repair in the infarcted heart. Cardiovasc. Res. **65**(4), 782–792 (2005)

158. Ishikawa, M., et al.: A heart-specific increase in cardiotrophin-1 gene expression precedes the establishment of ventricular hypertrophy in genetically hypertensive rats. J. Hypertens. **17**(6), 807–816 (1999)

159. Pan, J., et al.: Involvement of gp130-mediated signaling in pressure overload-induced activation of the JAK/STAT pathway in rodent heart. Heart Vessel. **13**(4), 199–208 (1998)

160. Talwar, S., Choudhary, S.K.: Tuberculous aneurysms of the aorta. J. Thorac. Cardiovasc. Surg. **125**(5), 1184 (2003)

161. Lopez, B., et al.: Is plasma cardiotrophin-1 a marker of hypertensive heart disease? J. Hypertens. **23**(3), 625–632 (2005)

162. Richards, C.D.: The enigmatic cytokine oncostatin m and roles in disease. ISRN Inflamm. **2013**, 512103 (2013)

163. Argast, G.M., et al.: Cooperative signaling between oncostatin M, hepatocyte growth factor and transforming growth factor-beta enhances epithelial to mesenchymal transition in lung and pancreatic tumor models. Cells Tissues Organs. **193**(1–2), 114–132 (2011)

164. Deng, G., et al.: Unique methylation pattern of oncostatin m receptor gene in cancers of colorectum and other digestive organs. Clin. Cancer Res. **15**(5), 1519–1526 (2009)

165. Chollangi, S., et al.: A unique loop structure in oncostatin M determines binding affinity toward oncostatin M receptor and leukemia inhibitory factor

166. Junk, D.J., et al.: Oncostatin M promotes cancer cell plasticity through cooperative STAT3-SMAD3 signaling. Oncogene. **36**(28), 4001–4013 (2017)

167. Pradeep, A.R., et al.: Serum levels of oncostatin M (a gp 130 cytokine): an inflammatory biomarker in periodontal disease. Biomarkers. **15**(3), 277–282 (2010)

168. Hasegawa, M., et al.: Serum levels of interleukin 6 (IL-6), oncostatin M, soluble IL-6 receptor, and soluble gp130 in patients with systemic sclerosis. J. Rheumatol. **25**(2), 308–313 (1998)

169. Liang, H., et al.: Interleukin-6 and oncostatin M are elevated in liver disease in conjunction with candidate hepatocellular carcinoma biomarker GP73. Cancer Biomark. **11**(4), 161–171 (2012)

170. Robak, E., et al.: Circulating interleukin-6 type cytokines in patients with systemic lupus erythematosus. Eur. Cytokine Netw. **8**(3), 281–286 (1997)

171. Stephens, J.M., Elks, C.M.: Oncostatin M: potential implications for malignancy and metabolism. Curr. Pharm. Des. **23**(25), 3645–3657 (2017)

172. Miki, Y., et al.: Oncostatin M induces C2C12 myotube atrophy by modulating muscle differentiation and degradation. Biochem. Biophys. Res. Commun. **516**(3), 951–956 (2019)

173. Sands, B.E., et al.: Randomized, controlled trial of recombinant human interleukin-11 in patients with active Crohn's disease. Aliment. Pharmacol. Ther. **16**(3), 399–406 (2002)

174. Permyakov, E.A., Uversky, V.N., Permyakov, S.E.: Interleukin-11: a multifunctional cytokine with intrinsically disordered regions. Cell Biochem. Biophys. **74**(3), 285–296 (2016)

175. Matadeen, R., et al.: The dynamics of signal triggering in a gp130-receptor complex. Structure. **15**(4), 441–448 (2007)

176. Barton, V.A., et al.: Interleukin-11 signals through the formation of a hexameric receptor complex. J. Biol. Chem. **275**(46), 36197–36203 (2000)

177. Nguyen, P.M., Putoczki, T.L., Ernst, M.: STAT3-activating cytokines: a therapeutic opportunity for inflammatory bowel disease? J. Interf. Cytokine Res. **35**(5), 340–350 (2015)

178. Schwertschlag, U.S., et al.: Hematopoietic, immunomodulatory and epithelial effects of interleukin-11. Leukemia. **13**(9), 1307–1315 (1999)

179. Putoczki, T., Ernst, M.: More than a sidekick: the IL-6 family cytokine IL-11 links inflammation to cancer. J. Leukoc. Biol. **88**(6), 1109–1117 (2010)

180. Wan, B., et al.: Recombinant human interleukin-11 (IL-11) is a protective factor in severe sepsis with thrombocytopenia: a case-control study. Cytokine. **76**(2), 138–143 (2015)

181. Obana, M., et al.: Therapeutic administration of IL-11 exhibits the postconditioning effects against

receptor. J. Biol. Chem. **287**(39), 32848–32859 (2012)

ischemia-reperfusion injury via STAT3 in the heart. Am. J. Physiol. Heart Circ. Physiol. **303**(5), H569–H577 (2012)

182. Kimura, R., et al.: Identification of cardiac myocytes as the target of interleukin 11, a cardioprotective cytokine. Cytokine. **38**(2), 107–115 (2007)

183. Obana, M., et al.: Therapeutic activation of signal transducer and activator of transcription 3 by interleukin-11 ameliorates cardiac fibrosis after myocardial infarction. Circulation. **121**(5), 684–691 (2010)

184. Ernst, M., Putoczki, T.L.: Targeting IL-11 signaling in colon cancer. Oncotarget. **4**(11), 1860–1861 (2013)

185. Johnstone, C.N., et al.: Emerging roles for IL-11 signaling in cancer development and progression: focus on breast cancer. Cytokine Growth Factor Rev. **26**(5), 489–498 (2015)

186. Winship, A.L., et al.: Targeting interleukin-11 receptor-alpha impairs human endometrial cancer cell proliferation and invasion in vitro and reduces tumor growth and metastasis in vivo. Mol. Cancer Ther. **15**(4), 720–730 (2016)

187. Putoczki, T.L., et al.: Interleukin-11 is the dominant IL-6 family cytokine during gastrointestinal tumorigenesis and can be targeted therapeutically. Cancer Cell. **24**(2), 257–271 (2013)

188. Nara-Ashizawa, N., et al.: Lipolytic and lipoprotein lipase (LPL)-inhibiting activities produced by a human lung cancer cell line responsible for cachexia induction. Anticancer Res. **21**(5), 3381–3387 (2001)

189. Saitoh, M., et al.: Recombinant human interleukin-11 improved carboplatin-induced thrombocytopenia without affecting antitumor activities in mice bearing Lewis lung carcinoma cells. Cancer Chemother. Pharmacol. **49**(2), 161–166 (2002)

190. Saleh, A.Z., et al.: Binding of madindoline A to the extracellular domain of gp130. Biochemistry. **44**(32), 10822–10827 (2005)

191. Aparicio-Siegmund, S., et al.: Inhibition of protein kinase II (CK2) prevents induced signal transducer and activator of transcription (STAT) 1/3 and constitutive STAT3 activation. Oncotarget. **5**(8), 2131–2148 (2014)

192. Diegelmann, J., et al.: A novel role for interleukin-27 (IL-27) as mediator of intestinal epithelial barrier protection mediated via differential signal transducer and activator of transcription (STAT) protein signaling and induction of antibacterial and anti-inflammatory proteins. J. Biol. Chem. **287**(1), 286–298 (2012)

193. Aparicio-Siegmund, S., Garbers, C.: The biology of interleukin-27 reveals unique pro- and anti-inflammatory functions in immunity. Cytokine Growth Factor Rev. **26**(5), 579–586 (2015)

194. Hall, A.O., Silver, J.S., Hunter, C.A.: The immunobiology of IL-27. Adv. Immunol. **115**, 1–44 (2012)

195. Pflanz, S., et al.: IL-27, a heterodimeric cytokine composed of EBI3 and p28 protein, induces proliferation of naive CD4+ T cells. Immunity. **16**(6), 779–790 (2002)

196. Lu, D., et al.: Clinical implications of the interleukin 27 serum level in breast cancer. J. Investig. Med. **62**(3), 627–631 (2014)

197. Chiyo, M., et al.: Expression of IL-27 in murine carcinoma cells produces antitumor effects and induces protective immunity in inoculated host animals. Int. J. Cancer. **115**(3), 437–442 (2005)

198. Hisada, M., et al.: Potent antitumor activity of interleukin-27. Cancer Res. **64**(3), 1152–1156 (2004)

199. Salcedo, R., et al.: Immunologic and therapeutic synergy of IL-27 and IL-2: enhancement of T cell sensitization, tumor-specific CTL reactivity and complete regression of disseminated neuroblastoma metastases in the liver and bone marrow. J. Immunol. **182**(7), 4328–4338 (2009)

200. Salcedo, R., et al.: IL-27 mediates complete regression of orthotopic primary and metastatic murine neuroblastoma tumors: role for CD8+ T cells. J. Immunol. **173**(12), 7170–7182 (2004)

201. Li, Q., et al.: Increased interleukin-27 promotes Th1 differentiation in patients with chronic immune thrombocytopenia. Scand. J. Immunol. **80**(4), 276–282 (2014)

202. Fitzgerald, D.C., et al.: Suppression of autoimmune inflammation of the central nervous system by interleukin 10 secreted by interleukin 27-stimulated T cells. Nat. Immunol. **8**(12), 1372–1379 (2007)

203. Stumhofer, J.S., et al.: Interleukins 27 and 6 induce STAT3-mediated T cell production of interleukin 10. Nat. Immunol. **8**(12), 1363–1371 (2007)

204. Pot, C., et al.: Cutting edge: IL-27 induces the transcription factor c-Maf, cytokine IL-21, and the costimulatory receptor ICOS that coordinately act together to promote differentiation of IL-10-producing Tr1 cells. J. Immunol. **183**(2), 797–801 (2009)

205. Owaki, T., et al.: IL-27 induces Th1 differentiation via p38 MAPK/T-bet- and intercellular adhesion molecule-1/LFA-1/ERK1/2-dependent pathways. J. Immunol. **177**(11), 7579–7587 (2006)

206. Gwyer Findlay, E., et al.: IL-27 receptor signaling regulates CD4+ T cell chemotactic responses during infection. J. Immunol. **190**(9), 4553–4561 (2013)

207. Boumendjel, A., et al.: IL-27 induces the production of IgG1 by human B cells. Eur. Cytokine Netw. **17**(4), 281–289 (2006)

208. Larousserie, F., et al.: Differential effects of IL-27 on human B cell subsets. J. Immunol. **176**(10), 5890–5897 (2006)

209. Murakami, M., Kamimura, D., Hirano, T.: New IL-6 (gp130) family cytokine members, CLC/NNT1/BSF3 and IL-27. Growth Factors. **22**(2), 75–77 (2004)

210. Senaldi, G., et al.: Novel neurotrophin-1/B cell-stimulating factor-3: a cytokine of the IL-6 family.

Proc. Natl. Acad. Sci. USA. **96**(20), 11458–11463 (1999)

211. Shi, Y., et al.: Computational EST database analysis identifies a novel member of the neuropoietic cytokine family. Biochem. Biophys. Res. Commun. **262**(1), 132–138 (1999)

212. Benigni, F., et al.: Six different cytokines that share GP130 as a receptor subunit, induce serum amyloid A and potentiate the induction of interleukin-6 and the activation of the hypothalamus-pituitary-adrenal axis by interleukin-1. Blood. **87**(5), 1851–1854 (1996)

213. Vlotides, G., et al.: Novel neurotrophin-1/B cell-stimulating factor-3 (NNT-1/BSF-3)/cardiotrophin-like cytokine (CLC) – a novel gp130 cytokine with pleiotropic functions. Cytokine Growth Factor Rev. **15**(5), 325–336 (2004)

214. Plun-Favreau, H., et al.: The ciliary neurotrophic factor receptor alpha component induces the secretion of and is required for functional responses to cardiotrophin-like cytokine. EMBO J. **20**(7), 1692–1703 (2001)

215. Forger, N.G., et al.: Cardiotrophin-like cytokine/cytokine-like factor 1 is an essential trophic factor for lumbar and facial motoneurons in vivo. J. Neurosci. **23**(26), 8854–8858 (2003)

216. Uemura, A., et al.: Cardiotrophin-like cytokine induces astrocyte differentiation of fetal neuroepithelial cells via activation of STAT3. Cytokine. **18**(1), 1–7 (2002)

217. Schmidt-Ott, K.M., et al.: Novel regulators of kidney development from the tips of the ureteric bud. J. Am. Soc. Nephrol. **16**(7), 1993–2002 (2005)

218. Alexander, W.S., et al.: Suckling defect in mice lacking the soluble haemopoietin receptor NR6. Curr. Biol. **9**(11), 605–608 (1999)

219. Elson, G.C., et al.: Cytokine-like factor-1, a novel soluble protein, shares homology with members of the cytokine type I receptor family. J. Immunol. **161**(3), 1371–1379 (1998)

220. Vicent, S., et al.: Cross-species functional analysis of cancer-associated fibroblasts identifies a critical role for CLCF1 and IL-6 in non-small cell lung cancer in vivo. Cancer Res. **72**(22), 5744–5756 (2012)

221. Puppa, M.J., et al.: Skeletal muscle glycoprotein 130's role in Lewis lung carcinoma-induced cachexia. FASEB J. **28**(2), 998–1009 (2014)

222. *Clinicaltrials.gov.*

223. Naito, T.: Emerging treatment options for cancer-associated cachexia: a literature review. Ther. Clin. Risk Manag. **15**, 1253–1266 (2019)

224. Argilés, J.M., et al.: Therapeutic strategies against cancer cachexia. Eur J Transl Myol. **29**(1), 7960 (2019)

225. Belloum, Y., Rannou-Bekono, F., Favier, F.B.: Cancer-induced cardiac cachexia: Pathogenesis and impact of physical activity (Review). Oncol. Rep. **37**(5), 2543–2552 (2017)

226. Hain, B.A., et al.: Chemotherapy-induced loss of bone and muscle mass in a mouse model of breast cancer bone metastases and cachexia. JCSM Rapid Commun. **2**(1) (2019)

227. Bonetto, A., et al.: Differential bone loss in mouse models of colon cancer cachexia. Front. Physiol. **7**, 679 (2016)

228. Prado, B.L., Qian, Y.: Anti-cytokines in the treatment of cancer cachexia. Ann Palliat Med. **8**(1), 67–79 (2019)

229. Laird, B.J.A., Balstad, T.R., Solheim, T.S.: Endpoints in clinical trials in cancer cachexia: where to start? Curr. Opin. Support. Palliat. Care. **12**(4), 445–452 (2018)

230. McKeaveney, C., et al.: A critical review of multimodal interventions for cachexia. Adv. Nutr. (2020)

231. Roeland, E.J., et al.: Management of cancer cachexia: ASCO guideline. J. Clin. Oncol. **38**(21), 2438–2453 (2020)

232. Vaughan, V.C., Martin, P., Lewandowski, P.A.: Cancer cachexia: impact, mechanisms and emerging treatments. J. Cachexia. Sarcopenia Muscle. **4**(2), 95–109 (2013)

233. Rigas, J.R., et al.: Efect of ALD518, a humanized anti-IL-6 antibody, on lean body mass loss and symptoms in patients with advanced non-small cell lung cancer (NSCLC): results of a phase II randomized, double-blind safety and efficacy trial. J. Clin. Oncol. **28**(15_suppl), 7622–7622 (2010)

234. Hirata, H., et al.: Favorable responses to tocilizumab in two patients with cancer-related cachexia. J. Pain Symptom Manag. **46**(2), e9–e13 (2013)

235. Berti, A., et al.: Assessment of tocilizumab in the treatment of cancer cachexia. J. Clin. Oncol. **31**(23), 2970 (2013)

236. Chen, I., et al.: PACTO: a single center, randomized, phase II study of the combination of nab-paclitaxel and gemcitabine with or without tocilizumab, an IL-6R inhibitor, as first-line treatment in patients with locally advanced or metastatic pancreatic cancer. Ann. Oncol. **28**, v266 (2017)

237. Favalli, E.G.: Understanding the role of interleukin-6 (IL-6) in the joint and beyond: a comprehensive review of IL-6 inhibition for the management of rheumatoid arthritis. Rheumatol Ther. **7**(3), 473–516 (2020)

238. Heo, T.H., Wahler, J., Suh, N.: Potential therapeutic implications of IL-6/IL-6R/gp130-targeting agents in breast cancer. Oncotarget. **7**(13), 15460–15473 (2016)

239. Guo, D., et al.: Pantoprazole blocks the JAK2/STAT3 pathway to alleviate skeletal muscle wasting in cancer cachexia by inhibiting inflammatory response. Oncotarget. **8**(24), 39640–39648 (2017)

240. Reddel, C.J., et al.: Increased thrombin generation in a mouse model of cancer cachexia is partially interleukin-6 dependent. J. Thromb. Haemost. **15**(3), 477–486 (2017)

NF-kB Signaling in the Macroenvironment of Cancer Cachexia

Benjamin R. Pryce and Denis C. Guttridge

Abstract

The Nuclear Factor Kappa B (NF-κB) transcription factor is ubiquitously expressed and has been shown to control various cellular processes such as survival, proliferation, differentiation, and inflammation. While critical for cellular survival, dysregulation of NF-κB signaling has been linked to the progression of cancer. Although NF-κB has been shown to play a role in various cancers, this chapter will focus on pancreatic cancer and how canonical NF-κB signaling perturbs the macroenvironment of tumor–tissue interactions. This will include exploring how NF-κB can mediate various stages of disease development, beginning with tumor initiation and progression through interactions with the immune systems as well as promoting epithelial–mesenchymal transition (EMT), metastasis, and chemoresistance. All these aspects underscore NF-κB as being involved in various stages of pancreatic cancer development. We will also describe the role of NF-κB in normal skeletal muscle and how this is changed drastically in pancreatic cancer patients, leading to a muscle wasting disease known as cancer cachexia. These effects can be observed by NF-κB activity in both the myofiber and muscle stem cells, ultimately leading to muscle decline. Overall, this chapter will illustrate the central role that NF-κB signaling plays in the macroenvironment between pancreatic tumor cells and peripheral skeletal muscle.

Learning Objectives

1. Overview of the NF-κB signaling pathway
2. Examine the critical role of NF-κB signaling in pancreatic cancer
3. Explore the function of NF-κB signaling in muscle atrophy in cachexia
4. Understand the mechanism of muscle regeneration and how NF-κB dysregulates this process during cachexia

1 Introduction

The Nuclear Factor Kappa B (NF-κB) transcription factor was initially discovered in B cells, where it was found to bind to DNA enhancer elements regulating immunoglobulin light chain expression [1]. Since its discovery, NF-κB has been shown to be ubiquitously expressed and to

B. R. Pryce · D. C. Guttridge (✉)
Department of Pediatrics, and the Hollings Cancer Center, Medical University of South Carolina, Charleston, SC, USA
e-mail: guttridg@musc.edu

© Springer Nature Switzerland AG 2022
S. Acharyya (ed.), *The Systemic Effects of Advanced Cancer*,
https://doi.org/10.1007/978-3-031-09518-4_7

regulate numerous cell processes, such as cell survival, proliferation, differentiation, and immunity [2]. The activity of NF-κB, when dysregulated, has also been found to be tightly associated with various chronic diseases, including cancer [3]. In this chapter, we will focus on the NF-κB signaling pathway in the context of pancreatic cancer. We are particularly interested in this tumor type, not only because NF-κB has been found to play a role in the development and progression of pancreatic cancer, which typically presents in an advanced state, but also because this cancer has one of the highest incidences of the cachexia syndrome [4–8]. This gives us an opportunity in this chapter to dissect NF-κB's role in what we call the macroenvironment of pancreatic cancer-induced cachexia. We will also explore the mechanisms by which NF-κB contributes to muscle atrophy as well as its effect on muscle regeneration and how both processes influence skeletal muscle wasting in cancer cachexia.

1.1   NF-κB Regulation

The NF-κB family of transcription factors consists of five proteins; RelA, also known as p65, RelB, c-Rel, NF-κB 1 and NF-κB 2 (the latter two also known as p50 and p52, respectively), which form combination of homodimers and heterodimers to modulate gene expression [9]. All five proteins contain the Rel-homology domain (RHD) at the N-terminal end, which mediates both dimerization and DNA binding [10]. In the inactive state, NF-κB is predominantly maintained in the cytoplasm due to the binding of the inhibitor of kappa B protein (IκB) which prevents nuclear localization by blocking the NLS of NF-κB monomers [11]. The IκB protein family includes IκBα, IκBβ, IκBγ, IκBε, IκBζ, Bcl-3, p105, and p100 [12]. NF-κB activation is controlled by a signaling pathway, whose primary regulator is the IκB kinase (IKK) complex [11, 13]. The complex contains two catalytically active subunits, IKKα and IKKβ, as well as a regulatory subunit, IKKγ/NEMO. Under the correct extra- and intra-cellular signaling cues, IKK becomes activated, leading to multitude of cellular responses. These cellular responses occur in most tissues, since NF-κB is ubiquitously expressed [14].

1.2   Canonical and Non-canonical NF-κB Pathway

In the classical (canonical) NF-κB signaling pathway, extracellular factors activate IKK, which phosphorylates the IκB proteins [13] (Fig. 1). The phosphorylation leads to IκB ubiquitination and subsequent degradation, releasing NF-κB in the process. This unmasks the NLS domain of NF-κB heterodimers, such as p50/p65 or p50/c-Rel, and allows nuclear localization of NF-κB and target gene activation to proceed [15]. Based on genetic evidence, the canonical pathway is largely dependent on the catalytic activity of IKKβ and the regulatory activity of IKKγ but not IKKα [16]. In contrast, the non-canonical system is activated by a separate set of receptors and leads to the phosphorylation of p100 by IKKα homodimers in the absence of IKKγ, which causes processing of p100 to its p52 active form [2, 17] (Fig. 1). This then leads to dimerization of p52 with RelB and activation of non-canonical gene targets, which are distinct from those of the canonical pathway.

NF-κB is activated by a number of upstream receptors. At the forefront of these receptors are those involved in the inflammatory response, such as TNF, Toll-like receptor (TLR), and IL-1 [18]. Many other pro-inflammatory cytokines as well as reactive oxygen species (ROS) and immune stimulators lead to the activation of NF-κB [19, 20]. Conversely, the non-canonical system appears to be regulated by a distinct subset of receptors, related to lymphoid development, such as RANK and B-cell activating factor receptor (BAFF) [21, 22].

Classical Pathway   Alternative Pathway

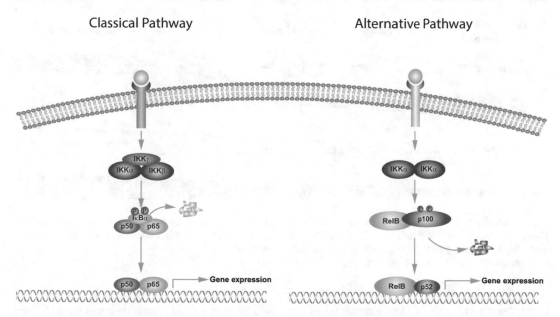

Fig. 1 Classical (Canonical) vs Alternative (Non-Canonical) NF-κB Signaling: In the Classical Pathway (left) extracellular signals lead to the activation of IKK, which subsequently phosphorylates IκBα leading to its ubiquitination and degradation. The degradation of IκBα unmasks the nuclear localization sequence of NF-κB heterodimers, which then enter the nucleus to regulate gene expression. The Alternative Pathway (right) leads to activation of IKKα homodimers, and phosphorylation of p100, leading to processing of p100 to the active p52. Localization of RelB/p52 heterodimers to the nucleus leads to alternative pathway gene expression

1.3 NF-κB Signaling in Healthy Skeletal Muscle

The canonical and non-canonical NF-κB signaling pathways play distinct roles in skeletal muscle. Early work on canonical signaling presented conflicting results, with studies demonstrating both positive and negative roles in muscle regeneration and myoblast differentiation [2]. However, data showed that NF-κB DNA binding and transactivation activities significantly decreased in differentiating myoblasts [23] (Fig. 2). Inhibiting NF-κB was also showed to accelerate myogenesis, increasing both myogenic gene expression and myotube formation *in vitro* [23]. These findings indicate that NF-κB must be downregulated for myogenesis to proceed. More recent work suggests that NF-κB functions to maintain muscle progenitor cells in an undifferentiated stem like state and that downregulation of NF-κB leads to a decrease in survival and proliferation as well as premature differentiation [24]. Therefore, NF-κB is considered necessary for muscle repair, as its activation prevents premature differentiation of MPCs, and its downregulation is required to allow terminal differentiation to proceed.

Contrary to the canonical pathway, evidence suggests that non-canonical NF-κB signaling is pro-myogenic. In one study, Enwere et al. demonstrated that activation of the non-canonical pathway, either by knockdown of cellular inhibitor of apoptosis (c-IAP), increased expression of p52, or treatment with TNF-like weak inducer of apoptosis (an activator of the non-canonical pathway, also known as TWEAK), resulted in an increase in the fusion of myoblasts into myotubes [25]. Whether the non-canonical pathway acts to promote myoblast differentiation during development or in injury-induced muscle repair remains unclear until further genetic studies are performed. Another potential function for the non-canonical signaling pathway of NF-κB relevant in myogenesis is its

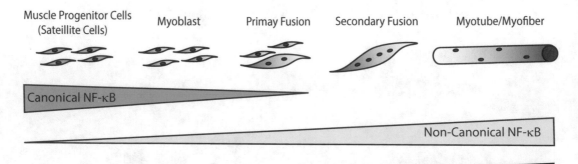

Fig. 2 Activity of Canonical and Non-Canonical NF-κB throughout Myogenic Differentiation: Canonical NF-κB signaling is active in proliferating muscle progenitor cells and maintains cells in an undifferentiated state. Downregulation of canonical NF-κB is correlated with increased differentiation and activation of the alternative pathway

ability to directly regulate the expression of the co-activator, PGC-1β, which is required to stimulate mitochondrial biogenesis and oxidative metabolism [26, 27]. Consistent with this role to regulate muscle metabolism, RelB and IKKα were found to work in concert with the myogenic transcription factor MyoD to stimulate the transcription of a host of oxidative metabolic genes, well beyond PGC-1β [28]. Together, these findings suggest that the canonical and non-canonical NF-κB signaling pathways play opposing yet necessary functions in myogenesis.

2   NF-κB Signaling in Pancreatic Cancer

Pancreatic Ductal Adenocarcinoma (PDAC) remains one of the deadliest cancers, with a current five-year survival rate of only 10% [29]. In addition to metastasis, PDAC is strongly associated with cachexia, with upwards of 80% of PDAC patients showing some signs of body weight decline, largely attributed to wasting of skeletal muscle [4]. Interestingly, NF-κB signaling is upregulated in a high proportion of PDAC and has been shown to regulate various cellular processes, such as proliferation, migration, metastasis, inflammation, apoptosis, and chemoresistance [30] (Fig. 3). In this section, we

will explore several effects downstream of NF-κB that mediate PDAC development and progression, as well as identifying the activators of NF-κB.

2.1   Role of NF-κB in PDAC Initiation and Progression

Oncogenic mutations in *Kras* coupled with the loss of tumor suppressors such as p53 are the primary drivers of PDAC [31]. Signaling downstream of Kras leads to a constitutive activation of NF-κB in pancreatic tumor cells. This is accomplished through a Kras dependent upregulation of IL-1α, which, when binding to its respective receptor on tumor cells, leads to degradation of IκB through the activation of IKK [32]. This results in NF-κB nuclear localization and target gene activation. A subsequent study found that blocking IL-1 receptor inhibited growth of PDAC cell lines [33]. The tumor suppressor p53 is responsible for inhibiting cell cycle progression in response to DNA repair, preventing chromosomal instability due to increasing mutations [34]. The activation of p53 target genes results in cell death, preventing a mutated cell from proliferating. Interestingly, IKKβ can phosphorylate p53 and lead to its ubiquitination and subsequent degradation, suggesting that NF-κB

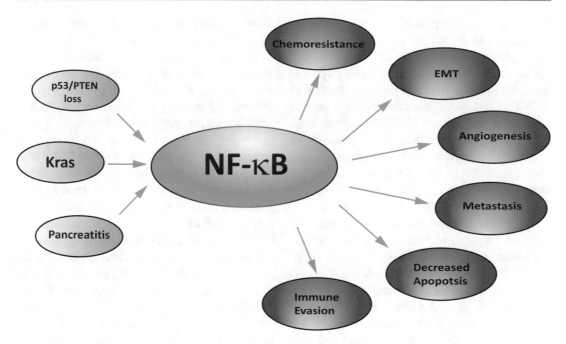

Fig. 3 Multiple Roles of NF-κB in Pancreatic Cancer: NF-κB activity can be upregulated in PDAC by multiple processes (blue). Increased NF-κB activity regulates multiple processes that lead to tumor progression and metastasis (red)

functions as an antagonist of p53 [35]. Furthermore, NF-κB can also block the transactivation of p53, preventing early DNA repair, which may be another mechanism by which NF-κB overcomes tumor repressors to promote the induction of PDAC [36]. There is also a body of evidence that tumor suppressors such as p53 and Pten are themselves inhibitors of NF-κB, further validating the role of NF-κB as a promotor of oncogenesis, at least in conditions of cellular immortalization due to the transformation activity of Kras [36, 37].

Whereas Kras induces NF-κB in developing tumor cells, persistent inflammation within the tumor microenvironment (TME) similarly acts to enhance tumor formation [38]. NF-κB activation by inflammatory cytokines is necessary in the context of acute infection, where an immune response restores tissue homeostasis and eliminates foreign pathogens. However, in the context of PDAC, increased inflammatory cytokines, such as TNF, IL-6, and IL-1, enhance tumorigenesis by activating survival mechanisms

within tumor cells, thus allowing these cells to evade p53 mediated apoptosis. Interestingly, patients with pre-existing pancreas inflammation, known as pancreatitis, are at higher risk of developing PDAC, suggesting that the inflammatory cascade and NF-κB signaling together contribute to PDAC formation [38]. This is corroborated by data that demonstrated that upregulation of NF-κB increases the expression of pro-proliferative genes, such as cyclin D1 to drive early tumor formation [23, 39].

In addition to being regulated by the TME, NF-κB also feeds back and enhances the expression of cytokines to alter the TME into a pro-tumorigenic state. A study by Ratnam et al. in our own laboratory determined that NF-κB activity in PDAC cells upregulated the expression of GDF15, which is a cytokine distantly belonging to the TGFβ superfamily. GDF15 was found to inhibit the upstream regulator of IKK, TAK1, in macrophages *in vitro* [40]. Inhibition of TAK1 decreased NF-κB activity, thereby reducing the anti-tumor effects within these macrophages. This signaling was considered to enhance tumorigenesis, as it reduces NF-κB activity in macrophages, leading to a reduction of anti-apoptotic factors, such as nitric oxide (NO) and TNF. The authors speculate that the reduction of both TNF and NO allows cancer cells to survive during the early phases of PDAC development. These findings demonstrate the complexity of immune responses in PDAC development, and how a reduction in NF-κB activity in a cell specific manner may also be beneficial for the initiating phases of PDAC development.

Apoptosis is a process of programmed cell death in response to both internal and external stress and is necessary for proper tissue development and homeostasis. Dysregulated apoptosis is a common feature of cancer and allows tumorigenic cells to expand rapidly, which is in part mediated by NF-κB [41]. Additionally, NF-κB inhibition can increase apoptosis in PDAC cells lines. NF-κB plays a predominant role in regulating the expression of several anti-apoptotic genes, such as BCL-2, Bcl-XL, and c-IAP [42]. As with increased proliferation, decreased apoptosis leads to an increase in the number of propagating cancer cells needed to boost tumorigenesis. Altogether, we can see that NF-κB signaling is activated by pro-oncogenic signaling and is necessary for tumorigenesis to proceed.

2.2 NF-κB Regulates Angiogenesis and EMT to Promote Metastasis

The highest percentage of cancer related deaths result from tumor metastases, which involves the migration of a primary tumor cells to a secondary lesion site. Several events must happen for metastases to colonize these secondary sites. The first event is the transition of the primary epithelial tumor cell into a more motile mesenchymal cell. Second, increased angiogenesis provides primary tumor cells a path to exit the primary tumor site. Finally, tumor cells must be able to survive outside of the TME to invade secondary tissues.

Epithelial–mesenchymal transition (EMT) is a prominent feature of metastasizing tumor cells and is caused in large degree by TGFβ, which activates the transcription of mesenchymal genes while also repressing epithelial genes [43]. This makes the cell more motile in part by breaking down cell–cell junctions. A study by Maier et al. indicated that the inhibition of NF-κB signaling was sufficient to block TGFβ mediated EMT in pancreatic cancer cells [44]. Additionally, activation of NF-κB induced EMT through the upregulation of ZEB1 and downregulation of E-cadherin. This effect was also observed in the absence of functional TGFβ activity, indicating that NF-κB can drive EMT independent of other signaling pathways.

Angiogenesis is induced by VEGF, IL-8, and MMP9, all of which have been shown to be regulated by NF-κB [45–48]. Therefore, the upregulation of NF-κB in cells within the TME can contribute to the induction of angiogenesis. Furthermore, NF-κB blockade was sufficient to prevent metastasis [47]. Furthermore, preventing angiogenesis also reduces primary tumor growth by depriving tumor cells the uptake of these nutrients found in the circulation [49].

Events such as the promotion of EMT and angiogenesis allow tumor cells to exit their primary lesion site and begin to migrate to distant sites. Both processes are regulated in part by NF-κB and blocking either of these processes has been shown to prevent metastasis. However, even after leaving the primary TME, circulating tumor cells continue to rely on NF-κB to survive and invade secondary sites. The upregulation of Sox9 was found to be dependent on NF-κB and was essential for invasion [50]. Additionally, the upregulation of HIF-1α by NF-κB promotes the expression of LASP1 and FSCN1, which can both increase metastasis and cell survival [51]. Therefore, NF-κB has been considered indispensable for cancer cells to exit their primary tumor site and metastasize.

2.3 NF-κB Increases Chemoresistance in PDAC

Although NF-κB can directly regulate survival and growth of pancreatic tumor cells, it is also activated in response to cellular stress. As chemotherapies cause significant cellular stress, it is not surprising that NF-κB is activated by such therapies [52, 53]. One major mechanism that has been proposed by which cytotoxic agents activate NF-κB is through DNA damage [54]. Following treatment with DNA double stranded break inducers, such as class II topoisomerase inhibitors, the ATM kinase is exported from the nucleus with IKKγ/NEMO, where they both activate IKKβ to promote NF-κB nuclear translocation. NF-κB activation is not limited to DSB inducing agents, as both vinblastine and vincristine, microtubule-disrupting compounds, lead to the activation of NF-κB [55]. How these non-chromatin damaging agents activate IKKβ is less clear. Interestingly, PDAC cells lines resistant to gemcitabine, a chemotherapeutic agent that was used until recently as a standard monotherapy for PDAC, exhibit elevated NF-κB activity [56]. Therefore, it is perhaps not surprising that a combinatorial treatment of gemcitabine with NF-κB inhibitors has been shown to be more efficacious than gemcitabine alone in a number of tumor cell lines [57–59].

Gemcitabine resistance is mediated in part by the expression the nucleoside transporters, hENT1, hCNT1, and hCNT3, as these transporters are essential in the uptake of gemcitabine by pancreatic cancer cells [60, 61]. Reductions in these transporters result in gemcitabine resistance [62, 63]. The expression of hCNT1 has been shown to be downregulated by Mucin-4, a glycoprotein whose expression is upregulated in PDAC [64]. Downregulation of hCNT1 by Mucin-4 was also found to be NF-κB dependent. Thus, it is quite possible that the connection between the elevation of NF-κB and downregulation of hCNT1 can be viewed as a plausible mechanism to explain chemoresistance in multiple tumor types [65, 66]. A separate study demonstrated that NF-κB was capable of mediating gemcitabine resistance in cooperation with STAT3 downstream of ROS [67]. However, even though NF-κB inhibition can prevent the resistance of PDAC to chemotherapy, translating this into the clinical setting has been challenging, with a number of combinatorial therapies yielding negative results [68, 69]. It is clear that more studies will be needed to fully understand the complexities underlying the functions of NF-κB in PDAC to generate novel targeted therapies.

3 NF-κB Dependent Myofiber Atrophy and Muscle Dysfunction in Cancer Cachexia

It is difficult to uncouple the effects of NF-κB on PDAC without thinking of NF-κB in its control of a larger PDAC microenvironment, since this signaling pathway has been strongly implicated in cachexia. At the heart of the cachexia is the catabolism of skeletal muscle, which contributes the greatest portion of weight loss in cancer patients [4]. Given the role of NF-κB in regulating skeletal muscle differentiation and regeneration it is logical to speculate that this signaling pathway might be involved in how PDAC regulates skeletal muscle wasting in cachexia [23, 24].

Skeletal muscle is remarkably adaptive, with the ability to undergo hypertrophy in response to resistance training, increasing in both size and strength due to upregulated protein synthesis [70]. However, the opposite also occurs, where underutilization leads to a significant decline in skeletal muscle activity [71]. While both effects can be observed in healthy individuals, the effect of inactivity can be devastating in patients that are immobilized, which is often observed during long periods of bed rest and the elderly [72]. Additionally, conditions such as muscular dystrophy, chronic inflammation, and cancer cachexia involve the activation of molecular signaling pathways that lead to muscle degradation [71]. In these conditions, NF-κB has been found to function in multiple ways to mediate muscle atrophy [71].

The following section will explore ways in which NF-κB signaling tips the scales toward protein breakdown in the wasting of skeletal muscle and explores the implications of these mechanisms in cancer cachexia.

3.1   NF-κB Activates the Ubiquitin Proteasome to Mediate Muscle Atrophy

The activation of pathways involved in protein breakdown is an active process mediated in by extracellular cues. The mediator of protein breakdown in skeletal muscle is the ubiquitin–proteasome system (UPS) [73]. The UPS system involves a series of three enzymes (termed E1, E2, and E3 ligases) that transfer a small protein, ubiquitin, from E1 to E2 and finally to the E3 ligase [74]. The E3 ligases then target proteins for degradation by the proteasome via the addition of a poly-ubiquitin chain onto specific lysine residues of substrate proteins, which in the context of muscle wasting, is primarily restricted to myofibrillar proteins which make up the bulk mass of skeletal muscle. Early work on muscle atrophy identified two E3 ligases that are responsible for the degradation of proteins in myofibers; atrogin-1 (MAFbx) and the muscle RING (really interesting new gene) finger protein 1 (Murf1) [75]. Both E3 ligases have been found to be upregulated in many wasting conditions, such as denervation and chronic inflammation, and their genetic deletion can block muscle wasting in animal models [75–77]. Additionally, other E3 ligases contribute to muscle atrophy. For example, the E3 ligases Fbxo40 is upregulated in denervation induced muscle atrophy and mediates the ubiquitination of the insulin receptor, IRS1, which blocks pro-growth signals, thus leading to decreased protein synthesis [78, 79].

The level of phosphorylated (active) p65 was found to be elevated in the muscle of gastric cancer patients [80]. Additionally, a decrease in IκBα expression was also observed, suggesting increased activation of NF-κB signaling. Interestingly, this occurred independent of muscle wasting, suggesting that NF-κB activation in

skeletal muscle occurs prior to muscle wasting in cancer cachexia. Similar findings were observed in our own laboratory in muscles from a mouse model of cancer cachexia. Evidence that NF-κB can upregulate the expression of the E3 ligases came from mouse models in which an active IKKβ transgene caused significant increases in the expression of MuRF1 [81]. Subsequent deletion of MuRF1 from these mice was sufficient to block this effect. The converse was also found to be true in a study that showed deletion of IKKβ prevented the expression of MuRF1, although the expression levels of atrogin-1 did not change in this study [82]. Further evidence revealed that NF-κB consensus binding sites are present in the MuRF1 promoter, supporting the notion that MuRF1 is a direct transcriptional target of NF-κB [83]. Atrogin-1 has also been shown to be regulated by NF-κB in a mouse model of fasting [84]. Such findings support the concept that one mechanism by which NF-κB promotes muscle atrophy is through the regulation of the E3 ubiquitin ligase genes (Fig. 4).

3.2   NF-κB and the UPS in Cancer Cachexia

Both MuRF1 and Atrogin-1 are increased in skeletal muscles in multiple animal models of cancer cachexia, with some, but less conclusive evidence from cachectic cancer patients [85, 86]. Human studies have shown that the upregulation of these ligases in patients is not consistent with the severity of the cachexia phenotype [87, 88]. Conversely, a study in pancreatic cancer patients found other components of the ubiquitin proteasome system to be upregulated only after the patient lost more the 10% of their pre-disease weight, but these components did not include either MuRF1 or Atrogin-1 [89]. Furthermore, analysis of rectus muscles from a cohort of GI cancer patients also failed to demonstrate a correlation between the expression of the E3 ligases and cachexia [90]. Similar findings were recently observed in our own laboratory where we were unable to show a significant induction of the E3 ligases in

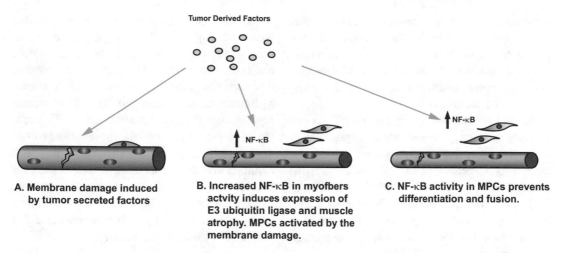

A. Membrane damage induced by tumor secreted factors

B. Increased NF-κB in myofbers actvity induces expression of E3 ubiquitin ligase and muscle atrophy. MPCs activated by the membrane damage.

C. NF-κB activity in MPCs prevents differentiation and fusion.

Fig. 4 Roles of NF-κB in Muscle Atrophy and Regeneration: Tumor derived factors result in muscle damage (**a**), which leads to subsequent activation of muscle progenitor cells and increased muscle atrophy (**b**). Sustained tumor derived signals lead to sustained activation of NF-κB in MPCs, preventing differentiation and fusion (**c**)

muscles from cachectic pancreatic cancer patients, whereas such regulation is easily observed in standard mouse models of cachexia [91]. These findings suggest that either greater care needs to be taken when comparing results from animal models to patient samples, or that further, more detailed human studies will be needed to fully appreciate the roles of the UPS in muscle wasting in cancer.

Finally, it is important to point out that in the C-26 mouse model of cachexia, investigators showed that p65 DNA binding activity did not readily change in cachectic muscles compared to controls [92]. However, in this same study, results also showed that cachexia was dependent on IKKβ. This implies the provocative notion that cancer-induced muscle wasting is independent of p65. These results challenge the hypothesis that NF-κB directly regulates the expression of E3 ubiquitin ligases in cachexia and suggests that a more complex mechanism orchestrates muscle wasting in cancer patients requiring IKKβ. In contrast to the function of the proteasome in tumor-induced muscle atrophy, how chemotherapy regulates muscle wasting is only beginning to be explored [93, 94]. In our own laboratory we showed that mice bearing C-26 tumors and treated with the chemotherapeutic agent, cisplatin, exhibit a predictable reduction of tumor burden, but not an associated rescue in muscle wasting [95]. Even without tumor burden, cisplatin administration was effective in causing muscle wasting in mice, showing that chemotherapy alone was sufficient to promote atrophy. Interestingly, in this study, cisplatin was unable to induce E3 ubiquitin ligases, reaffirming that the mechanisms driving cachexia might not be solely dependent on the UPS. However, we acknowledge that these conclusions have been disputed by alternative findings [93].

3.3 Beyond the Ubiquitin Proteasome System: Other Roles of NF-κB in Muscle Atrophy

As our understanding of the etiology of cancer cachexia grows, it becomes more apparent that multiple cellular signaling pathways converge to mediate muscle wasting. One such pathway is autophagy, which is a process of self-degradation by the cell, which is usually activated in times of nutrient stress [96]. Like apoptosis, autophagy is a stress response, but it acts as a survival mechanism rather than mediating cell death, although aberrant activation of autophagy can lead to cell death in some circumstances. Autophagy proceeds through the formation of double membrane phagophores that engulf cellular components that are degraded through the fusion of the phagophores to lysosomes. These components are then broken down to their essential subunits, such as amino acids, for export or recycling purposes by the cell.

It has been previously shown that NF-κB can regulate autophagy. For example, expression of constitutively active IKK was shown to stimulate autophagy, whereas genetic deletion of IKKβ prevented starvation induced autophagy [97]. This was also observed in another study, where IKKβ, but not the NF-κB transcription factors induced the expression of autophagy related genes, such as LC3 [98].

Tumor bearing mice show an increase in the expression of autophagy genes in skeletal muscle [99]. Similar regulation has been seen in other conditions of muscle atrophy, including denervation [100]. During denervation and fasting-induced atrophy, autophagy is thought to be necessary in order to maintain muscle mass, as the genetic deletion of autophagy genes worsened muscle pathology [101]. Conversely, it has been speculated that excessive autophagy in skeletal muscle of tumor bearing mice is a means of protein breakdown, which ultimately leads to muscle atrophy [99]. This was recently shown in the C-26 mouse model, where the activation of autophagy exacerbated disease pathology [102]. Interestingly, the inhibition of autophagy in this study was insufficient to completely reverse muscle wasting, but showed some beneficial effects. However, like the UPS, the regulation of autophagy genes is not consistently observed in patients with cancer cachexia [91]. Nevertheless, the dysregulation of this cellular process represents a relatively new concept to better understand the underlying mechanisms of muscle wasting in cancer.

4   NF-κB Role in Muscle Regeneration in Cancer Cachexia

The contractile apparatus within the myofiber is the functional unit of skeletal muscle. However, the stress of repeated overuse, such as those that occur during exercise, damages myofiber membranes and requires repair for normal muscle function to continue. Additionally, many chronic or acute conditions, such as muscular dystrophy or muscle contusions, require extensive muscle regeneration to regain normal function. To mediate this regeneration, skeletal muscle has evolved with a resident population of stem cells which facilitate muscle regeneration. In this section, we will provide an overview of muscle regeneration, as well as how this process accounts for the regulation of muscle atrophy and cancer cachexia that involve NF-κB signaling activity.

4.1   Events in Muscle Regeneration and Satellite Stem Cells

Quiescent muscle stem cells, known as satellite cells, reside between the basal lamina and the sarcolemma in healthy skeletal muscle [42]. Upon muscle damage, external signals, such as growth factors and cytokines, promote satellite cells activation. Once activated, these satellite cells, termed, muscle progenitor cells (MPCs) commit to repair by undergoing extensive proliferation and subsequently differentiating. During the differentiation process, MPCs mature and acquire the ability to fuse to other myogenic cells. The process of satellite cell activation is controlled by a cascade of transcription factors, starting with Pax7, which is critical for satellite cell specification as well as maintaining cells in a stem like state [103]. The transcription factors MyoD and Myf5 are expressed by MPCs and signify a commitment to differentiation [104]. In contrast to differentiating MPCs, a subset of cells will retain high levels of Pax7. This cell population expands and returns to quiescence to replenish the stem cells niche [105, 106]. In addition to satellite cells, growing evidence indicates that other cells within the muscle microenvironment (MME) can contribute to muscle regeneration, either by directly entering the myogenic lineage or by secreting pro-myogenic factors to facilitate repair [107, 108].

Perturbations in either the repopulation of the stem cell niche or in the capacity of MPCs to differentiate lead to significant skeletal muscle dysfunction. We now appreciate that external factors contribute to the fate of satellite cells, from activation to proliferation and differentiation. For example, Wnt signaling stimulates the expression of myogenic transcription factors in both development and muscle regeneration [109, 110]. Conversely, TGFβ negatively regulates myogenic differentiation by directly inhibiting MyoD transcriptional activity [111]. Additionally, the activity of Notch receptor signaling through the localization of the Notch inhibitor ligand, Numb, determines cell fate decision of satellite cells following their immediate activation [112].

4.2 NF-κB Prevents MPC Differentiation Leading to Muscle Wasting in Cachexia

Although the cachexia literature is dominated by studies that have focused on the atrophy of myofibers, a growing body of work has also implicated a role for the dysregulation of muscle repair and led us to propose the concept that muscle wasting in cancer involves not only the myofiber, but also encompasses the MME. A greater appreciation of this phenomenon was presented by He et al., where a correlation between myofiber membrane damage and the accumulation of interstitial cells within the MME was observed [113]. A large proportion of these cells stained positive for Pax7 in both the C-26 mouse model of cachexia and cachectic myofibers from pancreatic cancer patients.

These findings suggested that there was an increase in the number of satellite cells within the MME. These results were consistent with muscle damage in cancer cachexia, which results in the activation of satellite cells to repair injured skeletal muscle. However, when mice were challenged to repair damaged muscle, either by removing dystrophin (using the mdx mouse model) or injury in response to cardiotoxin, there was a significant decrease in the ability of

these cells to commit to the myogenic lineage, as evidenced by a decrease in differentiation markers and a decrease in centrally located nuclei, a hallmark of muscle regeneration. Previous studies have demonstrated that NF-κB has a significant role in preventing differentiation of MPC [23, 114]. Consistent with this, the authors found that media from cancer cells lines as well as cancer patient sera increased the expression of Pax7 as well as activated NF-κB in C2C12 myoblasts [113]. Inactivation of NF-κB, by a dominant negative IKKβ transgene in myoblasts, was enough to block the increase in Pax7. This process was reversible, as tumor resection was followed by a decrease in Pax7 levels. This was also observed by others in the C-26 tumor bearing mice, where an increase in the number of necrotic myofibers following cardiotoxin induced injury compared to control mice, suggesting a decrease in the capacity of muscle to regenerate [115] (Fig. 4). In addition, a similar perturbation in satellite cells and MPCs has been implicated in COPD and chronic kidney disease, whose conditions like cancer are well associated with cachexia [116, 117]. Interestingly, exercise was shown to reduce the expression of Pax7 and a decrease in NF-κB activity in C-26 tumor bearing mice, and increase myofiber diameter in a mouse model of cancer cachexia [118]. This suggests that exercise induced NF-κB downregulation may be a means to increase muscle regeneration in cancer cachexia and delay disease progression. Whether similar load bearing exercise protocols can be effectively translated to patients and other cachexia related disorders remains to be determined.

4.3 Multiple Mechanisms of NF-κB Mediated Inhibition of MPC Differentiation

Although it has been shown that the activation of NF-κB is increased in MPCs from cachectic muscle, a thorough analysis of the downstream mechanisms mediating the anti-myogenic effects of NF-κB in this context has not been explored. However, other studies have examined the role of

NF-κB in regulating the expression of many genes that inhibit MPC differentiation. Some of the earliest evidence of this was provided where increased NF-κB signaling was found to lead to the destabilization of MyoD mRNA [114]. Further work revealed that NF-κB upregulates the expression of cyclinD1 in proliferating myoblasts by directly binding to multiple elements on its promoter, preventing cell cycle exit and delaying differentiation [23]. Upon differentiation, NF-κB DNA binding activity and transcriptional activities were reduced, indicating that NF-κB downregulation correlated with a commitment to differentiate. Inactivation of NF-κB led to a significant decrease in cyclinD1 as well as an increase in differentiation. Further studies demonstrated that NF-κB was able to increase the expression of the transcriptional repressor YinYang (YY1) in MPCs [119]. Further work demonstrated that YY1 was bound to the promoter elements of myofibrillar genes, which would ultimately prevent MPCs from fully differentiating into functional skeletal muscle. Additionally, YY1 also silenced the expression of miR-29, which was found to function as a pro-myogenic factor, similar to miR-1 and miR-206, but without the same potency [120]. The overexpression of the transcription factor Twist1 in MPCs was recently found to induce a cachexia like phenotype [121]. Previous studies have demonstrated that NF-κB can directly upregulate Twist1 [122]. When expressed in muscle, Twist1 directly binds to MyoD, which prevents differentiation. Muscle wasting disorders are often accompanied by systemic inflammation, such as COPD, AIDS, and cancer cachexia. A multitude of pro-inflammatory cytokines have been shown to be upregulated in animal models of these disorders, as well as patients, including TNFα, IL-1β, and IL-6, all of which are potent activators of NF-κB as well as inhibitors of myogenic differentiation. Interestingly, these cytokines, as well as others, are also target genes of NF-κB. A review by Li et al. makes note of these facts and suggests that this cytokine regulation could lead to a feedforward loop in some disease context [71]. These examples demonstrate how NF-κB can signal at multiple nodes to reduce the myogenic potential of MPCs resulting in a decline of muscle regeneration in cancer cachexia.

5 Conclusions and Future Perspective

Cachexia has been recognized as a major contributor to increased morbidity and mortality in cancer patients for many years. However, standard of care for the management of cancer cachexia is still lacking due to the failure of many clinical trials. The failure of these clinical trials may be attributed in large part to the inability of current preclinical animal models to recapitulate cachexia observed in patients. For example, both the Lewis Lung Carcinoma (LLC) and colon-26 (C-26) xenograft cell models have been widely used to study the mechanisms of cancer cachexia, but have come under scrutiny due to the high tumor burden and acuteness of body wasting, which becomes evident within weeks after tumor cell injection [85, 123, 124]. More recently, studies have employed the use of genetically engineered mouse models as well the orthotopic injections of both mouse and human pancreatic cancer cells into a recipient mouse pancreas to study cachexia [91, 125–128]. Both approaches are thought to better replicate cachexia experienced by cancer patients and will be crucial in advancing our knowledge of disease pathology. Additionally, it has become increasingly clear that the causes of cancer cachexia are multimodal, with tumor secreted factors affecting multiple organs, which then indirectly contribute to muscle wasting [127, 129]. Therefore, clinical trials will need to be designed to account for multiple mechanisms underlying the disease pathology. Studies along these lines are already underway, with some interventions focusing on the anorexia associated cachexia through nutritional support in addition to traditional pharmacological intervention [130, 131].

The work presented in this chapter demonstrates the multiple nodes by which NF-κB is involved in a tumor macroenvironment, not only at the levels of tumor initiation, progression, metastasis, and chemotherapy resistance, but also as a mediator of muscle atrophy in cancer cachexia. Cumulatively, these findings indicate that NF-κB is a master regulator of the tumor macroenvironment, controlling both local tumor development as well as mediating peripheral tissue wasting. Given that NF-κB's role in these processes is essential for cancer and cachexia to develop, NF-κB remains a viable therapeutic target to alleviate these effects, blocking both tumor progression and preventing muscle wasting. However, to date, the use of NF-κB inhibitors for treatment of cancer and cachexia has remained elusive, as many clinical trials have been unsuccessful due to safety and efficacy failures [132]. A review by Prescott & Cook speculated that multiple therapies designed to target IKKβ may be susceptible to off target effects, as well as suffer from on-target toxicity, which may account for the lack of effectiveness in these trials. Additionally, as the NF-κB pathway is a stress response and survival pathway, its deletion would severely impair the ability of not only the tumor, but other tissues in the body to respond to stress, thus making them more susceptible to apoptosis, especially in patients undergoing chemotherapy. Finally, despite years of research on the subject, our understanding of NF-κB's role, as well factors mediating its activation, remains unclear in the context of the tumor macroenvironment in cancer cachexia. This is evident through the work disputing its role in regulating the UPS in cancer cachexia. More recent work on the subject suggests that NF-κB may mediate cachexia pathology by blocking muscle regeneration [113]. However, although downregulation of NF-κB activity in satellite cells was sufficient to promote differentiation and restore muscle mass in cachexia models [113], the long-term effects of this strategy remains to be further tested. Inhibition of NF-κB causes premature differentiation of MPCs [23]. Therefore, inhibiting NF-κB in satellite cells of cachectic muscle might be able to drive MPCs toward differentiation, but at the same time could also inhibit the replenishment of the satellite cell pool thus preventing future cycles of muscle regeneration to occur [24]. Furthering

our knowledge on NF-κB signaling in the tumor macroenvironment is the key to overcoming these challenges and will be essential to elucidate novel therapeutic approaches in PDAC-induced cancer cachexia.

Acknowledgements This work was supported by funding from the National Cancer Institute R01 CA180057 (DCG), and a Medical University of South Carolina Hollings Cancer Center Postdoctoral Fellowship (BRP).

References

1. Sen, R., Baltimore, D.: Inducibility of kappa immunoglobulin enhancer-binding protein Nf-kappa B by a posttranslational mechanism. Cell. **47**(6), 921–928 (1986)
2. Bakkar, N., Guttridge, D.C.: NF-kappaB signaling: a tale of two pathways in skeletal myogenesis. Physiol. Rev. **90**(2), 495–511 (2010)
3. Li, L., et al.: Nuclear factor-kappaB and IkappaB kinase are constitutively active in human pancreatic cells, and their down-regulation by curcumin (diferuloylmethane) is associated with the suppression of proliferation and the induction of apoptosis. Cancer. **101**(10), 2351–2362 (2004)
4. Baracos, V.E., et al.: Cancer-associated cachexia. Nat. Rev. Dis. Primers. **4**, 17105 (2018)
5. Sun, L., Quan, X.Q., Yu, S.: An epidemiological survey of cachexia in advanced cancer patients and analysis on its diagnostic and treatment status. Nutr. Cancer. **67**(7), 1056–1062 (2015)
6. Karin, M.: Nuclear factor-kappaB in cancer development and progression. Nature. **441**(7092), 431–436 (2006)
7. Wang, D.J., et al.: NF-kappaB functions in tumor initiation by suppressing the surveillance of both innate and adaptive immune cells. Cell Rep. **9**(1), 90–103 (2014)
8. Wang, W., et al.: The nuclear factor-kappa B RelA transcription factor is constitutively activated in human pancreatic adenocarcinoma cells. Clin. Cancer Res. **5**(1), 119–127 (1999)
9. Hayden, M.S., Ghosh, S.: Signaling to NF-kappaB. Genes Dev. **18**(18), 2195–2224 (2004)
10. Baldwin Jr., A.S.: The NF-kappa B and I kappa B proteins: new discoveries and insights. Annu. Rev. Immunol. **14**, 649–683 (1996)
11. Hayden, M.S., Ghosh, S.: Shared principles in NF-kappaB signaling. Cell. **132**(3), 344–362 (2008)
12. Ghosh, S., May, M.J., Kopp, E.B.: NF-kappa B and Rel proteins: evolutionarily conserved mediators of immune responses. Annu. Rev. Immunol. **16**, 225–260 (1998)

13. Zhang, Q., Lenardo, M.J., Baltimore, D.: 30 Years of NF-kappaB: a blossoming of relevance to human pathobiology. Cell. **168**(1–2), 37–57 (2017)

14. Judge, A.R., et al.: Role for IkappaBalpha, but not c-Rel, in skeletal muscle atrophy. Am. J. Physiol. Cell Physiol. **292**(1), C372–C382 (2007)

15. Smale, S.T.: Dimer-specific regulatory mechanisms within the NF-kappaB family of transcription factors. Immunol. Rev. **246**(1), 193–204 (2012)

16. Attar, R.M., et al.: Genetic approaches to study Rel/NF-kappa B/I kappa B function in mice. Semin. Cancer Biol. **8**(2), 93–101 (1997)

17. Xiao, G., Fong, A., Sun, S.C.: Induction of p100 processing by NF-kappaB-inducing kinase involves docking IkappaB kinase alpha (IKKalpha) to p100 and IKKalpha-mediated phosphorylation. J. Biol. Chem. **279**(29), 30099–30105 (2004)

18. Beg, A.A., et al.: Tumor necrosis factor and interleukin-1 lead to phosphorylation and loss of I kappa B alpha: a mechanism for NF-kappa B activation. Mol. Cell. Biol. **13**(6), 3301–3310 (1993)

19. Wang, L., et al.: IL-6 induces NF-kappa B activation in the intestinal epithelia. J. Immunol. **171**(6), 3194–3201 (2003)

20. Schreck, R., Rieber, P., Baeuerle, P.A.: Reactive oxygen intermediates as apparently widely used messengers in the activation of the NF-kappa B transcription factor and HIV-1. EMBO J. **10**(8), 2247–2258 (1991)

21. Claudio, E., et al.: BAFF-induced NEMO-independent processing of NF-kappa B2 in maturing B cells. Nat. Immunol. **3**(10), 958–965 (2002)

22. Novack, D.V., et al.: The IkappaB function of NF-kappaB2 p100 controls stimulated osteoclastogenesis. J. Exp. Med. **198**(5), 771–781 (2003)

23. Guttridge, D.C., et al.: NF-kappaB controls cell growth and differentiation through transcriptional regulation of cyclin D1. Mol. Cell. Biol. **19**(8), 5785–5799 (1999)

24. Straughn, A.R., et al.: Canonical NF-kappaB signaling regulates satellite stem cell homeostasis and function during regenerative myogenesis. J. Mol. Cell Biol. **11**(1), 53–66 (2019)

25. Enwere, E.K., et al., TWEAK and cIAP1 regulate myoblast fusion through the noncanonical NF-kappaB signaling pathway. Sci. Signal., 2012. 5(246): p. ra75

26. Bakkar, N., et al.: IKK/NF-kappaB regulates skeletal myogenesis via a signaling switch to inhibit differentiation and promote mitochondrial biogenesis. J. Cell Biol. **180**(4), 787–802 (2008)

27. Bakkar, N., et al.: IKKalpha and alternative NF-kappaB regulate PGC-1beta to promote oxidative muscle metabolism. J. Cell Biol. **196**(4), 497–511 (2012)

28. Shintaku, J., et al.: MyoD regulates skeletal muscle oxidative metabolism cooperatively with alternative NF-kappaB. Cell Rep. **17**(2), 514–526 (2016)

29. Rawla, P., Sunkara, T., Gaduputi, V.: Epidemiology of pancreatic cancer: global trends, etiology and risk factors. World J. Oncol. **10**(1), 10–27 (2019)

30. Pramanik, K.C., et al.: Advancement of NF-kappaB signaling pathway: a novel target in pancreatic cancer. Int. J. Mol. Sci. **19**(12) (2018)

31. Arlt, A., Muerkoster, S.S., Schafer, H.: Targeting apoptosis pathways in pancreatic cancer. Cancer Lett. **332**(2), 346–358 (2013)

32. Ling, J., et al.: KrasG12D-induced IKK2/beta/NF-kappaB activation by IL-1alpha and p62 feedforward loops is required for development of pancreatic ductal adenocarcinoma. Cancer Cell. **21**(1), 105–120 (2012)

33. Zhuang, Z., et al.: IL1 receptor antagonist inhibits pancreatic cancer growth by abrogating NF-kappaB activation. Clin. Cancer Res. **22**(6), 1432–1444 (2016)

34. Zilfou, J.T., Lowe, S.W.: Tumor suppressive functions of p53. Cold Spring Harb. Perspect. Biol. **1**(5), a001883 (2009)

35. Xia, Y., et al.: Phosphorylation of p53 by IkappaB kinase 2 promotes its degradation by beta-TrCP. Proc. Natl. Acad. Sci. USA. **106**(8), 2629–2634 (2009)

36. Webster, G.A., Perkins, N.D.: Transcriptional cross talk between NF-kappaB and p53. Mol. Cell. Biol. **19**(5), 3485–3495 (1999)

37. Vasudevan, K.M., Gurumurthy, S., Rangnekar, V.M.: Suppression of PTEN expression by NF-kappa B prevents apoptosis. Mol. Cell. Biol. **24**(3), 1007–1021 (2004)

38. Shi, J., Xue, J.: Inflammation and development of pancreatic ductal adenocarcinoma. Chin. Clin. Oncol. **8**(2), 19 (2019)

39. Hinz, M., et al.: NF-kappaB function in growth control: regulation of cyclin D1 expression and G0/G1-to-S-phase transition. Mol. Cell. Biol. **19**(4), 2690–2698 (1999)

40. Ratnam, N.M., et al.: NF-kappaB regulates GDF-15 to suppress macrophage surveillance during early tumor development. J. Clin. Invest. **127**(10), 3796–3809 (2017)

41. Papademetrio, D.L., et al.: Inhibition of survival pathways MAPK and NF-kB triggers apoptosis in pancreatic ductal adenocarcinoma cells via suppression of autophagy. Target. Oncol. **11**(2), 183–195 (2016)

42. Xia, Y., Shen, S., Verma, I.M.: NF-kappaB, an active player in human cancers. Cancer Immunol. Res. **2**(9), 823–830 (2014)

43. Xu, J., Lamouille, S., Derynck, R.: TGF-beta-induced epithelial to mesenchymal transition. Cell Res. **19**(2), 156–172 (2009)

44. Maier, H.J., et al.: NF-kappaB promotes epithelial-mesenchymal transition, migration and invasion of pancreatic carcinoma cells. Cancer Lett. **295**(2), 214–228 (2010)

45. Matsuo, Y., et al.: Proteasome inhibitor MG132 inhibits angiogenesis in pancreatic cancer by

blocking NF-kappaB activity. Dig. Dis. Sci. **55**(4), 1167–1176 (2010)

46. Elliott, C.L., et al.: Nuclear factor-kappa B is essential for up-regulation of interleukin-8 expression in human amnion and cervical epithelial cells. Mol. Hum. Reprod. **7**(8), 787–790 (2001)

47. Xie, T.X., et al.: Constitutive NF-kappaB activity regulates the expression of VEGF and IL-8 and tumor angiogenesis of human glioblastoma. Oncol. Rep. **23**(3), 725–732 (2010)

48. Rhee, J.W., et al.: NF-kappaB-dependent regulation of matrix metalloproteinase-9 gene expression by lipopolysaccharide in a macrophage cell line RAW 264.7. J. Biochem. Mol. Biol. **40**(1), 88–94 (2007)

49. Folkman, J.: Role of angiogenesis in tumor growth and metastasis. Semin. Oncol. **29**(6 Suppl 16), 15–18 (2002)

50. Sun, L., et al.: Epigenetic regulation of SOX9 by the NF-kappaB signaling pathway in pancreatic cancer stem cells. Stem Cells. **31**(8), 1454–1466 (2013)

51. Zhao, X., et al.: Hypoxia-inducible factor-1 promotes pancreatic ductal adenocarcinoma invasion and metastasis by activating transcription of the actin-bundling protein fascin. Cancer Res. **74**(9), 2455–2464 (2014)

52. Samuel, T., et al.: Variable NF-kappaB pathway responses in colon cancer cells treated with chemotherapeutic drugs. BMC Cancer. **14**, 599 (2014)

53. Arora, S., et al.: An undesired effect of chemotherapy: gemcitabine promotes pancreatic cancer cell invasiveness through reactive oxygen species-dependent, nuclear factor kappaB- and hypoxia-inducible factor 1alpha-mediated up-regulation of CXCR4. J. Biol. Chem. **288**(29), 21197–21207 (2013)

54. Wu, Z.H., et al.: Molecular linkage between the kinase ATM and NF-kappaB signaling in response to genotoxic stimuli. Science. **311**(5764), 1141–1146 (2006)

55. Das, K.C., White, C.W.: Activation of NF-kappaB by antineoplastic agents. Role of protein kinase C. J. Biol. Chem. **272**(23), 14914–14920 (1997)

56. Arlt, A., et al.: Role of NF-kappaB and Akt/PI3K in the resistance of pancreatic carcinoma cell lines against gemcitabine-induced cell death. Oncogene. **22**(21), 3243–3251 (2003)

57. Guo, Q., et al.: Evodiamine inactivates NF-kappaB and potentiates the antitumor effects of gemcitabine on tongue cancer both in vitro and in vivo. Onco. Targets. Ther. **12**, 257–267 (2019)

58. Waters, J.A., et al.: Targeted nuclear factor-kappaB suppression enhances gemcitabine response in human pancreatic tumor cell line murine xenografts. Surgery. **158**(4), 881–8 (2015); discussion 888–9

59. Uwagawa, T., et al.: Combination chemotherapy of nafamostat mesilate with gemcitabine for pancreatic cancer targeting NF-kappaB activation. Anticancer Res. **29**(8), 3173–3178 (2009)

60. Mackey, J.R., et al.: Functional nucleoside transporters are required for gemcitabine influx and manifestation of toxicity in cancer cell lines. Cancer Res. **58**(19), 4349–4357 (1998)

61. Amrutkar, M., Gladhaug, I.P.: Pancreatic cancer chemoresistance to gemcitabine. Cancers (Basel). **9**, 11 (2017)

62. Spratlin, J., et al.: The absence of human equilibrative nucleoside transporter 1 is associated with reduced survival in patients with gemcitabine-treated pancreas adenocarcinoma. Clin. Cancer Res. **10**(20), 6956–6961 (2004)

63. Bhutia, Y.D., et al.: CNT1 expression influences proliferation and chemosensitivity in drug-resistant pancreatic cancer cells. Cancer Res. **71**(5), 1825–1835 (2011)

64. Skrypek, N., et al.: The MUC4 mucin mediates gemcitabine resistance of human pancreatic cancer cells via the Concentrative Nucleoside Transporter family. Oncogene. **32**(13), 1714–1723 (2013)

65. Momeny, M., et al.: Blockade of nuclear factor-kappaB (NF-kappaB) pathway inhibits growth and induces apoptosis in chemoresistant ovarian carcinoma cells. Int. J. Biochem. Cell Biol. **99**, 1–9 (2018)

66. Hung, S.W., et al.: Defective hCNT1 transport contributes to gemcitabine chemoresistance in ovarian cancer subtypes: overcoming transport defects using a nanoparticle approach. Cancer Lett. **359**(2), 233–240 (2015)

67. Zhang, Z., et al.: Gemcitabine treatment promotes pancreatic cancer stemness through the Nox/ROS/NF-kappaB/STAT3 signaling cascade. Cancer Lett. **382**(1), 53–63 (2016)

68. Epelbaum, R., et al.: Curcumin and gemcitabine in patients with advanced pancreatic cancer. Nutr. Cancer. **62**(8), 1137–1141 (2010)

69. Wang, H., Cao, Q., Dudek, A.Z.: Phase II study of panobinostat and bortezomib in patients with pancreatic cancer progressing on gemcitabine-based therapy. Anticancer Res. **32**(3), 1027–1031 (2012)

70. Wackerhage, H., et al.: Stimuli and sensors that initiate skeletal muscle hypertrophy following resistance exercise. J. Appl. Physiol. 2019. **126**(1), 30–43 (1985)

71. Li, H., Malhotra, S., Kumar, A.: Nuclear factor-kappa B signaling in skeletal muscle atrophy. J. Mol. Med. (Berl). **86**(10), 1113–1126 (2008)

72. Gao, Y., et al.: Muscle atrophy induced by mechanical unloading: mechanisms and potential countermeasures. Front. Physiol. **9**, 235 (2018)

73. Glass, D.J.: Signaling pathways perturbing muscle mass. Curr. Opin. Clin. Nutr. Metab. Care. **13**(3), 225–229 (2010)

74. Buetow, L., Huang, D.T.: Structural insights into the catalysis and regulation of E3 ubiquitin ligases. Nat. Rev. Mol. Cell Biol. **17**(10), 626–642 (2016)

75. Bodine, S.C., et al.: Identification of ubiquitin ligases required for skeletal muscle atrophy. Science. **294**(5547), 1704–1708 (2001)

76. Gomes, M.D., et al.: Atrogin-1, a muscle-specific F-box protein highly expressed during muscle atrophy. Proc. Natl. Acad. Sci. USA. **98**(25), 14440–14445 (2001)

77. Cao, P.R., Kim, H.J., Lecker, S.H.: Ubiquitin-protein ligases in muscle wasting. Int. J. Biochem. Cell Biol. **37**(10), 2088–2097 (2005)

78. Shi, J., et al.: The SCF-Fbxo40 complex induces IRS1 ubiquitination in skeletal muscle, limiting IGF1 signaling. Dev. Cell. **21**(5), 835–847 (2011)

79. Ye, J., et al.: FBXO40, a gene encoding a novel muscle-specific F-box protein, is upregulated in denervation-related muscle atrophy. Gene. **404**(1–2), 53–60 (2007)

80. Rhoads, M.G., et al.: Expression of NF-kappaB and IkappaB proteins in skeletal muscle of gastric cancer patients. Eur. J. Cancer. **46**(1), 191–197 (2010)

81. Cai, D., et al.: IKKbeta/NF-kappaB activation causes severe muscle wasting in mice. Cell. **119**(2), 285–298 (2004)

82. Mourkioti, F., et al.: Targeted ablation of IKK2 improves skeletal muscle strength, maintains mass, and promotes regeneration. J. Clin. Invest. **116**(11), 2945–2954 (2006)

83. Wu, C.L., et al.: NF-kappaB but not FoxO sites in the MuRF1 promoter are required for transcriptional activation in disuse muscle atrophy. Am. J. Physiol. Cell Physiol. **306**(8), C762–C767 (2014)

84. Lee, D., Goldberg, A.L.: Muscle wasting in fasting requires activation of NF-kappaB and inhibition of AKT/Mechanistic Target of Rapamycin (mTOR) by the protein acetylase, GCN5. J. Biol. Chem. **290**(51), 30269–30279 (2015)

85. Acharyya, S., et al.: Cancer cachexia is regulated by selective targeting of skeletal muscle gene products. J. Clin. Invest. **114**(3), 370–378 (2004)

86. Yuan, L., et al.: Muscle-specific E3 ubiquitin ligases are involved in muscle atrophy of cancer cachexia: an in vitro and in vivo study. Oncol. Rep. **33**(5), 2261–2268 (2015)

87. D'Orlando, C., et al.: Gastric cancer does not affect the expression of atrophy-related genes in human skeletal muscle. Muscle Nerve. **49**(4), 528–533 (2014)

88. Williams, A., et al.: The expression of genes in the ubiquitin-proteasome proteolytic pathway is increased in skeletal muscle from patients with cancer. Surgery. **126**(4): p. 744–9 (1999); discussion 749–50

89. Khal, J., et al.: Increased expression of proteasome subunits in skeletal muscle of cancer patients with weight loss. Int. J. Biochem. Cell Biol. **37**(10), 2196–2206 (2005)

90. Stephens, N.A., et al.: Using transcriptomics to identify and validate novel biomarkers of human skeletal muscle cancer cachexia. Genome Med. **2**(1), 1 (2010)

91. Talbert, E.E., et al.: Modeling human cancer-induced cachexia. Cell Rep. **28**(6), 1612–1622 e4 (2019)

92. Cornwell, E.W., et al.: C26 cancer-induced muscle wasting is IKKbeta-dependent and NF-kappaB-independent. PLoS One. **9**(1), e87776 (2014)

93. Sakai, H., et al.: Mechanisms of cisplatin-induced muscle atrophy. Toxicol. Appl. Pharmacol. **278**(2), 190–199 (2014)

94. Barreto, R., et al.: Cancer and chemotherapy contribute to muscle loss by activating common signaling pathways. Front. Physiol. **7**, 472 (2016)

95. Damrauer, J.S., et al.: Chemotherapy-induced muscle wasting: association with NF-kappaB and cancer cachexia. Eur. J. Transl. Myol. **28**(2), 7590 (2018)

96. Glick, D., Barth, S., Macleod, K.F.: Autophagy: cellular and molecular mechanisms. J. Pathol. **221**(1), 3–12 (2010)

97. Criollo, A., et al.: The IKK complex contributes to the induction of autophagy. EMBO J. **29**(3), 619–631 (2010)

98. Comb, W.C., et al.: IKK-dependent, NF-kappaB-independent control of autophagic gene expression. Oncogene. **30**(14), 1727–1732 (2011)

99. Penna, F., et al.: Autophagic degradation contributes to muscle wasting in cancer cachexia. Am. J. Pathol. **182**(4), 1367–1378 (2013)

100. O'Leary, M.F., Hood, D.A.: Denervation-induced oxidative stress and autophagy signaling in muscle. Autophagy. **5**(2), 230–231 (2009)

101. Masiero, E., et al.: Autophagy is required to maintain muscle mass. Cell Metab. **10**(6), 507–515 (2009)

102. Penna, F., et al.: Autophagy exacerbates muscle wasting in cancer cachexia and impairs mitochondrial function. J. Mol. Biol. **431**(15), 2674–2686 (2019)

103. Seale, P., et al.: Pax7 is required for the specification of myogenic satellite cells. Cell. **102**(6), 777–786 (2000)

104. Sabourin, L.A., Rudnicki, M.A.: The molecular regulation of myogenesis. Clin. Genet. **57**(1), 16–25 (2000)

105. Yin, H., Price, F., Rudnicki, M.A.: Satellite cells and the muscle stem cell niche. Physiol. Rev. **93**(1), 23–67 (2013)

106. Lukjanenko, L., et al.: Aging disrupts muscle stem cell function by impairing matricellular WISP1 secretion from fibro-adipogenic progenitors. Cell Stem Cell. **24**(3), 433–446 e7 (2019)

107. Joe, A.W., et al.: Muscle injury activates resident fibro/adipogenic progenitors that facilitate myogenesis. Nat. Cell Biol. **12**(2), 153–163 (2010)

108. Dellavalle, A., et al.: Pericytes of human skeletal muscle are myogenic precursors distinct from satellite cells. Nat. Cell Biol. **9**(3), 255–267 (2007)

109. Rudolf, A., et al.: beta-Catenin activation in muscle progenitor cells regulates tissue repair. Cell Rep. **15**(6), 1277–1290 (2016)

110. Ikeya, M., Takada, S.: Wnt signaling from the dorsal neural tube is required for the formation of the medial dermomyotome. Development. **125**(24), 4969–4976 (1998)

111. Liu, D., Black, B.L., Derynck, R.: TGF-beta inhibits muscle differentiation through functional repression of myogenic transcription factors by Smad3. Genes Dev. **15**(22), 2950–2966 (2001)

112. Conboy, I.M., Rando, T.A.: The regulation of Notch signaling controls satellite cell activation and cell fate determination in postnatal myogenesis. Dev. Cell. **3**(3), 397–409 (2002)

113. He, W.A., et al.: NF-kappaB-mediated Pax7 dysregulation in the muscle microenvironment promotes cancer cachexia. J. Clin. Invest. **123**(11), 4821–4835 (2013)

114. Guttridge, D.C., et al.: NF-kappaB-induced loss of MyoD messenger RNA: possible role in muscle decay and cachexia. Science. **289**(5488), 2363–2366 (2000)

115. Inaba, S., et al.: Muscle regeneration is disrupted by cancer cachexia without loss of muscle stem cell potential. PLoS One. **13**(10), e0205467 (2018)

116. Theriault, M.E., et al.: Satellite cells senescence in limb muscle of severe patients with COPD. PLoS One. **7**(6), e39124 (2012)

117. Zhang, L., et al.: Satellite cell dysfunction and impaired IGF-1 signaling cause CKD-induced muscle atrophy. J. Am. Soc. Nephrol. **21**(3), 419–427 (2010)

118. Coletti, D., et al.: Spontaneous physical activity downregulates Pax7 in cancer cachexia. Stem Cells Int. **2016**, 6729268 (2016)

119. Wang, H., et al.: NF-kappaB regulation of YY1 inhibits skeletal myogenesis through transcriptional silencing of myofibrillar genes. Mol. Cell. Biol. **27**(12), 4374–4387 (2007)

120. Wang, H., et al.: NF-kappaB-YY1-miR-29 regulatory circuitry in skeletal myogenesis and rhabdomyosarcoma. Cancer Cell. **14**(5), 369–381 (2008)

121. Parajuli, P., et al.: Twist1 activation in muscle progenitor cells causes muscle loss akin to cancer cachexia. Dev. Cell. **45**(6), 712–725 e6 (2018)

122. Li, C.W., et al.: Epithelial-mesenchymal transition induced by TNF-alpha requires NF-kappaB-mediated transcriptional upregulation of Twist1. Cancer Res. **72**(5), 1290–1300 (2012)

123. Bonetto, A., et al.: The Colon-26 carcinoma tumor-bearing mouse as a model for the study of cancer cachexia. J. Vis. Exp. **117** (2016)

124. Zhang, G., et al.: Toll-like receptor 4 mediates Lewis lung carcinoma-induced muscle wasting via coordinate activation of protein degradation pathways. Sci. Rep. **7**(1), 2273 (2017)

125. Nosacka, R.L., et al.: Distinct cachexia profiles in response to human pancreatic tumours in mouse limb and respiratory muscle. J. Cachexia. Sarcopenia Muscle. (2020)

126. Judge, S.M., et al.: MEF2c-dependent downregulation of myocilin mediates cancer-induced muscle wasting and associates with cachexia in patients with cancer. Cancer Res. **80**(9), 1861–1874 (2020)

127. Burfeind, K.G., et al.: Circulating myeloid cells invade the central nervous system to mediate cachexia during pancreatic cancer. elife. **9** (2020)

128. Suzuki, T., Von Haehling, S., Springer, J.: Promising models for cancer-induced cachexia drug discovery. Expert Opin. Drug Discovery. **15**(5), 627–637 (2020)

129. Das, S.K., et al.: Adipose triglyceride lipase contributes to cancer-associated cachexia. Science. **333**(6039), 233–238 (2011)

130. Naito, T.: Emerging treatment options for cancer-associated cachexia: a literature review. Ther. Clin. Risk Manag. **15**, 1253–1266 (2019)

131. Solheim, T.S., et al.: A randomized phase II feasibility trial of a multimodal intervention for the management of cachexia in lung and pancreatic cancer. J. Cachexia. Sarcopenia Muscle. **8**(5), 778–788 (2017)

132. Prescott, J.A., Cook, S.J.: Targeting IKKbeta in cancer: challenges and opportunities for the therapeutic utilisation of IKKbeta inhibitors. Cell. **7**(9) (2018)

Part III

Therapy-Induced Muscle Wasting

Therapy-Induced Toxicities Associated with the Onset of Cachexia

Joshua R. Huot, Fabrizio Pin, and Andrea Bonetto

Abstract

Cancer is the second-leading cause of death worldwide. Notably, 40% of all patients suffer from cachexia, a debilitating muscle wasting syndrome that represents one of the primary causes of death in advanced cancer patients. Despite their known side effects, such as vomiting, anorexia, mucositis, neuropathy, musculoskeletal and metabolic alterations, chemotherapy treatments are the most effective and extensively used strategies against cancer. Moreover, chemotherapeutic agents may participate in and even exacerbate the pathophysiological alterations driving cachexia. Indeed, skeletal muscle atrophy and cardiac abnormalities occurring as a consequence of chemotherapy administration can worsen the response to anticancer therapies, reduce physical activity and severely impede the quality of life in cancer patients. Hence, chemotherapy-induced cachexia has recently emerged as a critical clinical problem impacting outcomes and overall survival in cancer patients. Here, we review some of the mechanism(s) through which chemotherapeutics induce musculoskeletal deficits related to cachexia. Lastly, we discuss different strategies currently under consideration to protect muscle tissue from chemotherapy-related toxicities, including nutritional interventions, mitochondria-targeted molecules, ghrelin, bisphosphonates, ACVR2B antagonists, and physical exercise.

Joshua R. Huot and Fabrizio Pin contributed equally with all other contributors.

J. R. Huot
Department of Surgery, Indiana University School of Medicine, Indianapolis, IN, USA

Department of Anatomy, Cell Biology and Physiology, Indiana University School of Medicine, Indianapolis, IN, USA

F. Pin
Department of Anatomy, Cell Biology and Physiology, Indiana University School of Medicine, Indianapolis, IN, USA

A. Bonetto (✉)
Department of Surgery, Indiana University School of Medicine, Indianapolis, IN, USA

Department of Anatomy, Cell Biology and Physiology, Indiana University School of Medicine, Indianapolis, IN, USA

Department of Otolaryngology – Head and Neck Surgery, Indiana University School of Medicine, Indianapolis, IN, USA

Simon Comprehensive Cancer Center, Indiana University School of Medicine, Indianapolis, IN, USA

Indiana Center for Musculoskeletal Health, Indiana University School of Medicine, Indianapolis, IN, USA

Department of Pathology, University of Colorado Anschutz Medical Campus, Aurora, CO, USA
e-mail: abonetto@iu.edu; andrea.bonetto@cuanschutz.edu

© Springer Nature Switzerland AG 2022
S. Acharyya (ed.), *The Systemic Effects of Advanced Cancer*,
https://doi.org/10.1007/978-3-031-09518-4_8

Learning Objectives

- Anticancer drugs have specific negative off-target effects related to cachexia.
- Commonly used chemotherapies induce skeletal muscle atrophy and weakness.
- Atrophy and weakness induced by anticancer drugs are linked to several perturbations in normal muscle homeostasis.
- Novel strategies being employed to counteract skeletal muscle atrophy and weakness induced by anticancer drugs.

1 Introduction

Based on the most recent statistics, over 1.9 million individuals will be diagnosed with cancer in 2022 in the United States alone, and about 609,000 patients will succumb to the disease [1]. Up to 40% of all individuals are expected to develop a tumor during their lifetime, making cancer the second-leading cause of death in the United States and worldwide [2]. Interestingly, despite a 33% increase in cancer diagnoses over the past few years, there has been a 29% decline in mortality rates among cancer patients [1], likely due to the introduction of more effective drugs, innovative therapeutic strategies, and early medical interventions [3]. Whenever possible, cancer treatments aim to cure the disease or, at least, limit suffering and prolong life when the disease is too advanced (i.e., palliative cure) [4]. The most common therapeutic options for cancer treatment are surgery, chemotherapy, and radiotherapy.

Surgical cancer treatment was initially characterized by radical and aggressive resection of the tumor, thus representing an invasive approach. With the introduction in 1992 of laparoscopic technology survival among patients being surgically treated for their cancer improved significantly, primarily due to the reduction of the complications associated with the procedure [5]. Another option for treatment of localized cancers, such as primary tumors or regional lymph nodes, is radiotherapy [6]. Many technological advancements, such as the accurate radiation delivery, the dose of radiation in proportion to the tumor mass, and the limitation of exposure to surrounding normal tissue, have markedly improved the benefits of radiation oncology in recent years, thus also reducing the associated side effects [7].

Surgery and radiotherapy work in a localized manner, but in order to treat cancer that has spread or metastasized to distant organs, additional systemic therapy is required. In these cases, the use of cytotoxic drugs (i.e., chemotherapy) remains one of the most effective and extensively used treatment approaches. Alternative approaches, such as hormonal and immune therapy, have also been introduced and today are often offered as first choice of treatment for many tumor types [5]. In particular, hormonal therapy uses exogenous hormones or analogs and inhibitors of hormone synthesis or receptors to interfere with tumor growth. Additionally, cancer immunotherapy exploits the relationship between cancer and immune system, using the latter to stimulate an anti-tumor response against the cancer [8, 9]. Several different immunotherapies for cancer therapy are now FDA-approved and include cytokines, cancer vaccines, immune checkpoint inhibitors and chimeric antigen receptor (CAR) T cell therapy (CAR-T cells) [9]. Though the goal of this strategy is to specifically recognize and kill the tumor cells, this type of therapy has been shown to cause serious side effects in patients [10].

2 Chemotherapy

Chemotherapy drugs, targeting essential and common mechanisms essential for cell survival, are designed to kill or antagonize cancer cell division. Unfortunately, due to their relatively poor specificity, chemotherapeutic agents can induce cytotoxic effects, which can result into systemic complications in the host body. Several types of cytotoxic molecules are approved for cancer treatment, and are classified depending on different mechanisms of action as follows:

- Alkylating agents (e.g., carboplatin, cisplatin, cyclophosphamide, and oxaliplatin) inhibit the transcription of DNA into RNA by the formation of cross-links in the DNA chain [11].
- Antimetabolites (e.g., 5-fluorouracil) are structural analogs of the DNA-building molecules that act to interfere with DNA synthesis [12].
- Anti-tumor antibiotics anthracyclines (e.g., doxorubicin) and non-anthracyclines (e.g., bleomycin and mitomycin-C) bind to the DNA preventing cell division.
- Topoisomerase inhibitors I (e.g., irinotecan) or II (e.g., etoposide) interfere with the relaxation of the DNA supercoiling, thus inhibiting an essential step required for cell division [13].
- Mitotic inhibitors/taxanes (e.g., docetaxel and paclitaxel) and vinca alkaloids (e.g., vinblastine and vincristine) act on the mitotic spindle causing metaphase arrest through microtubule derangement [14, 15].

Chemotherapy treatments act systemically and can damage normal (healthy) tissue leading to severe side effects, thus also representing a serious problem and a source of concern for both patients and clinicians [16]. Side effects induced by chemotherapy are heterogeneous and depend on the type of drugs used. The most common side effects of chemotherapy include myelosuppression, nausea, vomiting, diarrhea, constipation, fatigue, muscle weakness, neuropathy, ecchymosis, and hair loss. Loss of body weight is also reported in many cases and can be related with a reduction in appetite and anorexia. Moreover, chemotherapy was shown to promote changes in how food tastes and smells, hence reducing meal appreciation and altering the dietary intake and nutritional status [17]. Notably, myelosuppression is among the most relevant dose-limiting adverse effects of chemotherapy, primarily due to the high turnover rate that characterizes marrow cells, similar to cancer cells [18]. This treatment can lead to severe bone marrow alterations such as anemia, neutropenia, and thrombocytopenia. These complications can be accompanied by fatigue, increased infection, and bleeding.

Chemotherapy-induced nausea and vomiting (CINV) is a complex reflex involving central and peripheral nervous systems [19]. Chemotherapy can induce the release of neurotransmitters that later stimulate receptors in various locations, such as the *nucleus tractus solitarius*, one of the most important areas regulating emesis [20]. Consistently, recent findings showed that cisplatin can cause emesis and contribute to chemotherapy-induced malaise by increasing the level of circulating MIC-1/GDF15 [21], previously shown to subvert a physiological pathway of appetite regulation by acting on hindbrain neurons [22]. Other common alterations resulting from chemotherapy treatment are mucositis, which can involve any portion of the gastrointestinal tract (i.e., oral, gastric, small intestine, and rectal mucosa) and lead to pain, ulceration, diarrhea, malabsorption, dysphagia, anorexia, as well as weight loss and increased risk of sepsis [23, 24]. Treatments can also induce the production of reactive oxygen species (ROS), with subsequent macrophage activation and release of pro-inflammatory cytokines such as TNF-α, IL-6, and COX-2, previously shown to alter basal epithelial layer, submucosa, and endothelium [23].

Chemotherapy agents are also known to damage the nervous system and induce different types of neuropathy, the most common being chemotherapy-induced peripheral neuropathy (CIPN) [25]. While less frequent, chemotherapy can also decrease motor and/or autonomic functions. Patients experience different types of symptoms such as hyperalgesia, pain, numbness, tingling, altered touch sensation, impaired vibration, as well as mechanical or thermal allodynia. In the most serious cases, it can induce balance disturbances, impaired movements, and severe disability [26]. The pathogenetic mechanism(s) of CIPN is multifactorial and includes inflammation, mitochondrial dysfunction, increased oxidative stress, dysregulation of Ca^{2+} hemostasis, changes in peripheral nerve excitability and axonal degeneration [25]. Finally, the central nervous system can be affected by chemotherapy primarily by inducing memory disruption, impaired attention, processing speed, executive function [27], as well as emotional problem like anxiety and depression [28].

Other important alterations driven by anticancer therapy concern the musculoskeletal system.

Muscle wasting and bone loss can occur after prolong/chronic chemotherapy administration leading to increased risk of fracture and reduced mobility and muscle weakness [29]. In particular, the effects of antineoplastic therapy on bone homeostasis have been known for quite some time, including the role of corticosteroids in inducing osteoporosis [30]. Other studies showed that the increase in pro-inflammatory cytokines released following chemotherapy administration may participate in increasing bone resorption and bone loss [31]. Different chemotherapeutic agents can also affect the gonads, hence affecting the hormonal status and inducing endocrine and bone alterations. As an example, in premenopausal breast cancer patients, chemotherapy causes ovarian failure, thus resulting in rapid bone loss and increased bone fracture rates [32].

Recently, we and others have found that anticancer regimens such as Folfiri (a combination of 5-fluorouracil, leucovorin, and irinotecan), as well as platinum-based drugs such as carboplatin and cisplatin can induce severe depletion of bone tissue [33–35]. Furthermore, our work highlighted the evidence that circulating factors released by bone during resorption can also contribute to skeletal muscle alterations [35, 36]. Along this line, several observations showed that the anticancer therapies may participate in causing skeletal muscle atrophy and weakness in patients with different types of tumors such as lung, breast, colorectal, prostate, and non-small cell lung cancer [37–39]. In this regard, the first evidence showing a mechanistic effect of chemotherapeutic agents on muscle mass was associated with the detection of negative nitrogen balance in rats receiving 5-fluorouracil, cisplatin, or methotrexate [40]. However, the mechanism of action of chemotherapy on skeletal and cardiac muscle homeostasis is poorly understood.

3   Chemotherapy-Associated Cachexia: Phenotype and Mechanisms

Anticancer therapies have been linked to the occurrence of cachexia [39, 41]. Cachexia is defined as the progressive loss of skeletal muscle mass, with or without fat loss, which cannot be fully recovered by conventional nutritional support [42–44]. Additionally, cachexia is thought to be a multi-organ wasting syndrome occurring in the majority of cancer patients. Notably, such comorbidity of cancer can be exacerbated by the anticancer drugs, thereby leading to worsened quality of life and poorer outcomes. Though several organ systems are negatively impacted by chemotherapy treatments, here we will focus primarily on the effects on skeletal muscle mass. Indeed, we and others have repeatedly demonstrated that chemotherapeutics directly induce skeletal muscle wasting and weakness [33, 34, 36, 39, 45–47]. While the effects of anticancer drugs are restricted largely to animal and cell culture models, several clinical studies have demonstrated that the use of chemotherapeutics in colorectal, gastric, breast, and lung cancer patients may also accelerate muscle wasting and weakness [48–54].

For example, Folfiri, routinely prescribed to treat solid tumors such as colon cancer, has shown to induce muscle wasting in different experimental settings. Indeed, Folfiri treatment of C2C12 myotubes, an *in vitro* model widely used to study the regulation of muscle size, resulted in the appearance of severe muscle atrophy [45]. When administered to animals, Folfiri induced severe weight loss, which was accompanied by skeletal muscle wasting and reduced myofiber size [33, 36, 45, 47]. The loss of muscle mass induced by Folfiri was also accompanied by reduced whole body grip strength and specific force of the EDL (*extensor digitorum longus*) muscle [45]. Similarly, the commonly used platinum-based therapies cisplatin, carboplatin, and oxaliplatin are known to induce a cachectic response in skeletal muscle. We and others have demonstrated that cisplatin administration causes C2C12 myotube atrophy and induces severe loss of muscle mass, fiber size, and grip strength in mice [35, 38, 55, 56]. Our data are corroborated by similar findings showing effects in response to both oxaliplatin and carboplatin [34, 57–59], also reflected by reduced walking speed and rearing time in oxaliplatin-treated mice [57]. Additionally, the commonly used chemotherapeutic doxorubicin has consistently shown to induce muscle wasting and weakness in mice as well as to induce myotube atrophy [60–64]. In more recent years, new waves of second-line chemotherapeutics have been developed in hopes of halting tumor progression and metastases, some of which include multi-kinase inhibitors such as sorafenib and regorafenib. However, as we have recently shown, these newly found treatment options, which are widely used in treating hepatocellular carcinoma, colorectal, and gastric cancers, can also induce skeletal muscle wasting and weakness [46, 65–67].

In an attempt to identify strategies to prevent chemotherapeutics from promoting the onset of muscle disorders, our group and others have investigated some of the mechanisms responsible for chemotherapy-driven effects on muscle. In particular, investigations on the role of muscle mitochondria in cancer or following chemotherapy treatment have received much attention. In this regard, we demonstrated that administration of Folfiri is accompanied by reduced levels of skeletal muscle PGC-1α, PGC-1β, and cytochrome-C, proteins involved in the control of mitochondrial biogenesis, as well as impaired succinate dehydrogenase (SDH), an enzyme participating in oxidative phosphorylation, and reduced number and size of skeletal muscle mitochondria [45]. Additional studies have since followed further implicating mitochondrial impairments with Folfiri and other chemotherapy regimens, including cisplatin and doxorubicin [36, 68, 69].

Dysfunctions of the neuromuscular junctions (NMJs), known to contribute to muscle weakness, have also received much attention over the past few years [70]. Indeed, some chemotherapeutics are known to induce mitochondrial abnormalities, whereas mitochondrial dysfunction has previously been implicated in maintenance, formation, and fragmentation of NMJs [71–74]. Furthermore, the widely studied E3-ubiquitin ligase MuRF-1, and the myogenic regulatory factor myogenin, both known to play a role in NMJ remodeling, were previously found dysregulated by anticancer drugs [75, 76]. Moreover, oxaliplatin has shown to induce motor alterations of diaphragm muscle as well as enteric neuropathy; however, whether this effect extends to limb musculature has not been reported [77, 78]. More recently, doxorubicin was shown to alter several NMJ-associated proteins of mouse soleus muscles further supporting a possible mechanism by which anticancer drugs directly affect muscle force [79]. Interestingly, it was recently reported that cancer patients affected with weight loss and cachexia did not present with perturbed NMJ morphology in the *rectus abdominis* muscle [80]. However, the assessment of NMJ morphology was performed 4–6 weeks after cessation of chemotherapy. Therefore whether NMJ integrity in cancer patients is actually directly compromised by the administration of anticancer drugs, especially in limb skeletal muscles, is yet to be determined.

Recent studies also highlighted how anticancer treatments alter pivotal regulators of muscle size, thus impairing muscle anabolic and catabolic processes and ultimately resulting in skeletal muscle depletion. Certainly, consistent with markedly reduced muscle mass Folfiri was shown to activate MAPKs, including MEK1/2, ERK1/2, and p38, which have been previously implicated in mediating muscle wasting in rodent models of cachexia, while also reducing the anabolic mediator AKT within the skeletal muscle [45, 81–83]. Similarly, cisplatin was described in association with increased levels of known mediators of skeletal muscle atrophy, including Atrogin-1, MuRF1, p38, myostatin, and NF-KB, as well as with reduced activation of the anabolic AKT/IGF-1 axis [35, 38, 76, 84, 85]. In line with such anabolic suppression by chemotherapy, oxaliplatin was reported along with reduced p70S6K and rpS6 protein content within skeletal muscle, thereby suggesting suppression of protein synthesis, while doxorubicin treatment was shown to promote direct suppression of muscle protein synthesis [58, 62]. Additionally, experimental evidence also suggested that doxorubicin and oxaliplatin administration can elevate skeletal muscle Atrogin-1 [57, 61]. Of note, second-line chemotherapeutics, such as regorafenib and sorafenib, though not affecting anabolism, were shown to increase the levels of markers of autophagy within skeletal muscle. These findings further strengthen the idea that abnormal muscle homeostasis may represent a common feature in chemotherapy-induced muscle loss [46].

Though this chapter is largely focused on skeletal muscle abnormalities induced by anticancer treatments, it is noteworthy that also the cardiac muscle can be severely impacted by chemotherapeutics. In particular, doxorubicin has been widely studied with respect to cardiotoxicity, and its administration has been consistently linked to negative cardiac effects from elevating cardiomyocyte senescence, apoptosis, necroptosis, as well as impairing cardiac function [86–88]. Similarly, cisplatin and carboplatin were both shown to induce cardiac toxicities, whereas the multi-kinase inhibitors regorafenib and sorafenib were shown to cause reductions in left ventricular mass [46, 89, 90]. Though some studies have implicated mitochondrial dysfunction and oxidative stress in cardiotoxicities observed with the use of doxorubicin and platinum-based chemotherapies, further studies are required to understand the negative toxicities anticancer drugs have on cardiac mass and function.

4   The Impact of Cancer and Chemotherapy on Muscle Metabolism

Cachexia frequently leads to progressive body wasting and functional impairment due to reduced energy intake, anorexia, enhanced hypercatabolism, and increased energy expenditure [42], thus also contributing to the definition of "energy-wasting syndrome" [91]. Indeed, cachexia is usually also accompanied by energy imbalance and significant reduction of mitochondrial ATP production, as shown in mice bearing the Lewis lung carcinoma model [92]. Consistent with the idea proposed by Warburg in 1956, tumors mainly rely on a glycolytic metabolism as a source of energy, thereby utilizing the aerobic glycolysis as a very inefficient way to generate ATP compared to normal tissue, which, conversely, relies on the more efficient oxidative phosphorylation [93]. In this context, tumor metabolism may play a direct role in the energetic alterations that characterize cancer patients [93–

95]. Abnormalities in energy expenditure during cancer cachexia can also derive from the overactivation of futile cycles, such as the Cori cycle, the recycling of protons at mitochondrial level, and the increased lipid turnover [91, 96–98]. These are energy-consuming cellular processes without anabolic or catabolic functions, which altogether can impact on the whole energy expenditure [95]. Further, tumor progression is also associated with a systemic inflammatory response. In this regard, the release of several humoral factors, such as pro-inflammatory cytokines, can directly target multiple tissues, including skeletal muscle, fat, liver, pancreas, and brain, ultimately leading to impaired body metabolism.

Though cancer- and chemotherapy-induced cachexia are both characterized by altered glucose metabolism, the primary mechanism responsible for this remains unclear. Cancer cachexia in both clinical and experimental settings can be associated with insulin resistance, which can lead to persistent hyperglycemia and altered glucose utilization by skeletal muscle [99, 100]. Also chemotherapy has been reported to induce transient hyperglycemia in up to 30% of the patients, often depending on treatment type, thus strengthening the recognition that some antineoplastic treatments can determine metabolic complications [101]. However, other studies reported that glycemia can be reduced in experimental cancer cachexia [47, 102], as well as in patients undergoing chemotherapy treatments [103]. These findings suggest that the hypoglycemia in this case may derive from elevated amounts of glucose utilized by the tumor, further suggested by elevations in circulating lactate. This idea is also corroborated by the fact that the expression levels of the phosphoenolpyruvate carboxykinase (PEPCK) in the liver of tumor hosts are significantly elevated, thus suggesting that lactate is then mostly uptaken by the liver and used as gluconeogenic substrate within the Cori cycle [47].

The progressive skeletal muscle wasting observed during cachexia is also associated with changes in the activity of the TCA cycle, which we showed to be reduced in both cancer- and chemotherapy-induced cachexia. These findings are in line with decreased activities of SDH and pyruvate dehydrogenase (PDH) in the skeletal muscle [47]. Specifically, PDH regulates the entry of pyruvate in the TCA cycle [104], whereas its function in the skeletal muscle is inhibited by the activity of the pyruvate dehydrogenase kinase 4 (PDK4). Of note, several chemotherapy agents such as Folfiri, carboplatin, and cisplatin also increase PDK4 expression in the skeletal muscle, further supporting the idea that anticancer drugs directly alter the muscle's energetic metabolism [105]. This deregulation also suggests a switch towards a more glycolytic metabolism in skeletal muscle, as also supported by various findings reporting an oxidative-to-glycolytic switch in muscle fiber composition in animals bearing tumors or exposed to anticancer drugs [45, 106]. Such reduced oxidative phosphorylation capacity and mitochondrial function can result from a direct effect of chemotherapy agents or tumor burden on the mitochondria, as supported by evidence that mitochondrial number or morphology are compromised in cachectic muscles [45, 106–108]. Moreover, the appearance of a more glycolytic muscle phenotype could partially explain the susceptibility to fiber atrophy, especially considering that muscle type II fibers (i.e., glycolytic) are more prone to develop cachexia during cancer [109].

Finally, oxidative stress has been described as an important player during cancer cachexia. Consistently, ROS levels were found elevated in the blood of patients affected with cancer, in line with markedly reduced antioxidant defenses [110]. Similarly, skeletal muscle in cachectic patients was previously shown associated with increased protein oxidation, suggesting a direct effect of ROS on muscle tissue [111]. Of note, some chemotherapy agents also stimulate the production of ROS as one of the mechanisms to counteract tumor growth [112], and this could ultimately also contribute to the occurrence of muscle weakness [47]. In line with this, we recently showed that mice bearing C26 tumors or treated with Folfiri have increase of ROS in the circulation, thus further suggesting a common mechanism of action that could participate in the onset of cachexia.

5 Strategies to Counteract Chemotherapy Side Effects

5.1 Mitochondria and Exercise

Skeletal muscle mitochondria are greatly impaired by several chemotherapies, and thus preservation of the mitochondrial pool and mitochondrial function are a current area of interest to sustain muscle integrity in the presence of chemotherapy [36, 45]. This is also due to the fact that correction or improvement of mitochondrial biogenesis maintains skeletal muscle mass in other models of muscle wasting, including hindlimb suspension, fasting, and denervation [113, 114]. Interestingly, in a model of cancer-induced cachexia, overexpression of mitochondrial proteins such as PGC1α or Mitofusin-2 to the extent of sustaining muscle mass has yielded mixed results [106, 115, 116]. Interestingly, the exercise mimetic trimetazidine, which stimulated PGC1α in tumor-bearing mice, was also able to improve muscle size [117]. However, whether this strategy of genetically or therapeutically targeting mitochondrial function in order to sustain muscle mass and function in the presence of chemotherapy has not yet been conclusively established.

A practical manner to improve mitochondrial content and function is via physical exercise. The use of aerobic exercise has been investigated and has demonstrated some efficacy in improving cachexia induced by cancer in rodent models. Indeed, low-intensity aerobic exercise training in combination with erythropoietin was able to mildly preserve skeletal muscle mass in C26 tumor-bearing mice, whereas moderate aerobic exercise training alone proved beneficial [106, 118]. Additionally, a combination of resistance and aerobic exercise training was able to preserve both muscle mass and strength in tumor hosts, while resistance training alone was able to preserve muscle mass in rats bearing breast cancer tumors [119, 120]. Of interest, exercise interventions have also shown to be beneficial in the presence of chemotherapy. It was recently shown that aerobic exercise could partially preserve both skeletal muscle mass and strength in tumor-bearing mice receiving chemotherapy

[121]. Aerobic exercise was shown to prevent muscle fatigue and muscle derangements induced by doxorubicin and cisplatin [60, 63, 79, 122]. Aerobic exercise training was also found beneficial in cancer patients receiving chemotherapy, with exercise often improving physical functioning and chemotherapy completion rates, as well as reducing chemotherapy-associated toxicity [123]. Additionally, resistance exercise was found safe in breast cancer patients receiving chemotherapy, thereby leading to increased muscle strength [124]. Further studies including resistance exercise alone and in conjunction with aerobic exercise should be completed to optimize exercise prescription for patients receiving anticancer drugs.

5.2 Activin Signaling

Another approach to conserve skeletal muscle mass, which continues to show beneficial evidence, is the targeting of the activin type 2B receptors (ACVR2B), which are able to bind TGFβ family ligands [125, 126]. Upregulation of activin signaling within skeletal muscle is known to induce wasting. ACVR2B/Fc is a synthetic peptide and receptor decoy designed to compete for binding to the ACVR2B. This peptide is able to fully preserve skeletal muscle mass in rodents in the presence of chemotherapeutics, including Folfiri, cisplatin, and doxorubicin [33, 62, 127]. In addition to serving as a skeletal muscle preservation strategy, targeting of the activin signaling also rescues cardiac dysfunction induced by doxorubicin, and prolongs survival in mouse models of colorectal and pancreatic cancer cachexia [86, 128–131]. Taken together, targeting of the activin signaling is an attractive approach and should receive further attention for preservation of muscle mass, especially in combination with anticancer drugs in an attempt to improve overall survival in cancer settings. It is also interesting that the use of ACVR2B/Fc was reported to preserve bone mass in mice chronically exposed to the chemotherapeutic regimen Folfiri, which may have further implications in the rapidly growing "muscle-bone" cross talk research area [33].

5.3 Bone-Muscle Cross Talk

The mechanical interaction that occurs between skeletal muscle and bone has been historically well established. However, over the past several years, investigations into the possible biochemical interaction between these two organs have received much attention [132]. For these reasons, the possibility that abnormal muscle and bone cross talk could participate in the pathogenesis of cachexia is now gaining increasing consideration, as both cancer and chemotherapy administration have shown to induce bone loss along with skeletal muscle wasting [33, 34, 59, 133–135]. In support of a muscle-bone cross talk in cachexia, Waning et al. demonstrated that excess TGFβ released from bone could mediate muscle weakness in several models of cancer-induced bone metastases, whereby blocking TGFβ or inhibiting bone resorption via the bisphosphonate zoledronic acid was sufficient to sustain muscle function [135]. Also, in the context of burn-induced cachexia, another bisphosphonate, pamidronate, which originally demonstrated to improve muscle and bone in burn patients, reduced TGFβ signaling and atrophy within skeletal muscle cells, thus further implicating the use of anti-bone resorptive drugs as a novel skeletal muscle preservation strategy in wasting conditions [136, 137].

Interestingly, only recently the use of bisphosphonates to preserve muscle and bone in the context of chemotherapy has been investigated. Work from our group and others initially showed that platinum-based chemotherapeutics, including cisplatin and carboplatin, induce both skeletal muscle and bone loss. In this setting, combined administration of zoledronic acid was sufficient not only to block bone loss induced by both chemotherapeutics, but also· to preserve skeletal muscle size and strength in the presence of cisplatin and skeletal muscle function following administration of carboplatin [35, 59]. As it is clear that bisphosphonates are able to preserve skeletal muscle mass and function in conjunction with chemotherapy, future research should examine other anti-bone resorptive treatments as a way to preserve and/or correct both skeletal muscle and bone integrity. In this regard, the receptor activator of nuclear factor-kappaB ligand (RANKL) may represent an ideal target because RANKL is a driver of bone resorption via osteoclast activation, which has proved effective in improving muscle strength in osteoporosis [138].

5.4 Nutrition

Ensuring adequate nutrition is another strong consideration for combating muscle wasting in the presence of cancer and chemotherapy. Sustaining weight in individuals relies on energy balance maintenance via nutrient intake, which is particularly important in cancer patients, especially since chemotherapy can promote taste alterations, lower appetite, and cause vomiting, thus causing a negative caloric state and exacerbating the cachectic potential in cancer patients [19, 139]. With respect to skeletal muscle mass maintenance, protein consumption is pivotal, as circulating amino acids, which stimulate muscle protein synthesis and can serve to prevent protein breakdown, have greater turnover in advanced cancer patients [139–141]. Moreover, it has been reported that cancer patients often do not meet protein intake guidelines (1.0–1.5 g/kg/day), and many even fall short of the recommended daily intake (0.8 g/kg/day) for healthy individuals, making protein intake a valid target for sustaining skeletal muscle mass [139, 142]. In fact, whey protein isolate was recently shown to increase fat-free mass index, body weight, and muscle strength in advanced cancer patients receiving chemotherapy [143], which is in line with prior studies demonstrating positive outcomes of whey protein on weight loss and exercise capacity in cancer patients treated with anticancer drugs [144, 145].

5.5 Amino Acids and Their Derivatives

Amino acids and their derivatives are starting to receive attention with respect to preserving muscle mass in the presence of chemotherapy.

Leucine, one of three branched-chain amino acids (along with isoleucine and valine), serves as the most potent stimulator of protein synthesis and was shown to improve muscle mass and attenuate cardiac failure in tumor-bearing rats [146, 147]. Given the cardiac and skeletal muscle impairments induced by chemotherapy, leucine supplementation may result beneficial to cancer patients, though studies are needed to confirm such hypothesis. Similarly, a metabolite of leucine, β-hydroxy-β-methylbutyrate (HMB), was shown to improve body weight and muscle mass in rats carrying hepatocarcinoma tumors [148]. HMB, when combined with arginine and glutamine also improves fat-free mass in advanced cancer patients. It is not yet known whether HMB can restore muscle integrity in the presence of chemotherapy [149]. Further, carnitine administration has also shown beneficial effects in preserving lean mass in advanced cancer patients, while both taurine and methionine were shown to reverse cisplatin-induced muscle atrophy in experimental models [56, 76, 150]. Interestingly, the commonly used supplement fish oil, which includes eicosapentaenoic acid, has been suggested to improve lean body mass in cancer patients. For instance, individuals receiving chemotherapy sustain and even gain skeletal muscle with fish oil consumption to a greater extent than the patients who do not consume the supplement [151].

5.6 Ghrelin

Major side effects of chemotherapy treatment are anorexia and altered appetite. Protein or various amino acids may be able to maintain or even improve skeletal muscle mass in conjunction with chemotherapy; however, for this to work, patients must maintain, or even increase, their calorie intake. Hormonal control of appetite is mediated via ghrelin, a ligand for the growth hormone secretagogue receptor (GHSR)-1a [84]. Ghrelin has been implicated as a possible avenue to increase appetite, and thus muscle mass, in various rodent models of cancer cachexia. In particular, ghrelin and ghrelin receptor agonists attenuated cachexia in the commonly used C26 and LLC mouse models of cachexia, as well as tumor-bearing rats [84, 152–155]. Moreover, targeting ghrelin was also sufficient to preserve cachexia in rodents exposed to cisplatin or Folfiri [84, 156]. There is also evidence that direct ghrelin administration is able to improve adverse events in patients undergoing chemotherapy [157]. Hence, ghrelin certainly represents an attractive approach to stimulating appetite and preserving body and muscle mass in patients receiving chemotherapy.

6   Conclusion and Perspectives

Chemotherapy is the preferred treatment for cancer. Although anticancer drugs are effective in combatting tumor growth in most cases, they also present with a multitude of negative off-target systemic toxicities, some of which directly target muscle mass and induce skeletal muscle wasting and weakness. Given that maintenance of skeletal muscle mass has proven critical for tolerance to treatments, as well as for daily function and overall survival in patients affected with cancer, research must continue to investigate the mechanisms underlying the impairment in muscle homeostasis and to identify novel countermeasures to preserve muscle mass and function. In this regard, here we discussed some of the mechanisms associated with the onset of cachexia resulting from administration of anticancer drugs, thereby contributing to highlight novel targets for future interventions and to suggest new combination therapies designed to improve outcomes and survival in cancer patients (Fig. 1).

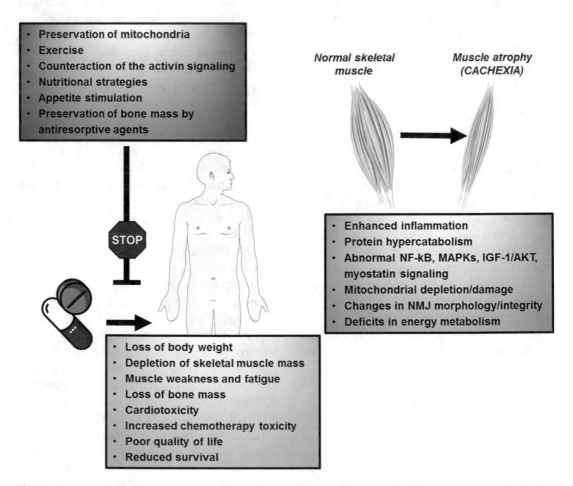

Fig. 1 Current strategies to preserve muscle mass and function in combination with anticancer drugs. Counteracting the mechanisms which drive chemotherapy-induced skeletal muscle perturbations contributes to reduced chemotherapy toxicity, improved quality of life and survival in cancer patients

Acknowledgments This study was supported by the Department of Surgery and the Department of Otolaryngology—Head & Neck Surgery at Indiana University, and by grants from the V Foundation for Cancer Research (V2017-021), the American Cancer Society (132013-RSG-18-010-01-CCG) and the Showalter Research Trust to AB. We would like to thank John Spence, PhD for his precious contribution in editing the chapter.

References

1. Siegel, R.L., et al.: Cancer statistics, 2022. CA Cancer J. Clin. **72**(1), 7–33 (2022)
2. Arem, H., Loftfield, E.: Cancer epidemiology: a survey of modifiable risk factors for prevention and survivorship. Am. J. Lifestyle Med. **12**(3), 200–210 (2018)
3. Falzone, L., Salomone, S., Libra, M.: Evolution of cancer pharmacological treatments at the turn of the third millennium. Front. Pharmacol. **9**, 1300 (2018)
4. Khan, F.A., Akhtar, S.S., Sheikh, M.K.: Cancer treatment – objectives and quality of life issues. Malays. J. Med. Sci. **12**(1), 3–5 (2005)
5. Arruebo, M., et al.: Assessment of the evolution of cancer treatment therapies. Cancers (Basel). **3**(3), 3279–3330 (2011)
6. Chen, H.H.W., Kuo, M.T.: Improving radiotherapy in cancer treatment: Promises and challenges. Oncotarget. **8**(37), 62742–62758 (2017)
7. Baumann, M., et al.: Radiation oncology in the era of precision medicine. Nat. Rev. Cancer. **16**(4), 234–249 (2016)
8. Jacinta Abraham, J.S.: Hormonal therapy for cancer. Medicine. **44**(1), 30–33 (2016)
9. Zhang, H., Chen, J.: Current status and future directions of cancer immunotherapy. J. Cancer. **9**(10), 1773–1781 (2018)
10. Zhao, Z., et al.: Delivery strategies of cancer immunotherapy: recent advances and future perspectives. J. Hematol. Oncol. **12**(1), 126 (2019)
11. Alkylating Agents, in LiverTox: clinical and research information on drug-induced liver injury. 2012: Bethesda (MD)
12. Lansiaux, A.: Antimetabolites. Bull. Cancer. **98**(11), 1263–1274 (2011)
13. Pommier, Y.: Topoisomerase I inhibitors: camptothecins and beyond. Nat. Rev. Cancer. **6**(10), 789–802 (2006)
14. Moudi, M., et al.: Vinca alkaloids. Int. J. Prev. Med. **4**(11), 1231–1235 (2013)
15. Oshiro, C., et al.: Taxane pathway. Pharmacogenet. Genomics. **19**(12), 979–983 (2009)
16. Nurgali, K., Jagoe, R.T., Abalo, R.: Editorial: adverse effects of cancer chemotherapy: anything new to improve tolerance and reduce sequelae? Front. Pharmacol. **9**, 245 (2018)
17. Boltong, A., Keast, R., Aranda, S.: Experiences and consequences of altered taste, flavour and food hedonics during chemotherapy treatment. Support Care Cancer. **20**(11), 2765–2774 (2012)
18. Carey, P.J.: Drug-induced myelosuppression: diagnosis and management. Drug Saf. **26**(10), 691–706 (2003)
19. Rapoport, B.L.: Delayed chemotherapy-induced nausea and vomiting: pathogenesis, incidence, and current management. Front. Pharmacol. **8**, 19 (2017)
20. Babic, T., Browning, K.N.: The role of vagal neurocircuits in the regulation of nausea and vomiting. Eur. J. Pharmacol. **722**, 38–47 (2014)
21. Borner, T., et al.: GDF15 Induces Anorexia through Nausea and Emesis. Cell Metab. (2020)
22. Tsai, V.W.W., et al.: The MIC-1/GDF15-GFRAL pathway in energy homeostasis: implications for obesity, cachexia, and other associated diseases. Cell Metab. **28**(3), 353–368 (2018)
23. Cinausero, M., et al.: New frontiers in the pathobiology and treatment of cancer regimen-related mucosal injury. Front. Pharmacol. **8**, 354 (2017)
24. Chaveli-Lopez, B., Bagan-Sebastian, J.V.: Treatment of oral mucositis due to chemotherapy. J Clin Exp Dent. **8**(2), e201–e209 (2016)
25. Zajaczkowska, R., et al.: Mechanisms of chemotherapy-induced peripheral neuropathy. Int. J. Mol. Sci. **20**(6) (2019)
26. Kolb, N.A., et al.: The association of chemotherapy-induced peripheral neuropathy symptoms and the risk of falling. JAMA Neurol. **73**(7), 860–866 (2016)
27. Dietrich, J., Prust, M., Kaiser, J.: Chemotherapy, cognitive impairment and hippocampal toxicity. Neuroscience. **309**, 224–232 (2015)
28. Li, M., Caeyenberghs, K.: Longitudinal assessment of chemotherapy-induced changes in brain and cognitive functioning: a systematic review. Neurosci. Biobehav. Rev. **92**, 304–317 (2018)
29. Sturgeon, K.M., et al.: Cancer- and chemotherapy-induced musculoskeletal degradation. JBMR Plus. **3**(3), e10187 (2019)
30. Baylink, D.J.: Glucocorticoid-induced osteoporosis. N. Engl. J. Med. **309**(5), 306–308 (1983)
31. Quach, J.M., et al.: Myelosuppressive therapies significantly increase pro-inflammatory cytokines and directly cause bone loss. J. Bone Miner. Res. **30**(5), 886–897 (2015)
32. Saarto, T., et al.: Chemical castration induced by adjuvant cyclophosphamide, methotrexate, and fluorouracil chemotherapy causes rapid bone loss that is reduced by clodronate: a randomized study in premenopausal breast cancer patients. J. Clin. Oncol. **15**(4), 1341–1347 (1997)
33. Barreto, R., et al.: ACVR2B/Fc counteracts chemotherapy-induced loss of muscle and bone mass. Sci. Rep. **7**(1), 14470 (2017)
34. Hain, B.A., et al.: Chemotherapy-induced loss of bone and muscle mass in a mouse model of breast

cancer bone metastases and cachexia. JCSM Rapid Commun. **2**(1) (2019)

35. Essex, A.L., et al.: Bisphosphonate treatment ameliorates chemotherapy-induced bone and muscle abnormalities in young mice. Front. Endocrinol. (Lausanne). **10**, 809 (2019)

36. Barreto, R., et al.: Cancer and chemotherapy contribute to muscle loss by activating common signaling pathways. Front. Physiol. **7**, 472 (2016)

37. Mock, V., et al.: Fatigue and quality of life outcomes of exercise during cancer treatment. Cancer Pract. **9**(3), 119–127 (2001)

38. Damrauer, J.S., et al.: Chemotherapy-induced muscle wasting: association with NF-kappaB and cancer cachexia. Eur. J. Transl. Myol. **28**(2), 7590 (2018)

39. Pin, F., Couch, M.E., Bonetto, A.: Preservation of muscle mass as a strategy to reduce the toxic effects of cancer chemotherapy on body composition. Curr. Opin. Support. Palliat. Care. **12**(4), 420–426 (2018)

40. Le Bricon, T., et al.: Negative impact of cancer chemotherapy on protein metabolism in healthy and tumor-bearing rats. Metabolism. **44**(10), 1340–1348 (1995)

41. Crawford, S.: Is it time for a new paradigm for systemic cancer treatment? Lessons from a century of cancer chemotherapy. Front. Pharmacol. **4**, 68 (2013)

42. Fearon, K., et al.: Definition and classification of cancer cachexia: an international consensus. Lancet Oncol. **12**(5), 489–495 (2011)

43. Siegel, R., Naishadham, D., Jemal, A.: Cancer statistics, 2012. CA Cancer J. Clin. **62**(1), 10–29 (2012)

44. Thoresen, L., et al.: Nutritional status, cachexia and survival in patients with advanced colorectal carcinoma. Different assessment criteria for nutritional status provide unequal results. Clin. Nutr. **32**(1), 65–72 (2013)

45. Barreto, R., et al.: Chemotherapy-related cachexia is associated with mitochondrial depletion and the activation of ERK1/2 and p38 MAPKs. Oncotarget. **7**(28), 43442–43460 (2016)

46. Huot, J.R., et al.: Chronic treatment with multi-kinase inhibitors causes differential toxicities on skeletal and cardiac muscles. Cancers (Basel). **11**(4) (2019)

47. Pin, F., et al.: Cachexia induced by cancer and chemotherapy yield distinct perturbations to energy metabolism. J. Cachexia. Sarcopenia Muscle. **10**(1), 140–154 (2019)

48. Naito, T., et al.: Skeletal muscle depletion during chemotherapy has a large impact on physical function in elderly Japanese patients with advanced non-small-cell lung cancer. BMC Cancer. **17**(1), 571 (2017)

49. Kasymjanova, G., et al.: Prognostic value of the six-minute walk in advanced non-small cell lung cancer. J. Thorac. Oncol. **4**(5), 602–607 (2009)

50. Kurk, S.A., et al.: Impact of different palliative systemic treatments on skeletal muscle mass in metastatic colorectal cancer patients. J. Cachexia. Sarcopenia Muscle. **9**(5), 909–919 (2018)

51. Klassen, O., et al.: Muscle strength in breast cancer patients receiving different treatment regimes. J. Cachexia. Sarcopenia Muscle. **8**(2), 305–316 (2017)

52. Freedman, R.J., et al.: Weight and body composition changes during and after adjuvant chemotherapy in women with breast cancer. J. Clin. Endocrinol. Metab. **89**(5), 2248–2253 (2004)

53. Awad, S., et al.: Marked changes in body composition following neoadjuvant chemotherapy for oesophagogastric cancer. Clin. Nutr. **31**(1), 74–77 (2012)

54. Poterucha, T., Burnette, B., Jatoi, A.: A decline in weight and attrition of muscle in colorectal cancer patients receiving chemotherapy with bevacizumab. Med. Oncol. **29**(2), 1005–1009 (2012)

55. Sakai, H., et al.: Mechanisms of cisplatin-induced muscle atrophy. Toxicol. Appl. Pharmacol. **278**(2), 190–199 (2014)

56. Stacchiotti, A., et al.: Taurine rescues cisplatin-induced muscle atrophy in vitro: a morphological study. Oxidative Med. Cell. Longev. **2014**, 840951 (2014)

57. Feather, C.E., et al.: Oxaliplatin induces muscle loss and muscle-specific molecular changes in Mice. Muscle Nerve. **57**(4), 650–658 (2018)

58. Sorensen, J.C., et al.: BGP-15 protects against oxaliplatin-induced skeletal myopathy and mitochondrial reactive oxygen species production in mice. Front. Pharmacol. **8**, 137 (2017)

59. Hain, B.A., et al.: Zoledronic acid improves muscle function in healthy mice treated with chemotherapy. J. Bone Miner. Res. (2019)

60. de Lima, E.A., et al.: Aerobic exercise, but not metformin, prevents reduction of muscular performance by AMPk activation in mice on doxorubicin chemotherapy. J. Cell. Physiol. **233**(12), 9652–9662 (2018)

61. Hulmi, J.J., et al.: Prevention of chemotherapy-induced cachexia by ACVR2B ligand blocking has different effects on heart and skeletal muscle. J. Cachexia. Sarcopenia Muscle. **9**(2), 417–432 (2018)

62. Nissinen, T.A., et al.: Systemic blockade of ACVR2B ligands prevents chemotherapy-induced muscle wasting by restoring muscle protein synthesis without affecting oxidative capacity or atrogenes. Sci. Rep. **6**, 32695 (2016)

63. Powers, S.K., et al.: Endurance exercise protects skeletal muscle against both doxorubicin-induced and inactivity-induced muscle wasting. Pflugers Arch. **471**(3), 441–453 (2019)

64. Nishiyama, K., et al.: Ibudilast attenuates doxorubicin-induced cytotoxicity by suppressing formation of TRPC3 channel and NADPH oxidase 2 protein complexes. Br. J. Pharmacol. **176**(18), 3723–3738 (2019)

65. Dewys, W.D., et al.: Prognostic effect of weight loss prior to chemotherapy in cancer patients. Eastern

Cooperative Oncology Group. Am. J. Med. **69**(4), 491–497 (1980)

66. Grothey, A., et al.: Regorafenib monotherapy for previously treated metastatic colorectal cancer (CORRECT): an international, multicentre, randomised, placebo-controlled, phase 3 trial. Lancet. **381**(9863), 303–312 (2013)

67. Shingina, A., et al.: In a 'real-world', clinic-based community setting, sorafenib dose of 400 mg/day is as effective as standard dose of 800 mg/day in patients with advanced hepatocellular carcimona, with better tolerance and similar survival. Can. J. Gastroenterol. **27**(7), 393–396 (2013)

68. Sirago, G., et al.: Growth hormone secretagogues hexarelin and JMV2894 protect skeletal muscle from mitochondrial damages in a rat model of cisplatin-induced cachexia. Sci. Rep. **7**(1), 13017 (2017)

69. Tarpey, M.D., et al.: Doxorubicin causes lesions in the electron transport system of skeletal muscle mitochondria that are associated with a loss of contractile function. J. Biol. Chem. **294**(51), 19709–19722 (2019)

70. Shigemoto, K., et al.: Muscle weakness and neuromuscular junctions in aging and disease. Geriatr Gerontol Int. **10**(Suppl 1), S137–S147 (2010)

71. De Vos, K.J., et al.: Familial amyotrophic lateral sclerosis-linked SOD1 mutants perturb fast axonal transport to reduce axonal mitochondria content. Hum. Mol. Genet. **16**(22), 2720–2728 (2007)

72. Magrane, J., et al.: Mitochondrial dynamics and bioenergetic dysfunction is associated with synaptic alterations in mutant SOD1 motor neurons. J. Neurosci. **32**(1), 229–242 (2012)

73. Malkki, H.: Neuromuscular disease: mitochondrial dysfunction could precipitate motor neuron loss in spinal muscular atrophy. Nat. Rev. Neurol. **12**(10), 556 (2016)

74. Ahn, B., et al.: Mitochondrial oxidative stress impairs contractile function but paradoxically increases muscle mass via fibre branching. J. Cachexia. Sarcopenia Muscle. **10**(2), 411–428 (2019)

75. Rudolf, R., et al.: Regulation of nicotinic acetylcholine receptor turnover by MuRF1 connects muscle activity to endo/lysosomal and atrophy pathways. Age (Dordr). **35**(5), 1663–1674 (2013)

76. Wu, C.T., et al.: D-methionine ameliorates cisplatin-induced muscle atrophy via inhibition of muscle degradation pathway. Integr. Cancer Ther. **18**, 1534735419828832 (2019)

77. McQuade, R.M., et al.: Role of oxidative stress in oxaliplatin-induced enteric neuropathy and colonic dysmotility in mice. Br. J. Pharmacol. **173**(24), 3502–3521 (2016)

78. Webster, R.G., et al.: Oxaliplatin induces hyperexcitability at motor and autonomic neuromuscular junctions through effects on voltage-gated sodium channels. Br. J. Pharmacol. **146**(7), 1027–1039 (2005)

79. Huertas, A.M., et al.: Modification of neuromuscular junction protein expression by exercise and doxorubicin. Med. Sci. Sports Exerc. (2020)

80. Boehm, I., et al.: Neuromuscular junctions are stable in patients with cancer cachexia. J. Clin. Invest. (2019)

81. Penna, F., et al.: Muscle wasting and impaired myogenesis in tumor bearing mice are prevented by ERK inhibition. PLoS One. **5**(10), e13604 (2010)

82. Liu, Z., et al.: p38beta MAPK mediates ULK1-dependent induction of autophagy in skeletal muscle of tumor-bearing mice. Cell Stress. **2**(11), 311–324 (2018)

83. Quan-Jun, Y., et al.: Selumetinib attenuates skeletal muscle wasting in murine cachexia model through ERK inhibition and AKT activation. Mol. Cancer Ther. **16**(2), 334–343 (2017)

84. Chen, J.A., et al.: Ghrelin prevents tumour- and cisplatin-induced muscle wasting: characterization of multiple mechanisms involved. J. Cachexia. Sarcopenia Muscle. **6**(2), 132–143 (2015)

85. Sakai, H., et al.: Dexamethasone exacerbates cisplatin-induced muscle atrophy. Clin. Exp. Pharmacol. Physiol. **46**(1), 19–28 (2019)

86. Magga, J., et al.: Systemic blockade of ACVR2B ligands protects myocardium from acute ischemia-reperfusion injury. Mol. Ther. **27**(3), 600–610 (2019)

87. Mitry, M.A., et al.: Accelerated cardiomyocyte senescence contributes to late-onset doxorubicin-induced cardiotoxicity. Am. J. Physiol. Cell Physiol. (2020)

88. Yu, X., et al.: Dexrazoxane ameliorates doxorubicin-induced cardiotoxicity by inhibiting both apoptosis and necroptosis in cardiomyocytes. Biochem. Biophys. Res. Commun. (2019)

89. Cheng, C.F., et al.: Pravastatin attenuates carboplatin-induced cardiotoxicity via inhibition of oxidative stress associated apoptosis. Apoptosis. **13**(7), 883–894 (2008)

90. Topal, I., et al.: The effect of rutin on cisplatin-induced oxidative cardiac damage in rats. Anatol. J. Cardiol. **20**(3), 136–142 (2018)

91. Argiles, J.M., et al.: Cancer cachexia: understanding the molecular basis. Nat. Rev. Cancer. **14**(11), 754–762 (2014)

92. Constantinou, C., et al.: Nuclear magnetic resonance in conjunction with functional genomics suggests mitochondrial dysfunction in a murine model of cancer cachexia. Int. J. Mol. Med. **27**(1), 15–24 (2011)

93. Warburg, O.: On the origin of cancer cells. Science. **123**(3191), 309–314 (1956)

94. Vander Heiden, M.G., Cantley, L.C., Thompson, C.B.: Understanding the Warburg effect: the metabolic requirements of cell proliferation. Science. **324**(5930), 1029–1033 (2009)

95. Rohm, M., et al.: Energy metabolism in cachexia. EMBO Rep. **20**, 4 (2019)

96. Argiles, J.M., et al.: Cachexia: a problem of energetic inefficiency. J. Cachexia. Sarcopenia Muscle. **5**(4), 279–286 (2014)

97. Kazak, L., et al.: A creatine-driven substrate cycle enhances energy expenditure and thermogenesis in beige fat. Cell. **163**(3), 643–655 (2015)

98. Mulligan, H.D., Beck, S.A., Tisdale, M.J.: Lipid metabolism in cancer cachexia. Br. J. Cancer. **66**(1), 57–61 (1992)

99. Dev, R., Bruera, E., Dalal, S.: Insulin resistance and body composition in cancer patients. Ann. Oncol. **29**(suppl_2), ii18–ii26 (2018)

100. Tomas, E., et al.: Hyperglycemia and insulin resistance: possible mechanisms. Ann. N. Y. Acad. Sci. **967**, 43–51 (2002)

101. Hwangbo, Y., Lee, E.K.: Acute hyperglycemia associated with anti-cancer medication. Endocrinol. Metab. (Seoul). **32**(1), 23–29 (2017)

102. Younes, R.N., Noguchi, Y.: Pathophysiology of cancer cachexia. Rev. Hosp. Clin. Fac. Med. Sao Paulo. **55**(5), 181–193 (2000)

103. Cho, E.M., et al.: Severe recurrent nocturnal hypoglycemia during chemotherapy with 6-mercaptopurine in a child with acute lymphoblastic leukemia. Ann. Pediatr. Endocrinol. Metab. **23**(4), 226–228 (2018)

104. Harris, R.A., et al.: Regulation of the activity of the pyruvate dehydrogenase complex. Adv. Enzym. Regul. **42**, 249–259 (2002)

105. Pin, F., et al.: PDK4 drives metabolic alterations and muscle atrophy in cancer cachexia. FASEB J. **33**(6), 7778–7790 (2019)

106. Pin, F., et al.: Combination of exercise training and erythropoietin prevents cancer-induced muscle alterations. Oncotarget. **6**(41), 43202–43215 (2015)

107. Shum, A.M., et al.: Disruption of MEF2C signaling and loss of sarcomeric and mitochondrial integrity in cancer-induced skeletal muscle wasting. Aging (Albany NY). **4**(2), 133–143 (2012)

108. Guigni, B.A., et al.: Skeletal muscle atrophy and dysfunction in breast cancer patients: role for chemotherapy-derived oxidant stress. Am. J. Physiol. Cell Physiol. **315**(5), C744–C756 (2018)

109. Wang, Y., Pessin, J.E.: Mechanisms for fiber-type specificity of skeletal muscle atrophy. Curr. Opin. Clin. Nutr. Metab. Care. **16**(3), 243–250 (2013)

110. Abrigo, J., et al.: Role of oxidative stress as key regulator of muscle wasting during cachexia. Oxidative Med. Cell. Longev. **2018**, 2063179 (2018)

111. Eley, H.L., Tisdale, M.J.: Skeletal muscle atrophy, a link between depression of protein synthesis and increase in degradation. J. Biol. Chem. **282**(10), 7087–7097 (2007)

112. Block, K.I., et al.: Impact of antioxidant supplementation on chemotherapeutic toxicity: a systematic review of the evidence from randomized controlled trials. Int. J. Cancer. **123**(6), 1227–1239 (2008)

113. Cannavino, J., et al.: PGC1-alpha over-expression prevents metabolic alterations and soleus muscle atrophy in hindlimb unloaded mice. J. Physiol. **592**(20), 4575–4589 (2014)

114. Sandri, M., et al.: PGC-1alpha protects skeletal muscle from atrophy by suppressing FoxO3 action and atrophy-specific gene transcription. Proc. Natl. Acad. Sci. USA. **103**(44), 16260–16265 (2006)

115. Wang, X., et al.: Increase in muscle mitochondrial biogenesis does not prevent muscle loss but increased tumor size in a mouse model of acute cancer-induced cachexia. PLoS One. **7**(3), e33426 (2012)

116. Xi, Q.L., et al.: Mitofusin-2 prevents skeletal muscle wasting in cancer cachexia. Oncol. Lett. **12**(5), 4013–4020 (2016)

117. Molinari, F., et al.: The mitochondrial metabolic reprogramming agent trimetazidine as an 'exercise mimetic' in cachectic C26-bearing mice. J. Cachexia. Sarcopenia Muscle. **8**(6), 954–973 (2017)

118. Ballaro, R., et al.: Moderate exercise improves experimental cancer cachexia by modulating the redox homeostasis. Cancers (Basel). **11**(3) (2019)

119. Padilha, C.S., et al.: Resistance exercise attenuates skeletal muscle oxidative stress, systemic pro-inflammatory state, and cachexia in Walker-256 tumor-bearing rats. Appl. Physiol. Nutr. Metab. **42**(9), 916–923 (2017)

120. Ranjbar, K., et al.: Combined exercise training positively affects muscle wasting in tumor-bearing mice. Med. Sci. Sports Exerc. **51**(7), 1387–1395 (2019)

121. Ballaro, R., et al.: Moderate exercise in mice improves cancer plus chemotherapy-induced muscle wasting and mitochondrial alterations. FASEB J. **33**(4), 5482–5494 (2019)

122. Sakai, H., et al.: Effect of acute treadmill exercise on cisplatin-induced muscle atrophy in the mouse. Pflugers Arch. **469**(11), 1495–1505 (2017)

123. Cave, J., et al.: A systematic review of the safety and efficacy of aerobic exercise during cytotoxic chemotherapy treatment. Support Care Cancer. **26**(10), 3337–3351 (2018)

124. Mijwel, S., et al.: Highly favorable physiological responses to concurrent resistance and high-intensity interval training during chemotherapy: the OptiTrain breast cancer trial. Breast Cancer Res. Treat. **169**(1), 93–103 (2018)

125. Lee, S.J., et al.: Regulation of muscle growth by multiple ligands signaling through activin type II receptors. Proc. Natl. Acad. Sci. USA. **102**(50), 18117–18122 (2005)

126. Tsuchida, K., et al.: Activin signaling as an emerging target for therapeutic interventions. Cell Commun. Signal. **7**, 15 (2009)

127. Hatakeyama, S., et al.: ActRII blockade protects mice from cancer cachexia and prolongs survival in the presence of anti-cancer treatments. Skelet. Muscle. **6**, 26 (2016)

128. Nissinen, T.A., et al.: Treating cachexia using soluble ACVR2B improves survival, alters mTOR localization, and attenuates liver and spleen responses. J. Cachexia. Sarcopenia Muscle. **9**(3), 514–529 (2018)

129. Zhong, X., et al.: The systemic activin response to pancreatic cancer: implications for effective cancer cachexia therapy. J. Cachexia. Sarcopenia Muscle. **10**(5), 1083–1101 (2019)

130. Benny Klimek, M.E., et al.: Acute inhibition of myostatin-family proteins preserves skeletal muscle in mouse models of cancer cachexia. Biochem. Biophys. Res. Commun. **391**(3), 1548–1554 (2010)

131. Zhou, X., et al.: Reversal of cancer cachexia and muscle wasting by ActRIIB antagonism leads to prolonged survival. Cell. **142**(4), 531–543 (2010)

132. Brotto, M., Bonewald, L.: Bone and muscle: Interactions beyond mechanical. Bone. **80**, 109–114 (2015)

133. Bonetto, A., et al.: Differential bone loss in mouse models of colon cancer cachexia. Front. Physiol. **7**, 679 (2016)

134. Pin, F., et al.: Growth of ovarian cancer xenografts causes loss of muscle and bone mass: a new model for the study of cancer cachexia. J. Cachexia. Sarcopenia Muscle. **9**(4), 685–700 (2018)

135. Waning, D.L., et al.: Excess TGF-beta mediates muscle weakness associated with bone metastases in mice. Nat. Med. **21**(11), 1262–1271 (2015)

136. Borsheim, E., et al.: Pamidronate attenuates muscle loss after pediatric burn injury. J. Bone Miner. Res. **29**(6), 1369–1372 (2014)

137. Pin, F., et al.: Molecular mechanisms responsible for the rescue effects of pamidronate on muscle atrophy in pediatric burn patients. Front. Endocrinol. (Lausanne). **10**, 543 (2019)

138. Bonnet, N., et al.: RANKL inhibition improves muscle strength and insulin sensitivity and restores bone mass. J. Clin. Invest. **129**(8), 3214–3223 (2019)

139. Prado, C.M., Purcell, S.A., Laviano, A.: Nutrition interventions to treat low muscle mass in cancer. J. Cachexia. Sarcopenia Muscle. (2020)

140. Phillips, S.M., Glover, E.I., Rennie, M.J.: Alterations of protein turnover underlying disuse atrophy in human skeletal muscle. J. Appl. Physiol. 2009. **107**(3), 645–654 (1985)

141. van der Meij, B.S., et al.: Increased amino acid turnover and myofibrillar protein breakdown in advanced cancer are associated with muscle weakness and impaired physical function. Clin. Nutr. **38**(5), 2399–2407 (2019)

142. Arends, J., et al.: ESPEN guidelines on nutrition in cancer patients. Clin. Nutr. **36**(1), 11–48 (2017)

143. Cereda, E., et al.: Whey protein isolate supplementation improves body composition, muscle strength, and treatment tolerance in malnourished advanced cancer patients undergoing chemotherapy. Cancer Med. **8**(16), 6923–6932 (2019)

144. Bumrungpert, A., et al.: Whey protein supplementation improves nutritional status, glutathione levels, and immune function in cancer patients: a randomized, double-blind controlled trial. J. Med. Food. **21**(6), 612–616 (2018)

145. Tozer, R.G., et al.: Cysteine-rich protein reverses weight loss in lung cancer patients receiving chemotherapy or radiotherapy. Antioxid. Redox Signal. **10**(2), 395–402 (2008)

146. Cruz, B., Oliveira, A., Gomes-Marcondes, M.C.C.: L-leucine dietary supplementation modulates muscle protein degradation and increases pro-inflammatory cytokines in tumour-bearing rats. Cytokine. **96**, 253–260 (2017)

147. Toneto, A.T., et al.: Nutritional leucine supplementation attenuates cardiac failure in tumour-bearing cachectic animals. J. Cachexia. Sarcopenia Muscle. **7**(5), 577–586 (2016)

148. Aversa, Z., et al.: beta-hydroxy-beta-methylbutyrate (HMB) attenuates muscle and body weight loss in experimental cancer cachexia. Int. J. Oncol. **38**(3), 713–720 (2011)

149. May, P.E., et al.: Reversal of cancer-related wasting using oral supplementation with a combination of beta-hydroxy-beta-methylbutyrate, arginine, and glutamine. Am. J. Surg. **183**(4), 471–479 (2002)

150. Gramignano, G., et al.: Efficacy of l-carnitine administration on fatigue, nutritional status, oxidative stress, and related quality of life in 12 advanced cancer patients undergoing anticancer therapy. Nutrition. **22**(2), 136–145 (2006)

151. Murphy, R.A., et al.: Nutritional intervention with fish oil provides a benefit over standard of care for weight and skeletal muscle mass in patients with nonsmall cell lung cancer receiving chemotherapy. Cancer. **117**(8), 1775–1782 (2011)

152. Borner, T., et al.: The ghrelin receptor agonist HM01 mimics the neuronal effects of ghrelin in the arcuate nucleus and attenuates anorexia-cachexia syndrome in tumor-bearing rats. Am. J. Physiol. Regul. Integr. Comp. Physiol. **311**(1), R89–R96 (2016)

153. Fujitsuka, N., et al.: Potentiation of ghrelin signaling attenuates cancer anorexia-cachexia and prolongs survival. Transl. Psychiatry. **1**, e23 (2011)

154. Villars, F.O., et al.: Oral treatment with the ghrelin receptor agonist HM01 attenuates cachexia in mice bearing Colon-26 (C26) tumors. Int. J. Mol. Sci. **18**(5) (2017)

155. Yoshimura, M., et al.: Z-505 hydrochloride, an orally active ghrelin agonist, attenuates the progression of cancer cachexia via anabolic hormones in Colon 26 tumor-bearing mice. Eur. J. Pharmacol. **811**, 30–37 (2017)

156. Shiomi, Y., et al.: Z-505 hydrochloride ameliorates chemotherapy-induced anorexia in rodents via activation of the ghrelin receptor, GHSR1a. Eur. J. Pharmacol. **818**, 148–157 (2018)

157. Hiura, Y., et al.: Effects of ghrelin administration during chemotherapy with advanced esophageal cancer patients: a prospective, randomized, placebo-controlled phase 2 study. Cancer. **118**(19), 4785–4794 (2012)

Bone-Muscle Crosstalk in Advanced Cancer and Chemotherapy

David L. Waning

Abstract

Advanced cancers metastasize to distant sites and bone is a common site for metastases from breast, lung, and prostate cancers. Breast and lung cancer bone metastases typically lead to osteolytic lesions. Osteolytic bone metastases lead to bone pain, nerve compression, hypercalcemia, increased risk of fractures from falls, and muscle weakness. Bone metastases are incurable, and therapies for osteolytic lesions are aimed at reducing tumor burden and limiting bone loss. In addition to growth of tumor cells in bone, systemic effects to the musculoskeletal system are important in the overall reduction in mobility and increased morbidity. The focus of this chapter will be on the process of breast cancer cell colonization of bone, and bone-muscle crosstalk in breast cancer bone metastases and chemotherapy-induced bone loss. Much progress has been made recently in our understanding of the interplay between bone and muscle in cancer and chemotherapy, and future therapeutic strategies will likely include considerations for both of these tissues in the context of reducing overall tumor burden in bone.

Learning Objectives

This chapter will provide an overview of the colonization of breast cancer cells to bone and systemic effects that degrade musculoskeletal function. Bone-muscle crosstalk that leads to muscle weakness will be highlighted. Therapeutic strategies that limit bone loss will be discussed in the context of improving skeletal muscle function.

1 Introduction

Bone is a common site for metastases from breast, lung, and prostate cancers. Breast and lung cancer bone metastases typically lead to osteolytic lesions, whereas prostate cancer bone metastases typically lead and abnormal deposition of new bone. In all types of bone metastases from solid tumors, there is evidence of increased bone resorption and new bone formation and it is the overall balance between these activities in bone that lead to the radiologic appearance. Bone metastases lead to significant bone pain, nerve compression, hypercalcemia, increased risk of fractures from falls, and muscle weakness. Once tumor cells colonize the bone, it is typically incurable and overall patient survival drops. In addition to growth of tumor cells in bone, systemic effects to the musculoskeletal system are

D. L. Waning (✉)
Penn State College of Medicine, Hershey, PA, USA

Penn State Cancer Institute, Hershey, PA, USA
e-mail: dwaning@psu.edu

© Springer Nature Switzerland AG 2022
S. Acharyya (ed.), *The Systemic Effects of Advanced Cancer*,
https://doi.org/10.1007/978-3-031-09518-4_9

important in the overall reduction in mobility and increased morbidity.

To generate effective therapies, a better understanding of the underlying mechanisms that drive the metastatic process are required. In addition, once tumor cells are present in the bone, treating the effects to the musculoskeletal system is an important aspect to maintaining mobility, quality of life, and a patient's ability to withstand antitumor therapy.

The focus for this chapter will be the process of breast cancer cell colonization of bone and the systemic musculoskeletal effects that occur as a result. This includes what is known about direct cell-to-cell interactions in the osteogenic niche as well as communication via cytokines and extracellular vesicles (carrying protein and regulatory RNAs). In addition, bone-muscle crosstalk in breast cancer bone metastases and in chemotherapy-induced bone loss will be described. Recent findings demonstrating the positive effects on muscle function by preventing bone loss will be highlighted.

2   Tumor Metastasis to Bone

The skeleton is a common site for metastasis from breast, lung, and prostate cancers. Patients with bone metastases report low quality of life compared to all other sites of metastasis. This is primarily due to reduced pain-free mobility in addition to fractures and hypercalcemia. Another consequence of bone metastases is that once cancer is present in the bone, the 5-year survival rate falls to under 10% and there is no cure although bone-saving treatments do offer good therapeutic strategies [1]. Solid tumor metastasis to bone is clinically diagnosed by radiography as either osteolytic (bone loss) or osteoblastic (abnormal bone formation), but regions of bone loss and bone formation are usually evident in all patients. Breast and lung cancer bone metastases are typically osteolytic, whereas prostate cancer bone metastases are typically osteoblastic. Osteolytic lesions are due to overstimulated osteoclast-mediated bone resorption in which osteoblasts cannot rebuild new bone, whereas osteoblastic

lesions are due to increased osteoblast-mediated bone deposition. In either case, there is an imbalance in the normal bone turnover resulting in net bone loss or gain [2–4].

Breast cancer colonization of bone is a complex process that is poorly understood but often leads to tumor cell dormancy for potentially years before causing clinical concern. When tumor cells enter the bone via circulation, they encounter small blood vessels of the venous sinusoids. The sinusoids are located in the epiphysis of long bones and this corresponds to the site of disseminated breast cancer metastases [5]. The epiphysis is the site of trabecular bone or cancellous bone. Trabecular bone aids in the transfer of mechanical loads from the articular surface and has a very high rate of normal bone turnover [6]. Trabecular bone has a high number of resident osteoblasts and osteoclasts to repair mechanically-induced microdamage. When tumor cells enter the venous sinusoids, they encounter bone stromal cells, osteoblasts, osteoclasts, osteocytes, and mesenchymal stem cells (MSCs). The interactions between tumor cells and cells present in bone can be via direct cell-to-cell contact and gap junctions, or via extracellular vesicles (EVs) and cytokines [7].

Breast cancer bone metastases proliferate via a well-known "vicious cycle" where invading tumor cells produce parathyroid-related hormone (PTHrP) that stimulates the overproduction of receptor activator of nuclear factor-κB ligand (RANK-L) on osteoblasts. RANK-L on osteoblasts bind to the receptor, RANK, on osteoclast precursor cells, stimulating osteoclastogenesis. Osteoclasts then secrete cathepsin K and other cysteine proteinases that help degrade the type I collagen of bone and drive bone resorption [3, 8]. Bone is a rich source of growth factors that are released upon bone resorption. These include insulin growth factor-1 (IGF-1) and transforming growth factor-beta (TGFβ), which stimulate invading breast cancer cells to make additional PTHrP [2, 3].

Communication between breast cancer cells and bone cells is critical for establishing bone metastases. Direct cell-to-cell contact is one important way invading cells communicate with bone cells (reviewed in [7]). Cancer cells induce changes in the bone microenvironment prior to the clinical appearance of bone lesions and before patients fracture. It has been shown that areas of cancer cell infiltration, there are more osteoclasts present with fewer osteoblasts compared to areas of the bone that do not have any visible cancer cells [9].

Invading cancer cells first contact the resident osteoblasts via N-cadherin/E-cadherin junctions. N-cadherins on the osteoblasts interacted with E-cadherins on the breast cancer cells and tumor proliferation occurs. One of the effects of these early direct cell-to-cell contact events leads to increased mTOR (mammalian target of rapamycin) signaling in the cancer cells, thus conferring strong growth advantages [10]. Further molecular insights into the process of tumor cell survival in the osteogenic niche point to a crucial need to extract calcium from resident bone cells. Invading breast cancer cells are capable of extracting calcium via connexin 43 (Cx43) gap junctions [11]. Connexins form intercellular channels via hexameric assemblies capable of allowing direct diffusion of ions and small molecules between cells [12]. In bone, Cx43 is the most prevalent connexin and this has been shown to increase cancer cell growth in the bone. However, in another study, gap junctions between breast cancer cells and stromal cells of the bone marrow decreased tumor cell growth [13], indicating a more complex process than what we currently understand. The expansion of micrometastases requires intracellular calcium signaling as calcium is an important second messenger for transcription, cellular proliferation, and migration. Calcium is also important for apoptotic signaling, yet cancer cells have developed novel mechanisms to prevent the negative effects that a surge of intracellular calcium creates and this resistance is a hallmark of cancer cells [14]. In fact, cancer cells have devised mechanisms of calcium and oxidant crosstalk to initiate pro-tumorigenic signaling pathways that blunt surges in calcium by modulating various calcium channels such as Bcl-2-IP3 receptor interactions ([15] and reviewed in [16]). Another example of direct cell-to-cell between breast cancer cells and

bone cells is via Jagged1 and Notch. Jagged1 on breast cancer cells interacts with Notch on osteoblasts to drive progression of cancer cell growth [17]. Interestingly, this same signaling pathway may actually also lead to chemoresistance to invading cancer cells in the bone microenvironment [18].

In addition to the role of osteoblasts in cancer cell colonization, osteoclasts have also been shown to play an important role in breast cancer cell progression. Dormant cancer cells can reawaken via direct cell-to-cell contact between breast cancer cells expressing vascular cell adhesion molecule 1 (VCAM-1) and osteogenic cells of the bone microenvironment. This recruits osteoclasts and leads to bone resorption and cancer cell proliferation and growth [19].

Cytokines present in circulation can also drive cancer cell growth in bone without making direct cell-to-cell contact. Bone is a rich storehouse of cytokines that are released during bone resorption. In addition, bone cells release factors that can drive growth and proliferation of cancer cells in bone. A well-characterized cytokine interaction in bone metastases is the SDF-1/CXCR4 axis. Osteoblasts produce stromal cell-derived factor 1 (SDF-1), and cancer cells express the receptor for SDF-1, C-X-C motif chemokine receptor 4 (CXCR4). In fact a neutralizing antibody to CXCR4 reduced bone metastases [20, 21]. Interleukin-8 (IL-8), made by breast cancer cells, helps drive osteoclastogenesis in the early stages of cancer cell colonization of bone [22]. In addition, interleukin-6 (IL-6), IL-8, C-X-C motif chemokine ligand 1 (CXCL1), C-C motif chemokine ligand 2 (CCL2), and vascular endothelial growth factor (VEGF), made by osteoblasts also enhance osteoclast formation and play a role in the initial steps of breast cancer cell colonization and growth in bone [23].

Recently, a very interesting study of osteoblasts in the metastatic niche showed that after prolonged exposure metastatic breast cancer cells, that osteoblasts were altered to produce a different suite of soluble factors compared to naive cells. This so-called process of osteoblast education, creates a subpopulation of osteoblasts that can modulate cancer cell growth and proliferation [24]. In vivo, "educated" osteoblasts led to a decrease in inflammatory cytokines and expressed high levels of runt-related transcription factor 2 (Runx2), osteocalcin (OCN), and osteopontin (OPN). Educated osteoblasts suppress both ER+ breast cancer and TNBC cells from proliferating in the bone [24].

Extracellular vesicles (EVs), which include exosomes and microvesicles, are membrane-bound structures that vary based in size and cellular origin but are capable of transporting RNA, DNA, and proteins and have the potential to mediate cell-to-cell communication by delivery of their contents to target cells. Bone cells produce EVs which have functional effects in bone but mounting evidence also suggests that shedding of these bone-derive EVs into circulation can alter distant tissues ([25–29] and reviewed in [30]). This raises the possibility that crosstalk between cancer cells and bone cells could influence metastasis to bone. Stromal cell-derived EVs tend to have anti-proliferative activity on cancer cells [31, 32]. Many of the EV-mediated effects are via small microRNAs (miRNA) that regulate target genes in target cells (reviewed in [7, 33]). Within these EVs, miRNAs that are critical post-transcriptional regulators of gene expression, can function in osteogenesis with influence on osteoblast and osteoclast-mediated bone remodeling. Of particular note are miRNA-218 that is active during osteoblastogenesis, and miRNA-148a that is active during osteoclastogenesis ([34, 35] and reviewed in [36]). miRNAs also play a role in development of bone metastases [37]. miRNAs that target Runx2 have been shown to reduce the progression of breast cancer bone metastases [38]. Bone metastatic breast cancer cells express genes normally associated with osteogenic cells, including the master regulator of osteoblast differentiation, Runx2. This provides cancer cells with the ability of osteomimicry, which is attenuated by miRNAs [38–40]. Antagonizing miRNA-218 also reduced the expression of RANK-L and PTHrP and reduced the development of osteolytic lesions in a preclinical model of breast cancer bone metastases [41].

3 Skeletal Muscle Weakness in Advanced Cancer with Bone Metastases

Until very recently, the connection between bone and muscle was thought to mainly be mechanical, which allowed for proper growth and development of these tissues. We now appreciate the fact that bone and muscle both act as endocrine organs that are capable and necessary for signaling to each other as well as distant organs. This is true of normal development and aging but also in disease insults to both tissues. Muscle and bone anabolism are tightly coupled during growth and development [42, 43] and muscle and bone catabolism occur during aging [44], and so it is not surprising that these tissues are tightly connected. These studies are guiding a new emphasis on the musculoskeletal system as a whole. It is critical to understand crosstalk between bone and muscle and in particular in diseases of the musculoskeletal system.

Muscle is an endocrine organ that can express many cytokines referred to as "myokines." The best-understood myokines are those that are secreted following bouts of exercise. Interleukin 6 (IL-6) is produced in muscle in response to exercise and regulates differentiation of the muscle stem cell population (satellite cells) to increase muscle mass [45, 46]. Other interleukins (IL-5, IL-7, IL-8) have been shown to play a critical role in metabolism following exercise [47]. Irisin, a hormone produced by muscle after exercise, is involved in "browning" of white fat, thereby affecting whole body metabolism. White fat browning increases energy expenditure in mice on high-fat diet [48] and in cancer cachexia [49, 50]. IL-6 has also been shown to cause white fat browning in cancer cachexia [51]. Muscle also secretes myostatin (GDF-8) which is a potent inhibitor of skeletal muscle growth and myostatin knockout mice have provided much information about muscle growth and regeneration [52, 53].

In addition to the myokines that affect muscle and metabolism, many factors secreted by muscle also act on bone. These include β-aminoisobutyric acid (L-BAIBA), insulin-like growth factor 1 (IGF-1), fibroblast growth factor 2 (FGF-2), myostatin (also called growth and differentiation factor 8 [GDF8]), and IL-6 [54, 55]. IGF-1 and FGF-2 have been shown to stimulate bone formation [56, 57]. IL-6 causes a defect in osteogenesis within the bone marrow compartment [58]. Myostatin influences bone density in diabetes mellitus type 2, inhibits osteoblastogenesis, and affects fracture callus size in bone repair [59–62]. L-BAIBA plays an important role in osteocyte survival under oxidative stress [55].

Bone-derived factors, also referred to as "osteokines," are capable of influencing muscle, both mass and functionally. Osteocytes produce fibroblast growth factor 23 (FGF23) that is important for phosphate metabolism [63]. FGF23 has also recently been reported to impact skeletal muscle [64] and causes hypertrophy in cardiac muscle [65]. Indian hedgehog (Ihh) promotes myogenesis in both mouse and chick embryos [42]. Osteocalcin, secreted by osteoblasts, has been shown to regulate glucose metabolism in muscle and increases during bouts of exercise and diminishes with aging coincident with decreases in exercise capacity and skeletal muscle mass and function [66, 67].

In advanced breast cancer that has metastasized to the bone, a significant co-morbidity is muscle weakness. Bone osteolytic lesions from breast cancer bone metastases lead to pathological signaling pathways of the normal bone-muscle crosstalk described above to impact the musculoskeletal system. The endocrine signals between bone and muscle are of great interest and likely play a large role in the overall health of the musculoskeletal system. In addition to the myokines and osteokines that may be activated in disease, the bone matrix itself can impact muscle and bone. Among these, several have known effects on skeletal muscle. Activin, transforming growth factor-β (TGFβ), IGF-1, and bone morphogenic protein 2 (BMP-2) are all stored in bone and have the potential to impact skeletal muscle [68, 69].

Bone is a large storehouse for TGFβ, which is deposited in the mineralized bone matrix by osteoblasts [70, 71], and bone-derived TGFβ plays a central role to promote tumor osteolysis [72–75]. TGFβ is released in high concentrations from the mineralized bone matrix during osteoclastic bone resorption [74]. Preclinical mouse models of human breast cancer from estrogen receptor-positive (ER+) breast cancer and triple-negative breast cancer (TNBC) with bone metastases and osteolytic lesions have shown decreased skeletal muscle function [76]. In addition, a syngeneic mouse model of breast cancer with osteolytic bone metastases also showed decreased skeletal muscle function [77]. Skeletal muscle weakness due to osteolytic bone metastases was shown to be due to excess bone-derived TGFβ signaling in skeletal muscle. TGFβ signaling leads to increased oxidative stress and ultimately sarcoplasmic reticulum (SR) calcium leak [76]. During muscle excitation-contraction coupling (E-C coupling), calcium release from the SR stores via the ryanodine receptor 1 (RyR1) calcium release channel, triggers contraction [78]. Oxidative stress that leads to oxidation and nitrosylation of RyR1 causes calcium leak and leads to skeletal muscle weakness [79, 80]. In the case of breast cancer bone metastases, oxidative stress via increased expression and binding of Nox4 to the RyR1 calcium channel causes calcium leak and muscle weakness. Nox4 is a constitutively active oxidase and also a TGFβ target gene. Nox4 generates reactive oxygen species (ROS) [81, 82] that modifies RyR1. Blocking bone loss using an anti-resorptive bisphosphonate (zoledronic acid (ZA)) that blocks osteoclast activity or directly blocking TGFβ using a pan-neutralizing antibody (1D11) both improved skeletal muscle calcium handling and function [75, 76, 83, 84]. The extent of bone metastatic lesion area correlated with a decrease in muscle function. These data and previous work in breast cancer bone metastases showed that the source of TGFβ is the mineralized matrix [74, 76]. Novel drugs that block Nox4 activity (GKT137831) or prevent RyR1 calcium leak (Rycal S107) improved calcium handling and skeletal muscle function that was independent of osteolytic lesion area [76, 77]. Interestingly, a biochemical signature of RyR1 calcium leak as well as SMAD2/3 phosphorylation, a downstream marker of TGFβ signaling were both present in skeletal muscle biopsies taken from human patients with breast cancer bone metastases which validates the importance of these findings [76, 77, 79, 80, 85]. The effect of osteolytic lesions on skeletal muscle was further supported by the finding that a tenfold larger inoculum of breast cancer cells into the mammary fat pad of mice did not cause skeletal muscle weakness, increased TGFβ signaling, oxidative stress, or RyR1 calcium leak in muscle [76]. In a direct assessment of the effect of TGFβ on muscle, contractility was decreased in mice that were exposed to recombinant TGFβ directly into the lower limb [86]. Increased TGFβ signaling in skeletal muscle also has the potential to impact function by inhibiting satellite cell activation. Satellite cells are the myocyte precursors that are activated for muscle repair and growth. TGFβ also impairs myocyte differentiation [87, 88] and is associated with skeletal muscle weakness in several models of muscular dystrophy [89, 90].

Other factors stored in bone may also impact skeletal muscle function. Activin receptor type 2B, ActRIIB, mediates signaling from activin, myostatin and growth and differentiation factor 11 (GDF-11) and has been shown to be a critical regulator of muscle mass [91]. Pharmacological blockade of ActRIIB using a soluble decoy receptor (ACVR2B/Fc) prevents muscle wasting, induces muscle satellite cell recruitment and differentiation, and prolongs survival in mouse models of muscle wasting [92]. In addition, ACVR2B/Fc improves muscle function in a Duchenne muscular dystrophy model (mdx mice) [93]. A caveat with these experiments is that it is not possible to determine the individual effects of blocking activin, myostatin, or GDF-11 due to receptor usage overlap. In contrast to negative effects on skeletal muscle from TGFβ, activin, and myostatin signaling, skeletal muscle hypertrophy has been observed following treatment of mice with IGF-1 and BMP-2 signaling [94–96]. IGF-1 causes myogenesis (proliferation and differentiation), and BMP causes muscle hypertrophy [94, 97].

In addition to factors released from the mineralized bone matrix during development of osteolytic lesions due to bone metastases, other factors that affect are capable of impacting skeletal muscle in cancer patients. Serum 25-hydroxyvitamin D levels are often low in breast cancer patients with bone metastases and who are placed on bisphosphonate therapy [98]. Mice with vitamin D receptor knockout (VDRKO) mice exhibit decreased skeletal muscle function measured using a forced swim test and hanging screen test [99, 100]. In humans, bone mineralization defects (rickets and osteomalacia) are often associated with muscle weakness. Patients showed reduced functional output in a timed up and go assay, 6-minute walk test, and stair climbing test [101, 102]. Myopathies due to vitamin D deficiency are complicated by presence of calcium and phosphate deficiencies that could complicate muscle functional assessments [64].

4   Chemotherapy-Induced Bone Loss and Muscle Weakness

Bone loss and muscle weakness are significant sequelae of cancers metastatic to bone as well as certain cancer chemotherapies, so-called cancer treatment-induced bone loss (CTIBL) [103, 104]. Overall cancer survivorship has increased in recent decades for several cancer types. Recently it was reported that there are currently over 8.1 female cancer survivors [105] and that breast cancer survivors make up greater than 40% of these survivors. Breast cancer survivors are among the highest risk group for developing bone metastases. In addition to the impact on bone from metastases, the choice of chemotherapy can lead to loss of bone mass with long-term consequences [106, 107]. In addition to acute effects on muscle and bone, chemotherapy causes chronic muscle weakness and exercise intolerance that can resolve quickly or last for years following cancer remission [108, 109]. The reduction in bone quality can be further exacerbated by inactivity that is often associated with cancer patients. This leads to a cycle of immobility, causing increased musculoskeletal impacts of reduced bone and loss of skeletal muscle mass. In severe cases, this can reduce treatment options in cancer patients, which impacts survival.

Current anti-cancer therapies for primary and advanced breast cancer include hormonal and non-hormonal depending on the tumor type. Many of these have the potential to cause bone loss [110]. These therapies include endocrine strategies for breast cancer which mitigate the effects of estrogen, and cytotoxic drugs such as platinum-derived compounds (cisplatin, carboplatin), alkylating agents (ifosfamide, cyclophosphamide, doxorubicin), anti-metabolites (methotrexate), and glucocorticoids. Other interventions for breast cancer, such as radiation therapy, and oophorectomy also result in bone loss [104], leading to osteopenia, osteoporosis, and increased risk of fractures and mortality [110].

There is much current research aimed at understanding CTIBL and best approaches to reduce, prevent, or reverse bone loss. Much of this work is also aimed at understanding changes in skeletal muscle mass and function. Anthracycline (doxorubicin), platinum-containing agents (cisplatin, carboplatin), and combination chemotherapy (Folfiri, 5-fluorouracil, leucovorin, and irinotecan) have been used to study musculoskeletal changes. Cisplatin has been shown to cause skeletal muscle atrophy due to activation of NF-κB and activation of the ubiquitin proteosomal system (UPS) [111]. These studies have revealed that these agents cause significant reduction in bone volume and muscle weakness [112–119].

Muscle wasting is a commonly observed phenomenon in the setting of cancer [120], but it is difficult to determine the etiology of this; immobility, chemotherapy, cancer progression. In addition to loss of muscle mass, patients experience muscle weakness and this is an equally important sequelae. Skeletal muscle function is gaining attention from clinicians as they are beginning to assess physical function and activity of cancer patients. Breast cancer patients have been reported to have impaired muscle function that affects quality of life [121]. In fact, the functional capacity of breast cancer patients has recently been determined using a stationary bicycle exercise to measure power output with results showing that loss of skeletal muscle mass was independent of loss of muscle mass [122]. These data support further clinical studies to determine the extent of muscle impairment.

5 Preventing Bone Loss Improve Muscle Function

Bone-muscle crosstalk plays an important role in the musculoskeletal system during development and also in disease. Bone metastases represent a unique insult to the musculoskeletal system, and the clinical aim is to prevent bone loss at the same time as treatment of the cancer (hormonal, cytotoxic, targeted therapy). Much recent work has shown that bone resorption can lead to skeletal muscle weakness and that therapies aimed at preventing bone loss improve muscle function [76, 77, 114, 117, 123].

Bone loss due to breast cancer bone metastases causes a cascade of effects in skeletal muscle in response to excess TGFβ signaling as described above [76]. In this study, muscle function was improved by using a bone anti-resorptive bisphosphonate strategy. Bisphosphonates are a class of drugs that prevent loss of bone density in osteoporosis and bone metastases. Bisphosphonates have a high-affinity mineralized bone matrix and are released during osteoclast-mediated bone resorption and cause osteoclast apoptosis, thus shutting down bone resorption [124]. In mice with breast cancer bone metastases, preventing bone resorption using the FDA-approved bisphosphonate zoledronic acid, prevented increased TGFβ signaling and effectively shutting down the signaling cascade at its origin in bone and improving skeletal muscle function [76]. More recently, another approach to preventing bone loss was employed in a mouse model of breast cancer bone metastases with very similar results. Sclerostin, an inhibitor of the canonical Wnt signaling pathway required for osteoblastogenesis, is expressed by metastatic breast cancer cells in bone [123]. Recently a fully humanized anti-sclerostin antibody (romosozumab) was approved by FDA to treat bone loss [125]. In mice with breast cancer bone metastases, a mouse anti-sclerostin antibody (setrusumab) prevented bone loss and improved skeletal muscle function [123].

In addition to bone metastases, chemotherapy alone can also cause reduced bone mass and skeletal muscle weakness. When chemotherapy is known to cause bone loss, a bone anti-resorptive is also commonly prescribed (bisphosphonate or denosumab) to reduce risk of fracture from falls [103, 126]. The role that bone-muscle crosstalk plays in the skeletal muscle weakness of chemotherapy patients has not been studied until recently. Studies of the platinum-containing chemotherapies, cisplatin and carboplatin have now been used to investigate strategies to improve skeletal muscle function in preclinical models of CTIBL [114, 117]. In mice with breast cancer bone metastases, carboplatin does have the desired anti-tumor effects of reduced tumor burden in bone, but mice still exhibit reduced skeletal muscle function [113]. In clinical practice, however, chemotherapy would be given in combination with a bone anti-resorptive. Therefore the effects of zoledronic acid on the musculoskeletal system of mice treated with cisplatin and carboplatin were investigated. Mice were treated with zoledronic acid and either cisplatin or carboplatin at the same time in order to mimic the clinical scenario of a cancer patient. In these studies, zoledronic acid was able to ameliorate skeletal muscle weakness in mice treated with cisplatin or carboplatin and these mice exhibited significantly reduced bone loss. These are exciting and entirely novel data showing that zoledronic acid is able to improve contractility in skeletal muscle using preclinical models of bone loss [114, 117]. These data suggest that preventing bone resorption, necessary to reduce fractures from falls in cancer patients, may have important positive effects beyond bone, that is improved muscle function and indeed overall musculoskeletal health.

6 Conclusions and Future Perspectives

Erosion of musculoskeletal function in advanced cancer patients is a major clinical concern that not only impacts quality of life, but also limits treatment strategies and can increase mortality. Insults to the musculoskeletal system come from: (1) direct cell-to-cell contacts between tumor cells and the bone, (2) endocrine signaling between bone and muscle, (3) bone-derived cytokines that are released during tumor growth in bone, and (4) therapies that are intended to reduce tumor burden but also have direct effects on bone and muscle.

In preclinical models of breast cancer bone metastases, loss of bone and muscle mass occurs along with significant skeletal muscle weakness. Inhibiting TGFβ signaling (1D11) or oxidative stress generated by TGFβ signaling in muscle via Nox4 (GKT137831) represent possible strategies to mitigate this weakness. In addition, preventing bone resorption using a bisphosphonate (zoledronic acid) or anti-sclerostin antibody (romosozumab) represents another strategy to prevent bone loss from tumor growth in bone as well as chemotherapy-induced bone loss. The potential of translational studies from these targets is high. ZA and romosozumab are already FDA-approved, and GKT137831 has recently completed phase 2 clinical trials for diabetic nephropathy (trial no. NCT02010242). Anti-TGFβ therapies have been tested in many diseases, including a recent phase 2 metastatic breast cancer trial by Eli Lilly (trial no. NCT02538471). Separate from pharmacological interventions, exercise is recommended for maintenance of bone and muscle in patients undergoing treatment for cancer [127, 128] and (reviewed in [129]). Nutrition supplementation could also help mitigate negative effects of bone loss and muscle loss in patients diagnosed with cancer. Calcium and vitamin D supplements are potential strategies to prevent bone loss in individuals undergoing treatment for cancer (reviewed in [129]). Indeed, combining these novel therapeutics could also lead to improved musculoskeletal outcomes for patients and will need many more clinical studies.

References

1. Siegel, R.L., Miller, K.D., Jemal, A.: Cancer statistics, 2018. CA Cancer J. Clin. **68**(1), 7–30 (2018)
2. Guise, T.A., Mundy, G.R.: Cancer and bone. Endocr. Rev. **19**(1), 18–54 (1998)
3. Weilbaecher, K.N., Guise, T.A., McCauley, L.K.: Cancer to bone: a fatal attraction. Nat. Rev. Cancer. **11**(6), 411–425 (2011)
4. Kolb, A.D., Bussard, K.M.: The bone extracellular matrix as an ideal milieu for cancer cell metastases. Cancers (Basel). **11**(7) (2019)
5. Phadke, P.A., Mercer, R.R., Harms, J.F., Jia, Y., Frost, A.R., Jewell, J.L., et al.: Kinetics of metastatic breast cancer cell trafficking in bone. Clin. Cancer Res. **12**(5), 1431–1440 (2006)
6. Burr, D.B., Akkus, O.: Bone morphology and organization. In: Burr, D.B., Allen, M.R. (eds.) Basic and applied bone biology, 2nd edn. Elsevier (2014)
7. Shupp, A.B., Kolb, A.D., Bussard, K.M.: Novel techniques to study the bone-tumor microenvironment. Adv. Exp. Med. Biol. **1225**, 1–18 (2020)
8. Shupp, A.B., Kolb, A.D., Mukhopadhyay, D., Bussard, K.M.: Cancer metastases to bone: concepts, mechanisms, and interactions with bone osteoblasts. Cancers (Basel). **10**(6) (2018)
9. Brown, H.K., Ottewell, P.D., Evans, C.A., Holen, I.: Location matters: osteoblast and osteoclast distribution is modified by the presence and proximity to breast cancer cells in vivo. Clin. Exp. Metastasis. **29**(8), 927–938 (2012)
10. Wang, H., Yu, C., Gao, X., Welte, T., Muscarella, A. M., Tian, L., et al.: The osteogenic niche promotes early-stage bone colonization of disseminated breast cancer cells. Cancer Cell. **27**(2), 193–210 (2015)
11. Wang, H., Tian, L., Liu, J., Goldstein, A., Bado, I., Zhang, W., et al.: The osteogenic niche is a calcium reservoir of bone micrometastases and confers unexpected therapeutic vulnerability. Cancer Cell. **34**(5), 823–39 e7 (2018)
12. Goodenough, D.A., Goliger, J.A., Paul, D.L.: Connexins, connexons, and intercellular communication. Annu. Rev. Biochem. **65**, 475–502 (1996)
13. Lim, P.K., Bliss, S.A., Patel, S.A., Taborga, M., Dave, M.A., Gregory, L.A., et al.: Gap junction-mediated import of microRNA from bone marrow stromal cells can elicit cell cycle quiescence in breast cancer cells. Cancer Res. **71**(5), 1550–1560 (2011)
14. Hanahan, D., Weinberg, R.A.: Hallmarks of cancer: the next generation. Cell. **144**(5), 646–674 (2011)
15. Rong, Y.P., Aromolaran, A.S., Bultynck, G., Zhong, F., Li, X., McColl, K., et al.: Targeting Bcl-2-IP3 receptor interaction to reverse Bcl-2's inhibition of apoptotic calcium signals. Mol. Cell. **31**(2), 255–265 (2008)
16. Hempel, N., Trebak, M.: Crosstalk between calcium and reactive oxygen species signaling in cancer. Cell Calcium. **63**, 70–96 (2017)
17. Sethi, N., Dai, X., Winter, C.G., Kang, Y.: Tumor-derived JAGGED1 promotes osteolytic bone metastasis of breast cancer by engaging notch signaling in bone cells. Cancer Cell. **19**(2), 192–205 (2011)
18. Zheng, H., Bae, Y., Kasimir-Bauer, S., Tang, R., Chen, J., Ren, G., et al.: Therapeutic antibody targeting tumor- and osteoblastic niche-derived Jagged1 sensitizes bone metastasis to chemotherapy. Cancer Cell. **32**(6), 731–47 e6 (2017)
19. Lu, X., Mu, E., Wei, Y., Riethdorf, S., Yang, Q., Yuan, M., et al.: VCAM-1 promotes osteolytic expansion of indolent bone micrometastasis of breast cancer by engaging alpha4beta1-positive osteoclast progenitors. Cancer Cell. **20**(6), 701–714 (2011)
20. Liang, Z., Wu, T., Lou, H., Yu, X., Taichman, R.S., Lau, S.K., et al.: Inhibition of breast cancer metastasis by selective synthetic polypeptide against CXCR4. Cancer Res. **64**(12), 4302–4308 (2004)
21. Mukherjee, D., Zhao, J.: The role of chemokine receptor CXCR4 in breast cancer metastasis. Am. J. Cancer Res. **3**(1), 46–57 (2013)
22. Bendre, M.S., Montague, D.C., Peery, T., Akel, N.S., Gaddy, D., Suva, L.J.: Interleukin-8 stimulation of osteoclastogenesis and bone resorption is a mechanism for the increased osteolysis of metastatic bone disease. Bone. **33**(1), 28–37 (2003)
23. Bussard, K.M., Venzon, D.J., Mastro, A.M.: Osteoblasts are a major source of inflammatory cytokines in the tumor microenvironment of bone metastatic breast cancer. J. Cell. Biochem. **111**(5), 1138–1148 (2010)
24. Kolb, A.D., Shupp, A.B., Mukhopadhyay, D., Marini, F.C., Bussard, K.M.: Osteoblasts are "educated" by crosstalk with metastatic breast cancer cells in the bone tumor microenvironment. Breast Cancer Res. **21**(1), 31 (2019)
25. Liu, X., Cao, M., Palomares, M., Wu, X., Li, A., Yan, W., et al.: Metastatic breast cancer cells overexpress and secrete miR-218 to regulate type I collagen deposition by osteoblasts. Breast Cancer Res. **20**(1), 127 (2018)
26. Li, D., Liu, J., Guo, B., Liang, C., Dang, L., Lu, C., et al.: Osteoclast-derived exosomal miR-214-3p inhibits osteoblastic bone formation. Nat. Commun. **7**, 10872 (2016)
27. Sun, W., Zhao, C., Li, Y., Wang, L., Nie, G., Peng, J., et al.: Osteoclast-derived microRNA-containing exosomes selectively inhibit osteoblast activity. Cell Discov. **2**, 16015 (2016)
28. Xie, Y., Chen, Y., Zhang, L., Ge, W., Tang, P.: The roles of bone-derived exosomes and exosomal microRNAs in regulating bone remodelling. J. Cell. Mol. Med. **21**(5), 1033–1041 (2017)

29. Cappariello, A., Loftus, A., Muraca, M., Maurizi, A., Rucci, N., Teti, A.: Osteoblast-derived extracellular vesicles are biological tools for the delivery of active molecules to bone. J. Bone Miner. Res. **33**(3), 517–533 (2018)

30. Li, Y., Yin, P., Guo, Z., Lv, H., Deng, Y., Chen, M., et al.: Bone-derived extracellular vesicles: novel players of interorgan crosstalk. Front. Endocrinol. (Lausanne). **10**, 846 (2019)

31. Bliss, S.A., Sinha, G., Sandiford, O.A., Williams, L. M., Engelberth, D.J., Guiro, K., et al.: Mesenchymal stem cell-derived exosomes stimulate cycling quiescence and early breast cancer dormancy in bone marrow. Cancer Res. **76**(19), 5832–5844 (2016)

32. Vallabhaneni, K.C., Penfornis, P., Xing, F., Hassler, Y., Adams, K.V., Mo, Y.Y., et al.: Stromal cell extracellular vesicular cargo mediated regulation of breast cancer cell metastasis via ubiquitin conjugating enzyme E2 N pathway. Oncotarget. **8**(66), 109861–109876 (2017)

33. Qin, W., Dallas, S.L.: Exosomes and extracellular RNA in muscle and bone aging and crosstalk. Curr. Osteoporos. Rep. **17**(6), 548–559 (2019)

34. Hassan, M.Q., Maeda, Y., Taipaleenmaki, H., Zhang, W., Jafferji, M., Gordon, J.A., et al.: miR-218 directs a Wnt signaling circuit to promote differentiation of osteoblasts and osteomimicry of metastatic cancer cells. J. Biol. Chem. **287**(50), 42084–42092 (2012)

35. Cheng, P., Chen, C., He, H.B., Hu, R., Zhou, H.D., Xie, H., et al.: miR-148a regulates osteoclastogenesis by targeting V-maf musculoaponeurotic fibrosarcoma oncogene homolog B. J. Bone Miner. Res. **28**(5), 1180–1190 (2013)

36. van Wijnen, A.J., van de Peppel, J., van Leeuwen, J. P., Lian, J.B., Stein, G.S., Westendorf, J.J., et al.: MicroRNA functions in osteogenesis and dysfunctions in osteoporosis. Curr. Osteoporos. Rep. **11**(2), 72–82 (2013)

37. Hesse, E., Taipaleenmaki, H.: MicroRNAs in bone metastasis. Curr. Osteoporos. Rep. **17**(3), 122–128 (2019)

38. Taipaleenmaki, H., Browne, G., Akech, J., Zustin, J., van Wijnen, A.J., Stein, J.L., et al.: Targeting of Runx2 by miR-135 and miR-203 impairs progression of breast cancer and metastatic bone disease. Cancer Res. **75**(7), 1433–1444 (2015)

39. Krzeszinski, J.Y., Wei, W., Huynh, H., Jin, Z., Wang, X., Chang, T.C., et al.: miR-34a blocks osteoporosis and bone metastasis by inhibiting osteoclastogenesis and Tgif2. Nature. **512**(7515), 431–435 (2014)

40. Croset, M., Pantano, F., Kan, C.W.S., Bonnelye, E., Descotes, F., Alix-Panabieres, C., et al.: miRNA-30 family members inhibit breast cancer invasion, osteomimicry, and bone destruction by directly targeting multiple bone metastasis-associated genes. Cancer Res. **78**(18), 5259–5273 (2018)

41. Taipaleenmaki, H., Farina, N.H., van Wijnen, A.J., Stein, J.L., Hesse, E., Stein, G.S., et al.: Antagonizing miR-218-5p attenuates Wnt signaling and reduces metastatic bone disease of triple negative breast cancer cells. Oncotarget. **7**(48), 79032–79046 (2016)

42. Bren-Mattison, Y., Hausburg, M., Olwin, B.B.: Growth of limb muscle is dependent on skeletal-derived Indian hedgehog. Dev. Biol. **356**(2), 486–495 (2011)

43. Rauch, F., Bailey, D.A., Baxter-Jones, A., Mirwald, R., Faulkner, R.: The 'muscle-bone unit' during the pubertal growth spurt. Bone. **34**(5), 771–775 (2004)

44. Greenlund, L.J., Nair, K.S.: Sarcopenia-consequences, mechanisms, and potential therapies. Mech. Ageing Dev. **124**(3), 287–299 (2003)

45. Steensberg, A., van Hall, G., Osada, T., Sacchetti, M., Saltin, B., Klarlund, P.B.: Production of interleukin-6 in contracting human skeletal muscles can account for the exercise-induced increase in plasma interleukin-6. J. Physiol. **529**(Pt 1), 237–242 (2000)

46. Serrano, A.L., Baeza-Raja, B., Perdiguero, E., Jardi, M., Munoz-Canoves, P.: Interleukin-6 is an essential regulator of satellite cell-mediated skeletal muscle hypertrophy. Cell Metab. **7**(1), 33–44 (2008)

47. The role of exercise-induced myokines in muscle homeostasis and the defense against chronic diseases, (2010)

48. Bostrom, P., Wu, J., Jedrychowski, M.P., Korde, A., Ye, L., Lo, J.C., et al.: A PGC1-alpha-dependent myokine that drives brown-fat-like development of white fat and thermogenesis. Nature. **481**(7382), 463–468 (2012)

49. Kir, S., White, J.P., Kleiner, S., Kazak, L., Cohen, P., Baracos, V.E., et al.: Tumour-derived PTH-related protein triggers adipose tissue browning and cancer cachexia. Nature. **513**(7516), 100–104 (2014)

50. Petruzzelli, M., Schweiger, M., Schreiber, R., Campos-Olivas, R., Tsoli, M., Allen, J., et al.: A switch from white to brown fat increases energy expenditure in cancer-associated cachexia. Cell Metab. **20**(3), 433–447 (2014)

51. Han, J., Meng, Q., Shen, L., Wu, G.: Interleukin-6 induces fat loss in cancer cachexia by promoting white adipose tissue lipolysis and browning. Lipids Health Dis. **17**(1), 14 (2018)

52. McPherron, A.C., Lawler, A.M., Lee, S.J.: Regulation of skeletal muscle mass in mice by a new TGF-beta superfamily member. Nature. **387**(6628), 83–90 (1997)

53. McFarlane, C., Plummer, E., Thomas, M., Hennebry, A., Ashby, M., Ling, N., et al.: Myostatin induces cachexia by activating the ubiquitin proteolytic system through an NF-kappaB-independent, FoxO1-dependent mechanism. J. Cell. Physiol. **209**(2), 501–514 (2006)

54. DiGirolamo, D.J., Kiel, D.P., Esser, K.A.: Bone and skeletal muscle: neighbors with close ties. J. Bone Miner. Res. **28**(7), 1509–1518 (2013)

55. Kitase, Y., Vallejo, J.A., Gutheil, W., Vemula, H., Jahn, K., Yi, J., et al.: beta-aminoisobutyric acid,

l-BAIBA, is a muscle-derived osteocyte survival factor. Cell Rep. **22**(6), 1531–1544 (2018)

56. Liang, H., Pun, S., Wronski, T.J.: Bone anabolic effects of basic fibroblast growth factor in ovariectomized rats. Endocrinology. **140**(12), 5780–5788 (1999)

57. Yakar, S., Rosen, C.J., Beamer, W.G., Ackert-Bicknell, C.L., Wu, Y., Liu, J.L., et al.: Circulating levels of IGF-1 directly regulate bone growth and density. J. Clin. Invest. **110**(6), 771–781 (2002)

58. Li, X., Zhou, Z.Y., Zhang, Y.Y., Yang, H.L.: IL-6 contributes to the defective osteogenesis of bone marrow stromal cells from the vertebral body of the glucocorticoid-induced osteoporotic mouse. PLoS One. **11**(4), e0154677 (2016)

59. Kellum, E., Starr, H., Arounleut, P., Immel, D., Fulzele, S., Wenger, K., et al.: Myostatin (GDF-8) deficiency increases fracture callus size, Sox-5 expression, and callus bone volume. Bone. **44**(1), 17–23 (2009)

60. Wallner, C., Jaurich, H., Wagner, J.M., Becerikli, M., Harati, K., Dadras, M., et al.: Inhibition of GDF8 (Myostatin) accelerates bone regeneration in diabetes mellitus type 2. Sci. Rep. **7**(1), 9878 (2017)

61. Kaji, H.: Effects of myokines on bone. Bonekey Rep. **5**, 826 (2016)

62. Hamrick, M.W., Samaddar, T., Pennington, C., McCormick, J.: Increased muscle mass with myostatin deficiency improves gains in bone strength with exercise. J. Bone Miner. Res. **21**(3), 477–483 (2006)

63. Liu, S., Zhou, J., Tang, W., Jiang, X., Rowe, D.W., Quarles, L.D.: Pathogenic role of Fgf23 in Hyp mice. Am. J. Physiol. Endocrinol. Metab. **291**(1), E38–E49 (2006)

64. Aono, Y., Hasegawa, H., Yamazaki, Y., Shimada, T., Fujita, T., Yamashita, T., et al.: Anti-FGF-23 neutralizing antibodies ameliorate muscle weakness and decreased spontaneous movement of Hyp mice. J. Bone Miner. Res. **26**(4), 803–810 (2011)

65. Faul, C., Amaral, A.P., Oskouei, B., Hu, M.C., Sloan, A., Isakova, T., et al.: FGF23 induces left ventricular hypertrophy. J. Clin. Invest. **121**(11), 4393–4408 (2011)

66. Mera, P., Laue, K., Ferron, M., Confavreux, C., Wei, J., Galan-Diez, M., et al.: Osteocalcin signaling in myofibers is necessary and sufficient for optimum adaptation to exercise. Cell Metab. **23**(6), 1078–1092 (2016)

67. Mera, P., Laue, K., Wei, J.W., Berger, J.M., Karsenty, G.: Osteocalcin is necessary and sufficient to maintain muscle mass in older mice. Mol. Metab. **5**(10), 1042–1047 (2016)

68. Sakai, R., Eto, Y.: Involvement of activin in the regulation of bone metabolism. Mol. Cell. Endocrinol. **180**(1–2), 183–188 (2001)

69. Wildemann, B., Kadow-Romacker, A., Haas, N.P., Schmidmaier, G.: Quantification of various growth factors in different demineralized bone matrix

preparations. J. Biomed. Mater. Res. A. **81**(2), 437–442 (2007)

70. Bonewald, L.F., Mundy, G.R.: Role of transforming growth factor-beta in bone remodeling. Clin. Orthop. Relat. Res. **250**, 261–276 (1990)

71. Dallas, S.L., Rosser, J.L., Mundy, G.R., Bonewald, L.F.: Proteolysis of latent transforming growth factor-beta (TGF-beta)-binding protein-1 by osteoclasts. A cellular mechanism for release of TGF-beta from bone matrix. J. Biol. Chem. **277**(24), 21352–21360 (2002)

72. Kang, Y., He, W., Tulley, S., Gupta, G.P., Serganova, I., Chen, C.R., et al.: Breast cancer bone metastasis mediated by the Smad tumor suppressor pathway. Proc. Natl. Acad. Sci. USA. **102**(39), 13909–13914 (2005)

73. Kang, Y., Siegel, P.M., Shu, W., Drobnjak, M., Kakonen, S.M., Cordon-Cardo, C., et al.: A multi-genic program mediating breast cancer metastasis to bone. Cancer Cell. **3**(6), 537–549 (2003)

74. Korpal, M., Yan, J., Lu, X., Xu, S., Lerit, D.A., Kang, Y.: Imaging transforming growth factor-beta signaling dynamics and therapeutic response in breast cancer bone metastasis. Nat. Med. **15**(8), 960–966 (2009)

75. Yin, J.J., Selander, K., Chirgwin, J.M., Dallas, M., Grubbs, B.G., Wieser, R., et al.: TGF-beta signaling blockade inhibits PTHrP secretion by breast cancer cells and bone metastases development. J. Clin. Invest. **103**(2), 197–206 (1999)

76. Waning, D.L., Mohammad, K.S., Reiken, S., Xie, W., Andersson, D.C., John, S., et al.: Excess TGF-beta mediates muscle weakness associated with bone metastases in mice. Nat. Med. **21**(11), 1262–1271 (2015)

77. Regan, J.N., Mikesell, C., Reiken, S., Xu, H., Marks, A.R., Mohammad, K.S., et al.: Osteolytic breast cancer causes skeletal muscle weakness in an immunocompetent syngeneic mouse model. Front. Endocrinol. (Lausanne). **8**, 358 (2017)

78. Santulli, G., Lewis, D.R., Marks, A.R.: Physiology and pathophysiology of excitation-contraction coupling: the functional role of ryanodine receptor. J. Muscle Res. Cell Motil. **38**(1), 37–45 (2017)

79. Andersson, D.C., Betzenhauser, M.J., Reiken, S., Meli, A.C., Umanskaya, A., Xie, W., et al.: Ryanodine receptor oxidation causes intracellular calcium leak and muscle weakness in aging. Cell Metab. **14**(2), 196–207 (2011)

80. Bellinger, A.M., Reiken, S., Carlson, C., Mongillo, M., Liu, X., Rothman, L., et al.: Hypernitrosylated ryanodine receptor calcium release channels are leaky in dystrophic muscle. Nat. Med. **15**(3), 325–330 (2009)

81. Ago, T., Kitazono, T., Ooboshi, H., Iyama, T., Han, Y.H., Takada, J., et al.: Nox4 as the major catalytic component of an endothelial NAD(P)H oxidase. Circulation. **109**(2), 227–233 (2004)

82. Carmona-Cuenca, I., Roncero, C., Sancho, P., Caja, L., Fausto, N., Fernandez, M., et al.: Upregulation of the NADPH oxidase NOX4 by TGF-beta in hepatocytes is required for its pro-apoptotic activity. J. Hepatol. **49**(6), 965–976 (2008)

83. Nyman, J.S., Merkel, A.R., Uppuganti, S., Nayak, B., Rowland, B., Makowski, A.J., et al.: Combined treatment with a transforming growth factor beta inhibitor (1D11) and bortezomib improves bone architecture in a mouse model of myeloma-induced bone disease. Bone. **91**, 81–91 (2016)

84. Dunn, L.K., Mohammad, K.S., Fournier, P.G., McKenna, C.R., Davis, H.W., Niewolna, M., et al.: Hypoxia and TGF-beta drive breast cancer bone metastases through parallel signaling pathways in tumor cells and the bone microenvironment. PLoS One. **4**(9), e6896 (2009)

85. Marx, S.O., Reiken, S., Hisamatsu, Y., Jayaraman, T., Burkhoff, D., Rosemblit, N., et al.: PKA phosphorylation dissociates FKBP12.6 from the calcium release channel (ryanodine receptor): defective regulation in failing hearts. Cell. **101**(4), 365–376 (2000)

86. Mendias, C.L., Gumucio, J.P., Davis, M.E., Bromley, C.W., Davis, C.S., Brooks, S.V.: Transforming growth factor-beta induces skeletal muscle atrophy and fibrosis through the induction of atrogin-1 and scleraxis. Muscle Nerve. **45**(1), 55–59 (2012)

87. Allen, R.E., Boxhorn, L.K.: Inhibition of skeletal muscle satellite cell differentiation by transforming growth factor-beta. J. Cell. Physiol. **133**(3), 567–572 (1987)

88. Allen, R.E., Boxhorn, L.K.: Regulation of skeletal muscle satellite cell proliferation and differentiation by transforming growth factor-beta, insulin-like growth factor I, and fibroblast growth factor. J. Cell. Physiol. **138**(2), 311–315 (1989)

89. Chen, Y.W., Nagaraju, K., Bakay, M., McIntyre, O., Rawat, R., Shi, R., et al.: Early onset of inflammation and later involvement of TGFbeta in Duchenne muscular dystrophy. Neurology. **65**(6), 826–834 (2005)

90. Kollias, H.D., McDermott, J.C.: Transforming growth factor-beta and myostatin signaling in skeletal muscle. J. Appl. Physiol. 2008, **104**(3), 579–587 (1985)

91. Lee, S.J., Reed, L.A., Davies, M.V., Girgenrath, S., Goad, M.E., Tomkinson, K.N., et al.: Regulation of muscle growth by multiple ligands signaling through activin type II receptors. Proc. Natl. Acad. Sci. USA. **102**(50), 18117–18122 (2005)

92. Zhou, X., Wang, J.L., Lu, J., Song, Y., Kwak, K.S., Jiao, Q., et al.: Reversal of cancer cachexia and muscle wasting by ActRIIB antagonism leads to prolonged survival. Cell. **142**(4), 531–543 (2010)

93. Pistilli, E.E., Bogdanovich, S., Goncalves, M.D., Ahima, R.S., Lachey, J., Seehra, J., et al.: Targeting the activin type IIB receptor to improve muscle mass and function in the mdx mouse model of Duchenne muscular dystrophy. Am. J. Pathol. **178**(3), 1287–1297 (2011)

94. Sartori, R., Schirwis, E., Blaauw, B., Bortolanza, S., Zhao, J., Enzo, E., et al.: BMP signaling controls muscle mass. Nat. Genet. **45**(11), 1309–1318 (2013)

95. Serrano, A.L., Munoz-Canoves, P.: Regulation and dysregulation of fibrosis in skeletal muscle. Exp. Cell Res. **316**(18), 3050–3058 (2010)

96. Schiaffino, S., Mammucari, C.: Regulation of skeletal muscle growth by the IGF1-Akt/PKB pathway: insights from genetic models. Skelet. Muscle. **1**(1), 4 (2011)

97. Florini, J.R., Ewton, D.Z., Coolican, S.A.: Growth hormone and the insulin-like growth factor system in myogenesis. Endocr. Rev. **17**(5), 481–517 (1996)

98. Wang-Gillam, A., Miles, D.A., Hutchins, L.F.: Evaluation of vitamin D deficiency in breast cancer patients on bisphosphonates. Oncologist. **13**(7), 821–827 (2008)

99. Burne, T.H., Johnston, A.N., McGrath, J.J., Mackay-Sim, A.: Swimming behaviour and post-swimming activity in Vitamin D receptor knockout mice. Brain Res. Bull. **69**(1), 74–78 (2006)

100. Kalueff, A.V., Lou, Y.R., Laaksi, I., Tuohimaa, P.: Impaired motor performance in mice lacking neurosteroid vitamin D receptors. Brain Res. Bull. **64**(1), 25–29 (2004)

101. Russell, J.A.: Osteomalacic myopathy. Muscle Nerve. **17**(6), 578–580 (1994)

102. Schott, G.D., Wills, M.R.: Muscle weakness in osteomalacia. Lancet. **1**(7960), 626–629 (1976)

103. Garg, A., Leitzel, K., Ali, S., Lipton, A.: Antiresorptive therapy in the management of cancer treatment-induced bone loss. Curr. Osteoporos. Rep. **13**(2), 73–77 (2015)

104. Guise, T.A.: Bone loss and fracture risk associated with cancer therapy. Oncologist. **11**(10), 1121–1131 (2006)

105. Miller, K.D., Siegel, R.L., Lin, C.C., Mariotto, A.B., Kramer, J.L., Rowland, J.H., et al.: Cancer treatment and survivorship statistics, 2016. CA Cancer J. Clin. **66**(4), 271–289 (2016)

106. Jensen, A.O., Jacobsen, J.B., Norgaard, M., Yong, M., Fryzek, J.P., Sorensen, H.T.: Incidence of bone metastases and skeletal-related events in breast cancer patients: a population-based cohort study in Denmark. BMC Cancer. **11**, 29 (2011)

107. Saad, F., Clarke, N., Colombel, M.: Natural history and treatment of bone complications in prostate cancer. Eur. Urol. **49**(3), 429–440 (2006)

108. Nicolson, G.L., Conklin, K.A.: Reversing mitochondrial dysfunction, fatigue and the adverse effects of chemotherapy of metastatic disease by molecular replacement therapy. Clin. Exp. Metastasis. **25**(2), 161–169 (2008)

109. Gilliam, L.A., St Clair, D.K.: Chemotherapy-induced weakness and fatigue in skeletal muscle: the role of oxidative stress. Antioxid. Redox Signal. **15**(9), 2543–2563 (2011)

110. D'Oronzo, S., Stucci, S., Tucci, M., Silvestris, F.: Cancer treatment-induced bone loss (CTIBL):

pathogenesis and clinical implications. Cancer Treat. Rev. **41**(9), 798–808 (2015)

111. Damrauer, J.S., Stadler, M.E., Acharyya, S., Baldwin, A.S., Couch, M.E., Guttridge, D.C.: Chemotherapy-induced muscle wasting: association with NF-kappaB and cancer cachexia. Eur. J. Transl. Myol. **28**(2), 7590 (2018)

112. Rana, T., Chakrabarti, A., Freeman, M., Biswas, S.: Doxorubicin-mediated bone loss in breast cancer bone metastases is driven by an interplay between oxidative stress and induction of TGFbeta. PLoS One. **8**(10), e78043 (2013)

113. Hain, B.A., Xu, H., Wilcox, J.R., Mutua, D., Waning, D.L.: Chemotherapy-induced loss of bone and muscle mass in a mouse model of breast cancer bone metastases and cachexia. JCSM Rapid. Communications. **2**(1) (2019)

114. Hain, B.A., Jude, B., Xu, H., Smuin, D.M., Fox, E.J., Elfar, J.C., et al.: Zoledronic acid improves muscle function in healthy mice treated with chemotherapy. J. Bone Miner. Res. **35**(2), 368–381 (2020)

115. Barreto, R., Kitase, Y., Matsumoto, T., Pin, F., Colston, K.C., Couch, K.E., et al.: ACVR2B/Fc counteracts chemotherapy-induced loss of muscle and bone mass. Sci. Rep. **7**(1), 14470 (2017)

116. Barreto, R., Waning, D.L., Gao, H., Liu, Y., Zimmers, T.A., Bonetto, A.: Chemotherapy-related cachexia is associated with mitochondrial depletion and the activation of ERK1/2 and p38 MAPKs. Oncotarget. **7**(28), 43442–43460 (2016)

117. Essex, A.L., Pin, F., Huot, J.R., Bonewald, L.F., Plotkin, L.I., Bonetto, A.: Bisphosphonate treatment ameliorates chemotherapy-induced bone and muscle abnormalities in young mice. Front. Endocrinol. (Lausanne). **10**, 809 (2019)

118. Gilliam LA, Ferreira LF, Bruton JD, Moylan JS, Westerblad H, St Clair DK, et al. Doxorubicin acts through tumor necrosis factor receptor subtype 1 to cause dysfunction of murine skeletal muscle. J. Appl. Physiol. 2009, **107**(6), 1935–1942 (1985)

119. Gilliam, L.A., Moylan, J.S., Callahan, L.A., Sumandea, M.P., Reid, M.B.: Doxorubicin causes diaphragm weakness in murine models of cancer chemotherapy. Muscle Nerve. **43**(1), 94–102 (2011)

120. von Haehling, S., Morley, J.E., Anker, S.D.: An overview of sarcopenia: facts and numbers on prevalence and clinical impact. J. Cachexia. Sarcopenia Muscle. **1**(2), 129–133 (2010)

121. Christensen, J.F., Jones, L.W., Andersen, J.L., Daugaard, G., Rorth, M., Hojman, P.: Muscle dysfunction in cancer patients. Ann. Oncol. **25**(5), 947–958 (2014)

122. Ballinger, T.J., Reddy, A., Althouse, S.K., Nelson, E. M., Miller, K.D., Sledge, J.S.: Impact of primary breast cancer therapy on energetic capacity and body composition. Breast Cancer Res. Treat. (2018)

123. Hesse, E., Schroder, S., Brandt, D., Pamperin, J., Saito, H., Taipaleenmaki, H.: Sclerostin inhibition alleviates breast cancer-induced bone metastases and muscle weakness. JCI. Insight. **5** (2019)

124. Drake, M.T., Clarke, B.L., Khosla, S.: Bisphosphonates: mechanism of action and role in clinical practice. Mayo Clin. Proc. **83**(9), 1032–1045 (2008)

125. Bandeira, L., Lewiecki, E.M., Bilezikian, J.P.: Romosozumab for the treatment of osteoporosis. Expert. Opin. Biol. Ther. **17**(2), 255–263 (2017)

126. Lipton, A., Uzzo, R., Amato, R.J., Ellis, G.K., Hakimian, B., Roodman, G.D., et al.: The science and practice of bone health in oncology: managing bone loss and metastasis in patients with solid tumors. J. Natl. Compr. Cancer Netw. **7**(Suppl 7), S1–29 (2009); quiz S30

127. Swenson, K.K., Henly, S.J., Shapiro, A.C., Schroeder, L.M.: Interventions to prevent loss of bone mineral density in women receiving chemotherapy for breast cancer. Clin. J. Oncol. Nurs. **9**(2), 177–184 (2005)

128. Kohrt, W.M., Bloomfield, S.A., Little, K.D., Nelson, M.E., Yingling, V.R., American College of Sports M: American College of Sports Medicine Position Stand: physical activity and bone health. Med. Sci. Sports Exerc. **36**(11), 1985–1996 (2004)

129. Sturgeon, K.M., Mathis, K.M., Rogers, C.J., Schmitz, K.H., Waning, D.L.: Cancer- and chemotherapy-induced musculoskeletal degradation. JBMR Plus. **3**(3), e10187 (2019)

Part IV

Preventing and Targeting Cachexia in Cancer

New Developments in Targeting Cancer Cachexia

Janice Miller, Michael I. Ramage, and Richard J. E. Skipworth

Abstract

Within the cancer population, 50–80% of patients will develop cachexia, impacting negatively on their ability to tolerate or gain benefit from either curative or palliative treatment. To date, although collaborative management guidelines have been developed, there are no internationally standardised management programmes used in the clinical forum for patients with cancer cachexia. Furthermore, current available treatment strategies have limited efficacy. This chapter considers recent developments in the ability to target cachexia. Such "targeting" comes in two key forms: firstly, the ability to target and recognise new patients with, or at risk of, developing cancer cachexia. Some of the new developments in this field of study relate not just to improvements in patient targeting, but also a better understanding of the inherent pitfalls in this process. The second aspect of targeting relates to the identification of novel therapeutic biological targets for further clinical investigation.

Learning Objectives

1. Targeting patients with, or at risk of, developing cancer cachexia through:
   a. Diagnostic, screening, staging, severity, and phase criteria
   b. Patient-specific factors: sexual dimorphism and genetics
   c. Body composition analysis using CT, magnetic resonance imaging (MRI) and contrast-enhanced ultrasound (CEUS)
2. Identification of novel therapeutic biological targets through:
   a. Pathophysiological mechanisms—including imbalance between protein synthesis and degradation, autophagy and fat-muscle crosstalk
   b. Mediators—including parathyroid-hormone-related-protein (PTHrP), TNF-related weak inducer of apoptosis (TWEAK), Angiotensin-II (Ang-II), ZIP 14 and the emerging role of microRNAs (miRNA)
   c. Potential therapeutic options, including recent clinical trials of selective androgen receptor modulators (SARMs) and ghrelin agonists (Anamorelin).
   d. Future targets including tumour-associated macrophages, heat shock proteins and small bowel microbiota

J. Miller · M. I. Ramage · R. J. E. Skipworth (✉)
Clinical Surgery, Royal Infirmary of Edinburgh, Edinburgh, UK
e-mail: richard.skipworth@nhslothian.scot.nhs.uk

© Springer Nature Switzerland AG 2022
S. Acharyya (ed.), *The Systemic Effects of Advanced Cancer*,
https://doi.org/10.1007/978-3-031-09518-4_10

1 Targeting Patients with Cancer Cachexia

1.1 Diagnostic, Staging, Severity, and Phase Criteria

1.1.1 Diagnostic Definition

Within the cancer population, between 50–80% of patients will develop the wasting syndrome of cachexia [1, 2], with resulting negative impact on their ability to withstand either radical or palliative treatment [3]. The term *cachexia* originates from the Greek words "*kakos*" meaning "bad" and "*-hexis*" meaning habit or state. Despite the historical context of cancer cachexia (CC), it has only been in the last 11 years that a consensus definition of the condition was reached [4]. In the 2011 consensus statement on cachexia, the authors defined the condition as a "multifactorial syndrome characterised by an ongoing loss of skeletal muscle mass (with or without loss of fat mass) that cannot be fully reversed by conventional nutritional support and leads to progressive functional impairment" [4]. As the authors state, "The definition emphasises the key role of muscle loss in the development of frailty, ADL disability, and impaired quality and duration of life".

The diagnosis of cancer cachexia requires the fulfilment of one of three criteria:

- Weight loss >5% in the past 6 months without starvation and/or
- Weight loss >2% and BMI <20 and/or
- Weight loss >2% and sarcopenia (low muscle mass as evidenced by computed tomography [CT] or Dual X-ray Absorptiometry (DXA)

Using these criteria, the prevalence(s) of cachexia in an unselected cancer population was 51% in inpatients and 22% in outpatients [5]. Furthermore, this definition identified patients with reduced physical function and quality of life (QoL), and a more rapid rate of decline in these parameters over longitudinal assessment [2, 6]. These criteria were investigated and validated by Blum et al. in 2014 [7]. The validation showed that patients who fulfilled cachexia criteria had worsened performance status and

shorter survival compared with those who did not. More recently, however, a Belgian study [8] compared the Fearon et al. definition to an earlier cachexia definition [9], and found that the earlier criteria showed greater discrimination with regards to overall survival when applied to the same population.

Of note, the CT cutoff criteria used in the 2011 definition were developed by converting DXA cut-points to CT using work done by Mourtzakis et al. [10]; however, these have since been superseded by sex and body mass index (BMI)-stratified criteria developed by Martin et al. [11].

1.1.2 Phase of Cachexia

The cachexia syndrome, as described in the 2011 consensus statement, exists as a continuum extending from *no cachexia*, through *pre-cachexia*, (a potential early stage of cachexia characterised by symptomatology and metabolic disturbance preceding severe weight loss) into *cachexia*, and ultimately, *refractory cachexia* (patients in whom ongoing cancer treatment is futile, and who have a life expectancy of <3 months) and death [4]. Severity stratification as described above can be useful for prognostication and discussion with patients and carers. However, the use of the scale requires strict vigilance from attending physicians. At the severe end of the scale, the accurate diagnosis of refractory cachexia is reliant on monitoring the patient's overall clinical condition and the specific response of their tumour to treatment. At the early end of the scale, the difference between "no cachexia" and "pre-cachexia" is subtle and can be difficult to distinguish. Early intervention is advocated in the consensus statement, and to this end, the future identification of early biomarkers is important to aid clinicians in the diagnosis of "pre-cachexia".

Both the phase and the stage of cachexia can and should be used to identify and stratify patients for inclusion in randomised controlled trials (RCTs). The future clinical priority will be to identify patients with pre-cachexia in order to institute prophylactic interventions. However, previous trials in CC have had a tendency to recruit patients with refractory cachexia, and have been plagued by patient attrition rates as high as 50% [12]. This past habit is understandable as patients with refractory cachexia are easier to identify than those with pre-cachexia. Furthermore, one can appreciate that pharmaceutical companies are attracted to the notion of a drug that is capable of reversing significant weight loss. However, as previous trials have not confirmed a drug that is currently capable of improving both nutritional and functional status, a change in tack to the prevention of deterioration, rather than the restoration of loss, is indicated in future patient targeting.

1.1.3 Complexities in Terminology

When considering loss of skeletal muscle mass in cancer cachexia, it is important to differentiate between the generic use of the term *sarcopenia* (meaning lack of skeletal muscle) and the sarcopenia of old age (as described in the statement of the European Working Group on the Sarcopenia of Older People [EWGSOP]) [13]. The EWGSOP definition requires a reduction in muscle function as well as mass to hold true, whereas the more generic use of the term relates predominantly to muscle mass only. There are a number of other definitions of sarcopenia within the literature, as recently summarised by von Haehling et al. [14], all of which require both low muscle mass and reduced muscle function for the definition of sarcopenia to be met. The differing interpretations of the same term are adding to confusion within the field, and misunderstandings in regard to patient targeting, especially as most cancer patients are elderly, and therefore, they have multifactorial reasons for low muscle mass. Differentiating between the various interpretations may require the use of an alternative term relating solely to low muscle mass with negative clinical impact: *myopenia* has been suggested [15].

1.1.4 Screening Patients with Cachexia

Screening patients for cachexia is difficult. Ideally, patients would be picked up at an early stage before significant weight loss becomes apparent. Pre-cachexia is defined as weight loss <5%, anorexia and systemic inflammation. There is currently only one validated screening tool in existence for the assessment of cachexia, namely the CASCO [16]. The current European Society for Parenteral and Enteral Nutrition (ESPEN) advise that all patients should be screened early in the course of their cancer treatment, including identifying signs of cachexia, sarcopenia and malnutrition. All patients should undergo measures of muscle mass and systemic inflammation as well as have a measurement of resting energy expenditure. Following this, management should be instigated in the form of nutritional support and physical rehabilitation [17].

1.1.5 Cross-sectional Imaging

Within both the international consensus definition of cachexia [4] and the EWGSOP statement on age-related sarcopenia [13] are defined cutoffs below which the patient is considered to have low muscle mass. These cutoff values were determined by population studies using two standard deviations below the mean of young healthy males and females to define those with sarcopenia [18]. These cutoffs are expressed in terms familiar to the users of DXA. In choosing cross-sectional techniques to use for body composition analysis, it is important to recognise not only the clinical advantages inherent in each modality but also the limitations posed by differing techniques [19]. The original body composition analysis technique used to describe sarcopenia was dual-energy X-ray absorptiometry (DXA) [18]. This technique allows the determination of skeletal mass and lean mass (LBM), as well as fat mass. The technique correlated well with validated measures of body composition assessment such as total body potassium counting [20], which is an index of body cell mass, and relies on the composition of adipocytes being predominantly fat for accuracy. However, within recent years, CT has become the norm for assessing body composition in cancer cachexia.

Nutritional Assessment

Individual patients do not have the same body habitus at diagnosis. Demographics have changed since the original cachexia definition was developed with rates of obesity increasing worldwide

[21]. Clinically significant weight loss therefore is harder to gauge. Obesity is usually considered to have a disadvantage on survival [22], however, it has been shown to confer a survival advantage in patients with weight loss associated with heart and renal failure due to their larger energy reserve [23]. Diagnostic criteria for cachexia therefore need to include additional information beyond weight loss.

A study based on pooled recent data by Martin et al. [24] ($n = 8160$) evaluated the prognostic significance of weight loss in patients who initially had low, intermediate or high BMI. The authors showed that the patients with a lower starting BMI had a worse prognosis, with BMI and weight loss being independent predictors of survival regardless of cancer site and stage. They were therefore able to devise a cancer BMI/weight loss grading system that predicted survival. Grade 4 (the most severe) is of value in identifying individuals whose expected survival is too short to merit specific treatment plans or interventions. Finding that degree of weight loss affects survival lends credence to the suggestion that cachexia has degrees of severity, even within the distinct phases of the 2011 consensus criteria. In order to better quantify the severity of cachexia, scoring systems or reproducible cutoff values below which survival time is impacted upon are required. One technique to stage the severity of existing cachexia is the use of body composition analysis to quantify skeletal muscle mass.

Fat deposition within muscle is also important in cachexia, so-called myosteatosis. This occurs in two forms: Intramyocellular fat and intermuscular fat, although the latter is not often included in CT body composition analysis. Myosteatosis appears to modulate adipokine secretion and muscle blood flow [25]. It is therefore of importance in the aetiology of insulin resistance and type II diabetes [26]. MRI scanning has been previously used to demonstrate increased intermuscular fat in cancer patients [27]. At the molecular level, patients with CT confirmed cancer cachexia have also demonstrated increased intramyocellular lipid deposition compared with healthy controls

[28]. Its clinical significance, however, remains unclear.

CT

The advent of spiral computed tomography (CT) scanning led to it becoming an attractive measurement methodology in clinical practice due to its ready availability and reduced radiation exposure relative to older CT methodologies. Most cancer patients will have CT scans performed as part of routine staging investigations and so the use of such scans does not require further invasive testing or ionising radiation. Given the clinical increase in the use of CT, more recent work has been done to equate DXA body composition values with those more commonly found when using planimetry software to analyse CT scans [10]. This technique gives a measurement of cross-sectional area, which can then be indexed against height to give the skeletal muscle index (SMI) in cm^2/m^2, which is more readily comparable between patients. The population on which these values were based included 1473 patients with GI or respiratory tract cancer [11]. The cutoff values reported are those below which survival time was reduced in this cohort. Subsequent analysis of BMI and sex allowed the authors to account for the possibility of elevated muscle mass in the obese and in males. The study thus reveals muscle mass (stratified for BMI) thresholds as a single variable below which survival is compromised and without reference to muscle function. Importantly, however, despite the hope of using CT as a targeting methodology for trial patients, and as an outcome measure following trial intervention, CT variables do not necessarily correlate with muscle protein composition [29]. An example of CT body composition analysis is shown in Fig. 1.

Whereas, the current published literature is peppered with studies confirming the clinical relevance of low muscle mass in cancer, the prognostic significance of adiposity at the time of cancer diagnosis on survival is not clear. A study by Ebadi et al. [30] investigated the independent prognostic significance of adipose tissue in predicting mortality. Total adipose index (TATI), visceral adipose index (VATI) and

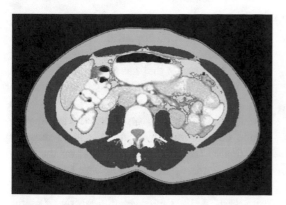

Fig. 1 CT body composition analysis. Cross-sectional CT image at L3 showing demarcation of skeletal muscle and adipose tissue using Slice-o-matic and ABACS software. Blue—subcutaneous fat, red—skeletal muscle, yellow—visceral fat, green—intramuscular fat

subcutaneous adipose index (SATI) were estimated for 1473 gastrointestinal and respiratory cancer patients. Low SATI was independently associated with increased mortality and shorter survival. In the presence of sarcopenia, however, the longest survival was observed in patients with high subcutaneous adiposity, so-called sarcopenic obesity. The relative importance of muscle mass variables compared with fat mass variables on trial inclusion and patient outcome will need to be the focus of future research in patient targeting.

MRI

Magnetic resonance imaging (MRI) is an increasingly attractive form of cross-sectional imaging, and is considered equivalent to CT as the gold standard imaging methodology for skeletal muscle mass analysis [4, 13, 31]. MRI scanning has been used in sports medicine [32] and in muscular dystrophy [33] to monitor the effect of injury or disease progression on the volume of scanned muscle. Recent advances in automation have shown this technique to be feasible in a whole-body context, allowing the definition and direct measurement of total skeletal muscle volume within a patient [34].

In contrast to the L3 method of CT scanning, which uses a single slice and algorithmic conversion to a height-based index [11], direct measurement using MRI has clear advantages. Firstly, the

entire volume of scanned muscle is measured directly rather than inferred or calculated. Secondly, there is no need for ionising radiation. Thirdly, it should be possible to use MRI to determine water and fat levels within a particular muscle, as myosteatosis has been shown to have an adverse prognostic effect on patients with pancreatic cancer [35]. Disadvantages to MRI scanning do exist, however, and include the labour-intensive nature of interpreting each scan; the cost in terms of the scanner and the staff required to run the scans; and the demand on patients—each scan takes 20–60 mins in a claustrophobic scanner environment with strict requirement for exact repositioning.

Contrast-Enhanced Ultrasound

Other imaging techniques are emerging that have been applied to explore the effect of cachexia on skeletal muscle morphology, metabolism and microcirculation. Weber et al. [36] investigated MRI, Magnetic Resonance Spectroscopy (MRS) and contrast-enhanced ultrasound (CEUS) in cachectic patients and healthy volunteers. CEUS is a relatively easy imaging technique that is underutilised in cachexia assessment. CEUS is able to quantify the skeletal muscle microcirculation by analysing the replenishment kinetics after destruction of intravenously injected microbubbles by ultrasonographic impulses [37]. It is sensitive enough to allow detection of low skeletal muscle perfusion (capillary blood flow at rest) and also allows for measurement of increased blood flow after exercise or during pathology, e.g. inflammatory myopathies.

MR spectroscopy (MRS) allows for non-invasive analysis of the chemical composition of tissues. It is able to quantify concentrations of several metabolites, in particular intramyocellular lipids and high energy phosphates, which are encountered in diseases leading to muscle degeneration [36]. In one published study of cachectic patients [36], capillary density (as determined by histology), microcirculation (measured in vivo by CEUS), and concentrations of metabolites (analysed in vivo by MRS) were the same in both patients with cachexia and healthy volunteers. Cancer cachexia was therefore associated with a loss of muscle volume but not of functionality. These data are indicative of sufficient oxidative phosphorylation potential, and oxygen availability, in the muscles of patients with cancer cachexia, suggesting therefore that cachectic muscles should respond to exercise programmes as they are apparently not irreversibly damaged.

Simple bedside ultrasound assessment of skeletal muscles is a relatively underutilised tool. It is of particular value in assessing patients in the intensive care setting and provides a non-invasive, simple and easily repeatable method. Previous studies have used this method to not only quantify initial muscle mass, but to evaluate the impact of different therapeutic strategies on wasting over time. Muscle mass measurement by ultrasonography has been shown to be reliable in patients even when oedema and fluid retention are present as well as in patients on renal replacement therapy [38]. US has also been shown to be able to demonstrate myofibre necrosis and fascial inflammation, and results correlate with a decrease in muscle fibre size and protein synthesis [39]. Different definitions of cancer cachexia that incorporate US cutoffs will be necessary for the widespread adoption of this technique in patient targeting.

1.2   Patient-Specific Factors

1.2.1   Sexual Dimorphism

Sex differences in the incidence of cancer are a critical issue in both cancer research and the development of therapeutics. Little is known, however, about the effect of gender on cachexia. Sexual dimorphism has been shown to exist in patient symptomatology, and in skeletal muscle mass, fibre type, and size in response to neoplasm. Physical function and fatigue scores, in particular, have been shown to be reduced in cachectic males compared to non-cachectic males but not in females [40]. Males with pancreatic cancer and systemic inflammation have a high prevalence of hypogonadism, and such

hypogonadal patients have worsened survival. In comparison, post-menopausal female patients have elevated rates of hyperoestrogenism, and it is these individuals that have shortened survival [41]. There is, therefore, a sex-specific relationship between the sex hormones and systemic inflammation (a key driver of cachexia) in pancreatic cancer. Stretch et al. [42] have recently investigated the influence of gender on the muscularity of cancer patients. Men were significantly more muscular. Molecular data were obtained from 41 K Agilent microarray analysis of rectus abdominus muscle and LC-MS analysis of urine in patients with cancer. This study was able to identify altered molecular pathways in men and women. Increasing muscularity was associated with JAK/STAT signalling pathways in men and growth hormone signalling pathways in women. It was negatively associated with mismatch DNA repair in men and with ER-mediated phagocytosis in women. The urinary proteome and muscle transcriptome suggested altered carbohydrate metabolism in women but not in men. Inflammation, however, was associated with low muscle mass independent of sex. These data suggest that future interventions and management of cancer cachexia would be best tackled with a sex-specific approach.

1.2.2   Genetics

Early studies that noted the association between certain single nucleotide polymorphisms (SNP) of pro-inflammatory cytokine genes and patient survival raised the intriguing possibility that SNP status may also be associated with patient development of cachexia. This supposition is supported by the observation that there is variation in the prevalence of cachexia between patients with the same type of cancer, e.g. up to 85% of patients with pancreatic cancer will present with cachexia, but 15% will not, possibly due to a differing genotype [43]. Candidate gene studies have shown that SNPs in the P-selectin, ACVR2B, ACE, LEPR and TNF genes were associated with low muscle mass in patients with cancer [44]. A small number of studies have related SNPs in a variety of pro-inflammatory cytokine genes to the presence of systemic inflammation, cachexia and poor survival in cancer. In particular the IL-6174CC polymorphism has been found to be associated with both systemic inflammation and high levels of intra-tumoural cytokines in patients with gastroesophageal cancer. The presence of systemic inflammation was also related to overall worsened survival [45]. On the contrary, the 308AA polymorphism in the TNFα gene was associated with poor outcome but did not relate to systemic inflammation or intra-tumoural cytokine levels, and the 511CC polymorphism on the IL-1β gene was not associated with tumour cytokine levels, systemic inflammation or prognosis [46, 47]. There are only a few studies that have investigated the genetic control of cachexia, but further studies in the future may allow us to target at-risk patients with a genetic predisposition to receive prophylactic nutritional intervention.

2 Identification of Novel Targets in Cachexia

The reasons why some cancers are more strongly associated with cachexia than others are not clear. It is true that some cancers are more likely to present earlier in the disease process than others, and therefore they are diagnosed before any significant weight loss is able to occur [48]. Alternative suggestions have involved the acute phase response and hypermetabolism being more common in certain cancers, therefore leading to a more rapid loss of skeletal muscle [49]. Whatever the case, this observation suggests that different tumours release either different types, or different concentrations, of cachectic mediators. Therefore, any newly identified targets will be of clinical interest, but may be disease-specific, and not of universal importance in all tumour types. Below, the authors describe mechanisms and mediators that have been of recent research interest in cancer cachexia.

2.1 Pathophysiological Mechanisms

2.1.1 Protein Metabolism

Previous studies on muscle protein synthesis and turnover have suggested that the presence of cancer significantly reduces the rate of protein synthesis, resulting in low muscle mass. Different organs have also been suggested to display differing rates of synthesis. In a study in pancreatic cancer patients, pancreatic tumour and skeletal muscle tissue demonstrated lower protein synthesis rates when compared with normal pancreatic and liver tissue [50]. The effect of feeding on rates of protein synthesis and breakdown were investigated using labelled amino acids in a study by van Dijk et al. [51]. This study showed that in pancreatic cancer patients with cachexia, basal protein turnover was elevated relative to healthy controls. Feeding appeared to improve the overall protein balance in cachectic patients through the mechanism of reduced protein breakdown, whilst protein synthetic rate remained unchanged. This contrasted with the healthy control group, in whom an equivalent relative increase in post-prandial anabolism took place, but in whom both protein synthetic and breakdown rates were increased in response to feeding. Thus, post-prandial anabolism was similar between cancer patients and controls but appeared to occur through different mechanisms. Additionally, Deutz et al. [52] found that the amino-acid and protein content of food preparations administered and ingested was important when considering muscle synthetic rates. Specifically formulated medical food with controlled protein, leucine, carbohydrate, and fat content increased anabolic rates in cancer patients in contrast to standard food preparations, suggesting a potential role for such supplements as part of a multimodal cachexia intervention.

Measurement of protein synthetic rates in previous studies has potentially been subject to confounding factors inherent in the isotope labelling process. For example infusions of labelled amino acids, specifically leucine, may increase protein synthetic rates due to the sudden increase in the presence of these amino acids if using a flooding dose technique. This is less of a problem using primed constant infusion tracer methods. However, in either technique, patients are frequently immobilised and fasted during the infusion and measurement process. This does not represent normal functioning and thus may not reflect normal protein metabolism. The main drawback to these measurements though is the short-term nature of the observation.

To counter this, MacDonald et al. [53] designed a human protocol for a method of radio-isotope labelling with heavy water to allow an analysis of protein synthetic rates, which more closely reflects the natural conditions of muscle metabolism. Following this method, the same group investigated the differences in myofibrillar protein synthesis rates between cancer patients and controls [54]. Perhaps surprisingly, the study performed did not show the expected differences in synthetic rates between the cancer and the control groups. Instead, they found a small but potentially important mismatch in protein synthetic and breakdown rates in the cancer group. In comparison with controls, in whom the rates of synthesis and breakdown were balanced, there was a 2.6% difference between the two in the cancer patients. This led the group to suggest that although the difference is small, it may be sufficient to account for the previously noted reduction in muscle mass and protein content in pancreatic cancer patients. The results of this study would suggest that the focus for therapeutic intervention in cancer cachexia should be on the reduction of protein breakdown rather than the promotion of protein synthesis.

2.1.2 Autophagy

Muscle loss in cachexia is associated with pathways dependent on ubiquitin-proteasome (UPP) and autophagy-lyosomes, resulting in increased protein breakdown [55]. Autophagy has recently been shown to be present in the pathogenesis of muscle wasting in myopathies as well as conditions such as sepsis, chronic obstructive pulmonary disease and cancer cachexia [56]. The process involves the sequestration of cytoplasm into double membrane vesicles (autophagosomes), which fuse with lysosomes, degrading the content [57]. Autophagy is highly selective and strictly regulated. All cells undergo degradation by autophagy but this is done at a basal speed. This can be accelerated or inhibited by different stimuli, e.g. starvation, where autophagy is increased to mobilise nutrients and essential amino acids. Growth of tumour is associated with a decreased availability of various nutrients. Tumour cells therefore adapt in order to increase their nutrient supply and survival. Previous studies have suggested that tumour cells can secrete substances that can accelerate autophagy in other cells in the tumour micro environment [58]. In cachectic patients with oesophageal and pancreatic cancers, the autophagy-lysosome system is induced as evidenced by the upregulation of beclin1 and Atg5—an autophagy activator and factor of phagophore formation [59].

2.1.3 Neural Innervation of Skeletal Muscle

Studies of related muscle wasting conditions have shown that pathological targeting of the neuromuscular junction (NMJ) may play a key role in the pathogenesis of cachexia. Similarly animal models of sarcopenia have suggested NMJ dysfunction and denervation to be a driver of muscle wasting. Recent detailed morphological analysis, however, has revealed that the NMJ remains stable in patients with cachexia who were suitable for curative surgery, suggesting that promotion of muscle hypertrophy using exercise and neural stimulation should remain therapeutic options for the future [60].

2.1.4 Fat-Muscle Crosstalk

Although the consensus definition of cancer cachexia [51] concentrates on the loss of lean tissue (particularly skeletal muscle) as a key identifier in the diagnosis of the condition, the loss of adipose tissue in cancer cachexia is increasingly thought to play important roles in the pathogenesis and negative clinical impact of cancer cachexia [61]. A better understanding of the physiological effects of cancer on adipose tissue would open new therapeutic avenues for treatment of cancer cachexia [4], resulting in improved outcomes following oncological treatments, better quality of life, and lengthened survival for patients.

Fat loss precedes muscle loss and is variable with respect to timing and intensity in various cancer populations. In cancer, the loss of adipose tissue is driven by lipolysis [62] rather than adipocyte apoptosis or necrosis. Mediators of lipolysis in cancer cachexia are largely unknown and currently no biomarker exists for fat wasting in cancer. Previously increased mRNA expression of zinc-α2-glycoprotein (ZAG), a proposed lipid mobilising factor, has been identified in fat samples from cachectic cancer patients; however, serum ZAG levels were unchanged from controls [63]. Microarray analysis has revealed that changes in the transcriptome of subcutaneous fat in cancer cachexia are opposite to those seen in obesity, underlining the importance of lipolysis. Visceral adipose tissue is lost more rapidly than subcutaneous adipose tissue during cachexia, suggesting differential tissue-dependent responses to the wasting process [64]. Bioinformatic analysis of mRNA expression in visceral (omental) and subcutaneous adipose depots in non-cachectic, obese endometrial cancer patients demonstrated nineteen shared biological pathways, eighteen of which were regulated in opposite directions between the fat depots [65]. Measuring changes in fat mass over time along with circulating levels of biomarkers would provide valuable information about the application of potential biomarkers throughout the disease trajectory. Comparison of healthy, weight stable and weight-losing cancer patients assessed by CT imaging should be considered in determining the prognostic ability of a biomarker of fat loss in cancer.

Recently, focus has concentrated on the concept of "fat-muscle crosstalk" in cancer cachexia [4]. Notably, genetic ablation of lipolytic pathways in adipocytes has been shown to protect against muscle mass loss in pre-clinical models of cancer cachexia [66], supporting the theory that the loss of visceral fat driven by lipolytic mechanisms is an early event in cancer cachexia and could have potential effects on skeletal muscle. The lipolytic ablation demonstrated further

the theory that the breakdown of fat precedes that of muscle loss and that signals generated during lipolysis may activate muscle proteolysis. The ablation of lipase was also associated with a lack of activation of the main pathway involved in muscle wasting, the UPP.

The infiltration of adipose tissue into muscle during the cachectic process may be another contributor to wasting. Stephens et al. [28] have demonstrated increasing intramyocellular lipid droplets in rectus abdominus muscles of cancer patients associated with increasing weight loss.

Understanding of the integrative physiology involved in cachexia may lead to the development of novel therapeutic approaches, particularly in patients suffering from sarcopenic obesity.

2.2 Mediators of Cachexia

Multiple possible mediators of cachexia have been proposed, falling into the general categories of

1. Systemic inflammation
2. Neuro-endocrine stress responses
3. Tumour products

Investigation of mediators within these categories using animal models, however, is subject to problems. Each animal model appears to have a different predominant mediator, meaning that generalisability is reduced. Large-scale human studies are lacking, and with incomplete crossover from animal models to humans, this is a hindrance to progress in the field. Additionally, (and as seen in studies of sepsis), the complexity and redundancy of mediator cascades have precluded cause and effect intervention studies. These mediators are summarised in Fig. 2.

2.2.1 PTHrP

The role of Parathyroid-Hormone (PTH) and Parathyroid-Hormone-related protein (PTHrP) in cachexia was described by Kir et al. [67] in 2014. In this paper, the team investigated the role of excessive thermogenesis and the browning of adipose tissue in cancer cachexia in a Lewis Lung Cancer (LLC) murine model. Using a

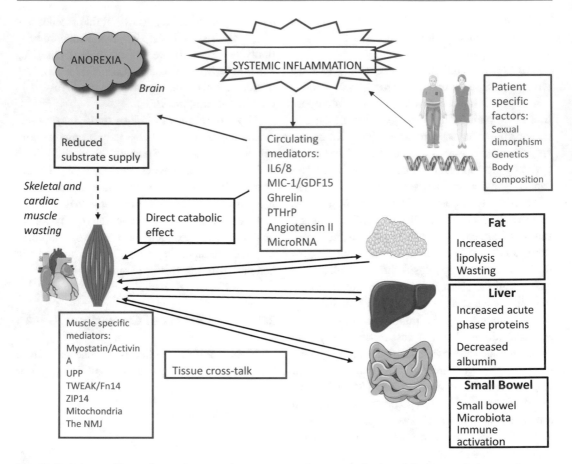

Fig. 2 Patient specific and novel targets in cancer cachexia. Targets of cancer cachexia are host and tumour specific. Differences in sex and genetics may predispose patients to cachexia, whilst some of the novel targets discussed act to stimulate excess catabolism or promote central anorectic effects

combination of cell culture and knockout methods, the group could show that inhibition of PTHrP in tumour-bearing mice markedly reduced the degree of muscle atrophy and fat browning in these mice compared to controls. Interestingly, whilst the injection of PTHrP into previously naïve tumour-bearing mice resulted in significant muscle atrophy, the injection of the protein into healthy mice did not. This led the team to conclude that whilst PTHrP has a role to play in the excessive thermogenesis and muscle atrophy of cancer cachexia in the LLC model, it must act with a co-factor as yet undetermined. The study also found a reduction in LBM in human cancer patients with increased PTHrP. This observation was unrelated to caloric intake or to systemic inflammation as measured by serum CRP concentrations. These results agree with previous work on human cancer patients [68], and the possibility of a tumour-specific co-factor requirement for cachexia development may explain why patients with primary hyperparathyroidism do not develop muscle wasting.

Following on from this initial study, the group then investigated the effect of Parathyroid-Hormone Receptor (PTHR) presence or absence in knockout mice undergoing 5/6 nephrectomy to simulate renal failure-associated cachexia [69]. They demonstrated that removal of PTHR from the adipose cells greatly reduced not only browning of the fat, but also the rate of muscle loss in these mice, despite being unable to fully deplete the receptor from the skeletal muscle. In further studies using the LLC model, the group similarly could demonstrate an apparent protection from the development of cachexia in mice whose adipose tissue was depleted of PTHR. Elevated serum levels of PTHrP have also been shown to be present in patients with gastroesophageal cancers in the absence of hypercalcaemia and were associated with systemic inflammation and adverse prognosis [68].

2.2.2 TWEAK

The discovery of TNF-related weak inducer of apoptosis (TWEAK) by Chicheportiche et al. [70], and subsequent work by Dogra et al. [71] on the effects of TWEAK on skeletal muscle, have prompted ongoing investigation into the precise role of TWEAK on muscle wasting. Discovery of a TWEAK receptor [72] and upregulation of TWEAK in both murine and human hepatocellular carcinoma [73] has led to increasing interest in the role of the TWEAK-Fn14 axis in human cancer and cachexia [74]. TWEAK through binding to Fn14 regulates many cellular responses, including proliferation, apoptosis and inflammation.

Studying healthy individuals, Raue et al. [75] investigated the induction of both TWEAK and Fn14 in human skeletal muscle following either resistance or running exercise. Taking serial muscle biopsies, the team measured TWEAK protein and mRNA levels and found them to be elevated in the recovery period following exercise. Interestingly, there was a rise in Fn14 mRNA and protein induction suggesting that, whilst the presence of TWEAK within skeletal muscle is involved in metabolism and turnover, the induction of its receptor Fn14 has more of an effect on skeletal muscle metabolism and turnover in the post-exercise period.

Using cultured myotubes, Bhatnagar et al. [76] investigated the effect of TWEAK on cell atrophy. They found that the addition of TWEAK to the cell cultures resulted in increased expression of factors previously found to be involved in muscle wasting, such as MuRF1 and Beclin 1. To confirm the action of TWEAK, the team used an autophagy inhibitor prior to TWEAK incubation and found reduced levels of autophagy. The group found activation of both caspases and NF-KB in response to TWEAK incubation, which led to their speculation that TWEAK induces autophagy through a number of possible pathways.

An interesting discovery of the effect of anti-Fn14 antibodies on tumour-bearing mice has additionally made the field slightly more complicated [77]. Using murine models to test the effects of anti-Fn14 antibodies on Fn14-bearing tumours, Johnston et al. noticed a reduction in the rate and severity of cachexia exhibited by these mice. Using knockout mice for both TWEAK and Fn14, the investigators found that administration of anti-TWEAK antibodies did not have the same reduction in the rate and severity of cachexia development, leading them to conclude that it is the expression of Fn14, rather than the secretion of TWEAK, which is more important in cachexia. These findings in both healthy humans and murine cancer models suggest that there is a close relationship between TWEAK and Fn14 and that Fn14 may be of increased relevance in skeletal muscle metabolism.

2.2.3 ZIP 14

Wang et al. [78] have recently identified the metal ion transporter ZRT and IRT like protein 14 (ZIP14). It has been shown to be upregulated in animal models and patients with metastatic cancer cachexia. It is stimulated by TNF-α and TGF-β and its muscle-specific depletion has been shown to markedly reduce muscle atrophy in cachectic rodents. ZIP14-mediated zinc uptake acts by inhibiting the expression of MyoD and Mef2c thereby blocking muscle cell differentiation and inducing myosin heavy chain loss. Its muscle-specific deletion has been shown to reduce muscle wasting in animal models of cachexia [78]. Similar findings have since been reproduced in experimental models of, and patients with, pancreatic cancer [79]. ZIP14 may therefore be a potential new target in the treatment of cancer cachexia.

2.2.4 Macrophage Inhibitory Cytokine1/Growth Differentiation Factor 15

Macrophage inhibitory cytokine1/growth differentiation factor 15 (MIC1/GDF15) has been consistently shown to induce cachexia in animal models and increased levels have been seen in patients with prostate cancer [80, 81]. The mechanism by which it acts is relatively unclear, but it is thought to act on the hypothalamus, thereby inducing anorexia and profound weight loss. Several studies have identified GFRAL, a member of the glial-derived neurotrophic factor receptor a family, as the receptor to which MIC-1/GDF15 binds. Ablation of MIC-1/GDF15 in animal models resulted in increased body weight, whereas overexpression led to lower body weight and fat mass [80]. A recent study has suggested that chemotherapy-induced sickness involves GD15 through the GDF15-GFRAL signalling pathway and that this nausea may preceed the onset of anorexia in animal models [82]. Human trials of anti-MIC-1/GDF15 are awaited.

2.2.5 Myostatin and Activin A

Myostatin is a member of the transforming growth factor-beta (TGFb) family with non-functional copies of the gene leading to muscle hypertrophy in both animals and humans [83]. Skeletal muscle is the main secretory source of myostatin with it thought to inhibit protein kinase B (Akt) and therefore the target of rapamycin complex 1 pathways, and in doing so enhance protein synthesis [84]. In addition to myostatin, other TGFb family members such as Activin A are upregulated in skeletal muscle. Activin A is increased following activation of the TNFa/transforming growth factor beta-activated kinase-1 signalling pathway. Many cancers lead to altered expression of activin A and are capable of inducing expression in muscle [85]. There is complex interaction between these pathways and therefore more than one target may need to be blocked in order to prevent muscle wasting. ActRIIB receptor has been hypothesised to be this target, with it being common to both myostatin and the activins. Indeed, in C26 mice,

its blockade reversed muscle wasting and prolonged survival [86].

Trials targeting Myostatin and the ActRIIB pathway have shown mixed results. One in patients with pancreatic cancer demonstrated an overall decreased survival in the treatment arm, whereas a study in elderly patients who had fallen, which demonstrated increased lean body mass and improved functional measures of muscle power albeit non-significantly [87, 88].

2.2.6 Angiotensin-II

In their investigation of the effects of Angiotensin-II (Ang-II) on insulin-like growth factor, Brink et al. noted a marked loss of weight in rats infused with Ang-II and that this effect was reduced in rats given losartan [89]. In humans, reduced weight loss in cardiac cachexia was also noted in a heart failure group given enalapril [90]. Following this, the research group including Tisdale et al. undertook multiple studies of the effects of angiotensin-converting-enzyme inhibitors (ACE-i), showing both reduced protein synthesis [91] and increased protein catabolism [92] in murine myotubes.

In their review of the effects of Ang-II, Yoshida et al. [93] discussed the effects of the hormone initially in the context of cachexia associated with congestive heart failure (CHF) and chronic kidney disease (CKD). It was noted within these populations that there were high levels of circulating Ang-II, and that those patients taking ACE-i did not lose as much weight as others in the group. Extrapolation from this non-cancer group encouraged investigation in a mouse model, again suggesting that Ang-II has a role in the propagation of systemic inflammation and the promotion of the cachectic state.

Following on from this, work carried out by Penafuerte et al. [94] used an innovative casual network analysis technique (CAN) to identify molecules likely to be acting as mediators or regulators of the cachectic process in patients with seven different cancer types. This involved the assessment of multiple cytokines in plasma, as well as their upstream regulators, and the mRNA segments related to these. The CAN techniques allow a modelling of complex, multi-faceted systems of interacting mediators to be reduced to more readily understandable schematics based on probabilities of association. This analysis gave Ang-II as a master upstream regulator of cachexia, and specifically pre-cachexia (where patients have not yet deteriorated sufficiently to meet previously-published cutoffs associated with worsened survival [11]). This prediction was validated in the patient cohort by showing a negative correlation between Ang-II levels and survival. Penafuerte et al. [94] hypothesised that Ang-II is among the major molecular mechanisms driving CC, and it is to be hoped that this may provide another possible therapeutic avenue for a commonly used medication in the form of the ACE-I.

Whilst not specifically a study of CC, a retrospective observational study observed that patients taking enalapril for heart failure were at lower risk of weight loss and demonstrated associated improved survival [95]. These results would seem to support related hypotheses regarding the potential clinical benefit of ACE-i. Although incompletely understood, the role of ACE and Ang-II in the molecular mechanism of CC remains relevant and an exciting possible target for intervention. Finally, at the genetic level, Johns et al. [44] investigated a cohort of cancer patients, looking at SNPs. Comparing patients with >2% weight loss and low SMI with weight-stable patients, there were significant correlations between SNPs coding for ACE and cachexia.

2.2.7 Micro-RNAs

The study of exosomal micro-RNAs (miRNAs) and muscle-specific miRNAs (myomiRs) is a novel focus of research for cancer cachexia. Both potentially play a role in the transduction of inflammatory signals and thus activate a catabolic status in muscles [96]. Exosomal-transported miRNAs are thought to play a key role in the propagation of inflammation in cancer cachexia. Exosomes are nanovesicles that promote tissue crosstalk in an autocrine, paracrine and endocrine manner. MiRNAs make this

communication more efficient. They have previously been implicated in the induction of metastasis and the regulation of protein pathways within skeletal muscle [97].

miRNA profile changes may indicate the presence or progression of muscular disease. An increase in miR-21 in particular has been highlighted in a variety of cancer types (colorectal, gastric, prostate and liver) [98, 99]. Its upregulation is thought to promote muscle atrophy by binding to the transcription factor YY1 as well as the translational initiation factor eIF4E3, and in doing so interferes with myogenesis [96]. By analysing rectus muscle biopsies, Narasimhan et al. [100] have demonstrated eight other miRNAs that are upregulated in cancer cachexia patients. They control appetite, interact with transferrin and insulin-like growth receptors, as well as down regulate genes involved in lipid biosynthesis and energy balance and include: hsa-let-7d-3p; hsa-miR-345-5p; hsa-miR-423-5p; hsa-miR-532-5p; hsa-miR-1296-5p; hsa-miR-3184-3p; hsa-miR-423-3p; and hsa-miR-199a-3p.

Cachexia-related inflammation is also regulated by miRNAs. He et al. [101] demonstrated that lung and pancreatic cells secrete exosomes containing miR-21. MiR-21 travels in the bloodstream and binds to toll-like receptor-9 (TLR-9) in human myoblasts therefore promoting apoptosis of muscle cells by activating the c-Jun N-terminal kinase pathway (JNK) [97].

As cachexia is a syndrome with systemic effects in which there is thought to be a large amount of tissue crosstalk, these particles are likely to play an important role. Advancing research into their characterisation and function could potentially lead to the discovery of a cachexia-related biomarker or a drug that could reduce the inflammatory effects of certain tumours.

2.2.8 Interleukins

Interleukins 6 and 8 have been implicated in the pathogenesis of cachexia. They are drivers of systemic inflammation and raised levels have been previously associated with cachexia [102, 103]. IL-6 signalling triggers the activation of STAT proteins which increase the transcription of genes involved in immune function, cell proliferation, differentiation and apoptosis [104]. IL-6 administration in humans is capable of inducing muscle wasting and deletion of the IL-6 gene in the APC$^{Min/fl}$ mouse prevented the development of cachexia [102]. IL-6 is also capable of increasing autophagy in myotubes when joined with soluble IL-6 receptor [58]. Raised IL-6 levels have been previously associated with weight loss and prognosis in lung cancer patients; however, a recent study in pancreatic cancer patients has shown an association with disease progression but not cachexia [105]. Trials of anti-IL-6 antibody have shown increases in LBM in patients with non-small cell lung cancer and rheumatoid arthritis;, however, concerns were raised about its cardiometabolic safety [106, 107].

A study profiling cytokines from conditioned medium from human pancreatic cancer cells, human tumour-associated stromal cells and their co-culture identified IL-8 as being released at high levels. Myotubes treated with IL-8 underwent atrophy through the activation of ERK1/2, STAT and Smad signalling. Treatment of mice with IL-8 also induced muscle wasting [108].

2.3 Mitochondria

The ability to target mitochondrial dysfunction in cancer cachexia offers a therapeutic opportunity to normalise energy metabolism. Exercise has long been known to regulate mitochondrial function and skeletal muscle metabolism but is often difficult to comply with in patients with advanced cancer [109]. The exercise mimetic 5-aminoimidazole-4-carboxamide-1-beta-D -ribofuranoside (AICAR), an adenosine monophosphate-activated protein kinase (AMPK) activator, has been investigated in mice.

It impairs mitochondrial oxidative respiration in myotubes through its interactions with IFNg/TNFa and in doing so enables a metabolic shift to aerobic glycolysis. AICAR has been shown to prevent muscle wasting in mouse models of cancer [110]. The disruption of fission and fusion process can also be due to mitochondrial dysfunction. Fusion is regulated by mitofusin 1 and 2 (MFN-1/2) and optic atrophy protein 1. Loss of mitochondrial fusion leads to muscle wasting in MFN-1/2 knockout mice, thus implicating itself as an important novel mediator in the future investigation of cancer-associated muscle wasting [111, 112].

2.3.1 Muscle Regeneration

It has been suggested that cachexia interrupts the regenerative ability of skeletal muscle. Cachexia muscle damage leads to activation of both muscle satellite and non-myogenic cells. In muscle samples from human patients, transcriptomic data give evidence for blockade of satellite cell maturation, upregulation of apoptosis and reduced oxidative defence in the muscle of cancer patients [113]. In comparison, the potential for differentiation and oxidative defence is maintained in muscle samples from the healthy elderly.

3 Target Lessons Learnt from Recent Cachexia Trials

The last few years have seen the results of at least five phase 3 randomised, double blind placebo-controlled trials in patients with stage 3 or 4 non-small cell lung cancer (NSCLC). Different targets were trialled in each but the lack of functional efficacy suggests the targets may not be the sole cause of cachexia or indeed the targeted endpoints may not be adequate. In the ROMANA 1 and 2 trials, anamorelin, an oral ghrelin receptor agonist (Helsinn) was able to increase LBM over a 12-week treatment period, but not the co-primary endpoint of handgrip strength (HGS) [114]. The results of the ROMANA 3 extension study (a further 12 weeks of treatment for those patients with a preserved Eastern Cooperative Oncology Group [ECOG] performance status ≤2) confirmed drug safety/tolerability and maintenance of weight/

symptom improvement, but once again did not demonstrate an improvement in HGS compared with placebo [115]. In comparison, in the POWER (*Prevention and treatment Of* muscle *Wasting in patients with cancER*) 1 and 2 studies, treatment with enobosarm, a non-steroidal Selective Androgen Receptor Modulator (SARM) (GTx), was associated with an increase in LBM, but not the functional co-primary endpoint of ≥10% improvement in stair climb power after 84 days [116]. These studies support the hypothesis that both ghrelin and SARMs may be beneficial for the amelioration of CC, but they also highlight the current difficulties in clinical trial design. Future trials must give consideration to patient targeting: inclusion criteria, by avoiding the recruitment of refractory patients but ensuring consistency in the identification of patients early in the disease continuum (ROMANA 1/2 required patients to be weight-losing or underweight, whereas POWER 1/2 did not); treatment protocols, including drug dosage and the adoption of multimodal interventions (the MENAC trial (Multimodal Exercise/Nutrition/Anti-inflammatory treatment for Cachexia trial) incorporating exercise, nutrition and anti-inflammatory medication is currently recruiting and aims to establish a standard of care for cachectic patients, [117]); and outcome measures/biomarkesr of therapeutic response, particularly with regards to functional improvement. Initial data from the MENAC precursor feasibility study [118] suggests that multimodal intervention is possible for these patients; however, conclusions on outcome data for secondary endpoints including muscle mass could not be drawn due to small sample size.

4   Future Developments

There are many under-investigated areas that offer promise for the future of cachexia research:

Improved patient targeting D_3-Creatine (D_3-Cr) dilution has been shown to provide a more accurate measurement of skeletal muscle mass. Studies in patients with sarcopenia have shown

to be more accurately associated with functional capacity and risk of disability. Studies in cancer patients are awaited [119].

The Microbiome The small bowel plays a large role in metabolic control, energy homeostasis and weight control through the microbiota-gut-brain axis. Satiety gut hormones are released when food comes into contact with the small bowel mucosa. This signalling is further enhanced by small bowel gut microbiota. It has been hypothesised that systemic inflammation may arise at least in part from failure in gut barrier function and changes in the microbiome resulting in persistent immune activation [120]. The gut microbiota interacts with other components of the gut barrier, including the epithelium, gut-associated lymphoid tissue, and the enteric nervous system. Several studies in patients with anorexia have linked low levels of bacteria that produce short chain fatty acids to the regulation of appetite, lipid and glucose metabolism as well as immune functions [121]. In this regard, they may contribute to the pathophysiology of cachexia.

Intra-Tumoural Aetiology of Systemic Inflammation, and the Role of Stromal Cells Tumour-associated macrophages (TAMs) increase tumour progression and suppress anti-tumour immune functions via infiltration of monocytes from the blood. This infiltrate contributes to tumour growth and the release of cytokines that promote the pro-cachectic environment. Despite tumour-induced immunosuppression being a well-researched area, little is known about immunosuppression in the development of cachexia [122]. There is also lack of understanding about how tumour cells and intra-tumoural immune cells contribute to wider systemic inflammation. In addition to cancer cells, supporting stromal cells in the tumour play a role in the establishment of cachexia [123].

Heat Shock proteins (Hsp) Hsp have recently become of interest. In particular Hsp70 and Hsp90 are implicated in the development and progression of muscle wasting through the activation of the UPP and autophagy-lysosome pathways in response to Toll-like receptor 4 (TLR4) activation. Neutralisation of extracellular HSP70 and 90 or silencing Hsp70 and 90 expression in tumour cells revoked muscle wasting in myotubes and in mice [124]. Studies in humans are awaited.

Central Regulation of Cachexia In this chapter, the authors have focussed predominantly on the targeting of cachexia in peripheral tissues, including muscle and fat. However, it is well-known that central anorexia and fatigue, in part driven by systemic inflammation and circulating pro-inflammatory cytokines, contribute to the overall wasting phenomenon in patients. Progressive weight loss is usually a stimulus to food intake in normal humans. The persistence of anorexia in cancer patients suggests a failure of this response. Complex interactions between nervous, endocrine and immune systems affect this response. Previous studies have described a

central nervous system-based mechanism of cachexia. Cytokines produced elsewhere in the body are increased and modified in the hypothalamus leading to diverging weight and activity modulating neurons [125]. The mechanisms via which peripheral inflammation is translated to central responses are still not well understood.

5 Conclusions and Future Perspectives

Cancer cachexia is a complex, multifactorial condition with detrimental physical and psychosocial effects. However, advances in our understanding of the pathophysiology, assessment and treatment of cachexia are being accrued steadily (Summarised in Fig. 3). Progress in treating the human form of cancer cachexia can only move forwards through carefully designed large randomised controlled clinical trials of specific therapies with validated biomarkers of relevance

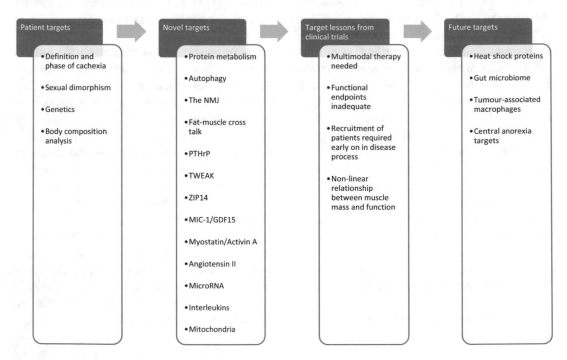

Fig. 3 Key learning points. There are many targets that have shown promise in providing a greater understanding of the pathophysiology of cancer cachexia. Results from clinical trials however, have failed to show in any increase

in functional endpoints. More research is required at the individual patient level, whilst considering novel options for the future to underlying mechanisms. The hope is that, with further advancements over the next few years, we will soon see the establishment of internationally accepted protocols for the successful management of cancer cachexia, leading to positive impacts on nutritional status, patient function, quality of life, and possibly survival. Themes for the future include the need to better quantify the severity of cachexia in order to develop scoring systems or cutoff values below which survival is impacted. Assessment of muscle mass should include the impact of muscle loss on function, whereas future trials and interventions need to consider sex-specific approaches and must try to avoid the recruitment of refractory patients.

References

1. Dewys, W.D., Begg, C., Lavin, P.T., Band, P.R., Bennett, J.M., Bertino, J.R., et al.: Prognostic effect of weight loss prior to chemotherapy in cancer patients. Eastern Cooperative Oncology Group. Am. J. Med. **69**(4), 491–497 (1980)
2. Wallengren, O., Lundholm, K., Bosaeus, I.: Diagnostic criteria of cancer cachexia: relation to quality of life, exercise capacity and survival in unselected palliative care patients. Support Care Cancer. **21**(6), 1569–1577 (2013)
3. Tan, B.H., Fearon, K.C.: Cachexia: prevalence and impact in medicine. Curr. Opin. Clin. Nutr. Metab. Care. **11**(4), 400–407 (2008)
4. Fearon, K., Strasser, F., Anker, S.D., Bosaeus, I., Bruera, E., Fainsinger, R.L., et al.: Definition and classification of cancer cachexia: an international consensus. Lancet Oncol. **12**(5), 489–495 (2011)
5. Vagnildhaug, O., Balstad, T., Almberg, S., Bruncli, C., Knudsen, A., Kaasa, S., et al.: A cross-sectional study examining the prevalence of cachexia and areas of unmet need in patients with cancer. Support Care Cancer. **26**(6), 1871–1880 (2017)
6. LeBlanc, T., Nipp, R., Rushing, C., Samsa, G., Locke, S., Kamal, A., et al.: Correlation between the international consensus definition of the Cancer Anorexia-Cachexia Syndrome (CACS) and patient-centered outcomes in advanced non-small cell lung cancer. J. Pain Symptom Manag. **49**(4), 680–689 (2014)
7. Blum, D., Stene, G.B., Solheim, T.S., Fayers, P., Hjermstad, M.J., Baracos, V.E., et al.: Validation of the Consensus-Definition for Cancer Cachexia and evaluation of a classification model—a study based on data from an international multicentre project (EPCRC-CSA). Ann. Oncol. **25**(8), 1635–1642 (2014)
8. Vanhoutte, G., van de Wiel, M., Wouters, K., Sels, M., Bartolomeeussen, L., Keersmaecker, S.D., et al.: Cachexia in cancer: what is in the definition? BMJ Open Gastroenterol. **3**(1), e000097 (2016)
9. Evans, W.J., Morley, J.E., Argilés, J., Bales, C., Baracos, V., Guttridge, D., et al.: Cachexia: a new definition. Clin. Nutr. **27**(6), 793–799 (2008)
10. Mourtzakis, M., Prado, C.M.M., Lieffers, J.R., Reiman, T., McCargar, L.J., Baracos, V.E.: A practical and precise approach to quantification of body composition in cancer patients using computed tomography images acquired during routine care. Appl. Physiol. Nutr. Metab. **33**(5), 997–1006 (2008)
11. Martin, L., Birdsell, L., MacDonald, N., Reiman, T., Clandinin, M.T., McCargar, L.J., et al.: Cancer cachexia in the age of obesity: skeletal muscle depletion is a powerful prognostic factor, independent of body mass index. J. Clin. Oncol. **31**(12), 1539–1547 (2013)
12. Fearon, K.C.H., Von Meyenfeldt, M.F., Moses, A.G.W., Van Geenen, R., Roy, A., Gouma, D.J., et al.: Effect of a protein and energy dense N-3 fatty acid enriched oral supplement on loss of weight and lean tissue in cancer cachexia: a randomised double blind trial. Gut. **52**(10), 1479–1486 (2003)
13. Cruz-Jentoft, A.J., Baeyens, J.P., Bauer, J.M., Boirie, Y., Cederholm, T., Landi, F., et al.: Sarcopenia: European consensus on definition and diagnosis Report of the European Working Group on Sarcopenia in Older People. Age Ageing. **39**(4), 412–423 (2010)
14. von Haehling, S., Ebner, N., dos Santos, M.R., Springer, J., Anker, S.D.: Muscle wasting and cachexia in heart failure: mechanisms and therapies. Nat. Rev. Cardiol. **14**(6), 323–341 (2017)
15. Fearon, K., Evans, W.J., Anker, S.D.: Myopenia—a new universal term for muscle wasting. J. Cachexia. Sarcopenia Muscle. **2**(1), 1–3 (2011)
16. Miller, J., Wells, L., Nwulu, U., Currow, D., Johnson, M., Skipworth, R.: Validated screening tools for the assessment of cachexia, sarcopenia, and malnutrition: a systematic review. Am. J. Clin. Nutr. **108**(6), 1196–1208 (2018)
17. Arends, J., Baracos, V., Bertz, H., Bozzetti, F., Calder, P., Deutz, N.E.P., et al.: ESPEN expert group recommendations for action against cancer-related malnutrition. Clin. Nutr. **36**(5), 1187–1196 (2017)
18. Baumgartner, R.N., Koehler, K.M., Gallagher, D., Romero, L., Heymsfield, S.B., Ross, R.R., et al.: Epidemiology of sarcopenia among the elderly in New Mexico. Am. J. Epidemiol. **147**(8), 755–763 (1998)
19. MacDonald, A.J., Greig, C.A., Baracos, V.: The advantages and limitations of cross-sectional body composition analysis. Curr. Opin. Support. Palliat. Care. **5**(4), 342–349 (2011)
20. Heymsfield, S.B., Smith, R., Aulet, M., Bensen, B., Lichtman, S., Wang, J., et al.: Appendicular skeletal

muscle mass: measurement by dual-photon absorptiometry. Am. J. Clin. Nutr. **52**(2), 214–218 (1990)

21. Nguyen, D.M., El-Serag, H.B.: The epidemiology of obesity. Gastroenterol Clin North Am. **39**(1), 1–7 (2010)

22. Flegal, K.M., Carroll, M.D., Kit, B.K., Ogden, C.L.: Prevalence of obesity and trends in the distribution of body mass index among US adults, 1999–2010. JAMA. **307**(5), 491–497 (2012)

23. Kalantar-Zadeh, K., Horwich, T.B., Oreopoulos, A., Kovesdy, C.P., Younessi, H., Anker, S.D., Morley, J.E.: Risk factor paradox in wasting diseases. Curr Opin Clin Nutr Metab Care. **10**(4), 433–442 (2007)

24. Martin, L., Senesse, P., Gioulbasanis, I., Antoun, S., Bozzetti, F., Deans, C., Strasser, F., Thoresen, L., Jagoe, R.T., Chasen, M., Lundholm, K., Bosaeus, I., Fearon, K.H., Baracos, V.E.: Diagnostic criteria for the classification of cancer-associated weight loss. J Clin Oncol. **33**(1), 90–99 (2015)

25. Delmonico, M., Harris, T., Visser, M., Park, S., Conroy, M., Velasquez-Mieyer, P., et al.: Health, Aging, and Body: longitudinal study of muscle strength, quality, and adipose tissue infiltration. Am. J. Clin. Nutr. **90**, 1579–1585 (2009)

26. Li, Y., Xu, S., Zhang, X., Yi, Z., Cichello, S.: Skeletal intramyocellular lipid metabolism and insulin resistance. Biophys. Rev. **1**, 90–98 (2015)

27. Gray, C., MacGillivray, T., Eeley, C., Stephens, N., Beggs, I., Fearon, K., et al.: Magnetic resonance imaging with k-means clustering objectively measures whole muscle volume compartments in sarcopenia/cancer cachexia. Clin. Nutr. **30**, 106–111 (2011)

28. Stephens, N., Skipworth, R., MacDonald, A., Greig, C., Ross, J., Fearon, K.: Intramyocellular lipid droplets increase with progression of cachexia in cancer patients. J. Cachexia. Sarcopenia Muscle. **2**, 111–117 (2011)

29. Ramage, M.I., Johns, N., Deans, C.D.A., Ross, J.A., Preston, T., Skipworth, R.J.E., et al.: The relationship between muscle protein content and CT-derived muscle radio-density in patients with upper GI cancer. Clin. Nutr. Edinb. Scotl. (2016 Dec 27)

30. Ebadi, M., Martin, L., Ghosh, S., Field, C.J., Lehner, R., Baracos, V.E., Mazurak, V.C.: Subcutaneous adiposity is an independent predictor of mortality in cancer patients. Br J Cancer. **117**(1), 148–155 (2017)

31. Yip, C., Dinkel, C., Mahajan, A., Siddique, M., Cook, G.J.R., Goh, V.: Imaging body composition in cancer patients: visceral obesity, sarcopenia and sarcopenic obesity may impact on clinical outcome. Insights Imaging. **6**(4), 489–497 (2015)

32. Mersmann, F., Bohm, S., Schroll, A., Boeth, H., Duda, G., Arampatzis, A.: Muscle shape consistency and muscle volume prediction of thigh muscles. Scand. J. Med. Sci. Sports. **25**(2), e208–e213 (2015)

33. Willis, T.A., Hollingsworth, K.G., Coombs, A., Sveen, M.L., Andersen, S., Stojkovic, T., et al.: Quantitative muscle MRI as an assessment tool for

monitoring disease progression in LGMD2I: a multicentre longitudinal study. PLoS One [Internet]. (2013 Aug 14) [cited 2016 May 18];**8**(8). Available from: http://www.ncbi.nlm.nih.gov/pmc/articles/PMC3743890/

34. Karlsson, A., Rosander, J., Romu, T., Tallberg, J., Grönqvist, A., Borga, M., et al.: Automatic and quantitative assessment of regional muscle volume by multi-atlas segmentation using whole-body water–fat MRI. J. Magn. Reson. Imaging. **41**(6), 1558–1569 (2015)

35. Rollins, K.E., Tewari, N., Ackner, A., Awwad, A., Madhusudan, S., Macdonald, I.A., et al.: The impact of sarcopenia and myosteatosis on outcomes of unresectable pancreatic cancer or distal cholangiocarcinoma. Clin. Nutr. Edinb. Scotl. **35**(5), 1103–1109 (2016)

36. Weber, M., Krakowski-Roosen, H., Schroder, L., Kinscherf, R., Krix, M., Kopp-Schneider, A., et al.: Morphology, metabolism, microcirculation, and strength of skeletal muscles in cancer-related cachexia. Acta Oncol. **48**(1), 116–124 (2009)

37. Krix, M., Krakowski-Roosen, H., Huttner, H., Delorme, S., Kauczor, H., Hildebrandt, W.: Assessment of skeletal muscle perfusion using contrast-enhanced ultrasonography. J. Ultrasound Med. **24**(4), 431–441 (2005)

38. Sabatino, A., Regolistil, G., Bozzoli, L., Fani, F., Antoniotti, R., Maggiore, U., et al.: Reliability of bedside ultrasound for measurement of quadriceps muscle thickness in critically ill patients with acute kidney injury. Clin. Nutr. **36**(6), 1710–1715 (2017)

39. Puthucheary, Z., Phadke, R., Rawal, J., McPhail, M., Sidhu, P., Rowlerson, A., et al.: Qualitative ultrasound in acute critical illness muscle wasting. Crit. Care Med. **43**(8), 1603–1611 (2015)

40. Stephens, N.A., Gray, C., MacDonald, A.J., Tan, B.H., Gallagher, I.J., Skipworth, R.J.E., et al.: Sexual dimorphism modulates the impact of cancer cachexia on lower limb muscle mass and function. Clin. Nutr. **31**(4), 499–505 (2012)

41. Skipworth, R.J.E., Moses, A., Sangster, K., Sturgeon, C., Voss, A., Fallon, M., et al.: Interaction of gonadal status with systemic inflammation and opioid use in determining nutritional status and prognosis in advanced pancreatic cancer. Support Care Cancer. **19**(3), 391–401 (2011)

42. Stretch, C., Wang, K., Rejtar, T., Reinker, S., Brachat, S., Badur, R., et al.: Sexual dimorphism in the skeletal muscle transcriptome and urinary proteome indicate sex specific pathways involved in regulation of muscularity in cancer patients. J. Cachexia. Sarcopenia Muscle. **9**(1), 183–212 (2018)

43. Tisdale, M.J.: Mechanisms of cancer cachexia. Physiol. Rev. **89**(2), 381–410 (2009)

44. Johns, N., Stretch, C., Tan, B.H.L., Solheim, T.S., Sørhaug, S., Stephens, N.A., et al.: New genetic signatures associated with cancer cachexia as defined

by low skeletal muscle index and weight loss. J Cachexia Sarcopenia Muscle. (2016 Jan 1);n/a–n/a

45. Deans, C., Rose-Zerilli, M., Wigmore, S., Ross, J., Howell, M., Jackson, A., et al.: Host cytokine genotype is related to adverse prognosis and systemic inflammation in gastro-oesophageal cancer. Ann. Surg. Oncol. **14**(2), 329–339 (2007)

46. Barber, M., Powell, J., Lynch, S., Fearon, K., Ross, J.: A polymorphism of the interleukin-1 beta gene influences survival in pancreatic cancer. Br. J. Cancer. **83**, 1443–1447 (2000)

47. Barber, M., Powell, J., Lynch, S., Gough, N., Fearon, K., Ross, J.: Two polymorphisms of the tumour necrosis factor gene do not influence survival in pancreatic cancer. Clin. Exp. Immunol. **117**, 425–429 (1999)

48. Mueller, T., Burmeister, M., Bachmann, J., Martignoni, M.: Cachexia and pancreatic cancer: are there treatment options? World J. Gastroenterol. **20**(28), 9361–9373 (2014)

49. Vazeille, C., Jouinot, A., Durand, J., Neveux, N., Boudou-Rouquette, P., Huillard, O., et al.: Relation between hypermetabolism, cachexia and survival in cancer patients: a prospective study in 390 cancer patients before initiation of anticancer therapy. Am. J. Clin. Nutr. **105**(5), 1139–1147 (2017)

50. van Dijk, D., Horstman, A., Smeets, J., den Dulk, M., Grabsch, H., Dejong, C., et al.: Tumour-specific and organ-specific protein synthesis rates in patients with pancreatic cancer. J. Cachexia. Sarcopenia Muscle. **10**(3) (2019)

51. van Dijk, D.P., van de Poll, M.C., Moses, A.G., Preston, T., Olde Damink, S.W., Rensen, S.S., et al.: Effects of oral meal feeding on whole body protein breakdown and protein synthesis in cachectic pancreatic cancer patients. J. Cachexia. Sarcopenia Muscle. **6**(3), 212–221 (2015)

52. Deutz, N.E.P., Safar, A., Schutzler, S., Memelink, R., Ferrando, A., Spencer, H., et al.: Muscle protein synthesis in cancer patients can be stimulated with a specially formulated medical food. Clin. Nutr. Edinb. Scotl. **30**(6), 759–768 (2011)

53. MacDonald, A.J., Small, A.C., Greig, C.A., Husi, H., Ross, J.A., Stephens, N.A., et al.: A novel oral tracer procedure for measurement of habitual myofibrillar protein synthesis. Rapid. Commun. Mass Spectrom. RCM. **27**(15), 1769–1777 (2013)

54. MacDonald, A.J., Johns, N., Stephens, N., Greig, C., Ross, J.A., Small, A.C., et al.: Habitual myofibrillar protein synthesis is normal in patients with upper GI cancer cachexia. Am. Assoc. Cancer Res. **21**(7), 1734–1740 (2015)

55. Sandri, M.: Protein breakdown in muscle wasting: role of autophagy-lysosome and ubiquitin-proteasome. Int. J. Biochem. Cell Biol. **45**(10) (2013)

56. Penna, F., Baccino, F., Costelli, P.: Coming back: autophagy in cachexia. Curr. Opin. Clin. Nutr. Metab. Care. **17**(3), 241–246 (2014)

57. Ryter, S., Mizumura, K., Choi, A.: The impact of autophagy on cell death modalities. Int. J. Cell Biol. (2014)

58. Pettersen, K., Andersen, S., Degen, S., Tadini, V., Grosjean, J., Hatakeyama, S., et al.: Cancer cachexia associates with a systemic autophagy-inducing activity mimicked by cancer cell-derived IL-6 trans-signalling. Sci. Rep. **7**, 2046 (2017)

59. Johns, N., Hatakeyama, S., Stephens, N.A., Degen, M., Degen, S., Frieauff, W., et al.: Clinical classification of cancer cachexia: phenotypic correlates in human skeletal muscle. PLoS One. **9**(1), e83618 (2014)

60. Boehm, I., Miller, J., Wishart, T., Wigmore, S., Skipworth, R., Jones, R., et al.: Neuromuscular junctions are stable in patients with cancer cachexia. J. Clin. Invest. (2019)

61. Ebadi, M., Mazurak, V.C.: Evidence and mechanisms of fat depletion in cancer. Nutrients. **6**(11), 5280–5297 (2014)

62. Ryden, M., Agustsson, T., Laurencikiene, J., Britton, T., Sjolin, E., Isaksson, B., et al.: Lipolysis; not inflammation, cell death, or lipogenesis is involved in adipose tissue loss in cancer cachexia. Cancer. **113**(7), 1695–1704 (2008)

63. Mracek, T., Stephens, N.A., Gao, D., Bao, Y., Ross, J.A., Rydén, M., et al.: Enhanced ZAG production by subcutaneous adipose tissue is linked to weight loss in gastrointestinal cancer patients. Br. J. Cancer. **104**(3), 441–447 (2011)

64. Ebadi, M., Baracos, V., Bathe, O., Robinson, L., Mazurak, V.: Loss of visceral adipose tissue precedes subcutaneous adipose tissue and associates with N-6 fatty acid content. Clin. Nutr. **35**(6), 1347–1353 (2016)

65. Modesitt, S.C., Hsu, J.Y., Chowbina, S.R., Lawrence, R.T., Hoehn, K.L.: Not all fat is equal: differential gene expression and potential therapeutic targets in subcutaneous adipose, visceral adipose, and endometrium of obese women with and without endometrial cancer. Int. J. Gynecol. Cancer Off. J. Int. Gynecol. Cancer Soc. **22**(5), 732–741 (2012)

66. Das, S.K., Eder, S., Schauer, S., Diwoky, C., Temmel, H., Guertl, B., et al.: Adipose triglyceride lipase contributes to cancer-associated cachexia. Science. **333**(6039), 233–238 (2011)

67. Kir, S., White, J.P., Kleiner, S., Kazak, L., Cohen, P., Baracos, V.E., et al.: Tumour-derived PTH-related protein triggers adipose tissue browning and cancer cachexia. Nature. **513**(7516), 100–104 (2014)

68. Deans, C., Wigmore, S., Paterson-Brown, S., Black, J., Ross, J., Fearon, K.C.H.: Serum parathyroid hormone-related peptide is associated with systemic inflammation and adverse prognosis in gastroesophageal carcinoma. Cancer. **103**(9), 1810–1818 (2005)

69. Kir, S., Komaba, H., Garcia, A.P., Economopoulos, K.P., Liu, W., Lanske, B., et al.: PTH/PTHrP receptor

mediates cachexia in models of kidney failure and cancer. Cell Metab. **23**(2), 315–323 (2016)

70. Chicheportiche, Y., Bourdon, P.R., Xu, H., Hsu, Y. M., Scott, H., Hession, C., et al.: TWEAK, a new secreted ligand in the tumor necrosis factor family that weakly induces apoptosis. J. Biol. Chem. **272**(51), 32401–32410 (1997)

71. Dogra, C., Changotra, H., Wedhas, N., Qin, X., Wergedal, J.E., Kumar, A.: TNF-related weak inducer of apoptosis (TWEAK) is a potent skeletal muscle-wasting cytokine. FASEB J. **21**(8), 1857–1869 (2007)

72. Wiley, S.R., Cassiano, L., Lofton, T., Davis-Smith, T., Winkles, J.A., Lindner, V., et al.: A novel TNF receptor family member binds TWEAK and is implicated in angiogenesis. Immunity. **15**(5), 837–846 (2001)

73. Feng, S.L.Y., Guo, Y., Factor, V.M., Thorgeirsson, S.S., Bell, D.W., Testa, J.R., et al.: The Fn14 immediate-early response gene is induced during liver regeneration and highly expressed in both human and murine hepatocellular carcinomas. Am. J. Pathol. **156**(4), 1253–1261 (2000)

74. Winkles, J.A.: The TWEAK–Fn14 cytokine–receptor axis: discovery, biology and therapeutic targeting. Nat. Rev. Drug Discov. **7**(5), 411–425 (2008)

75. Raue, U., Jemiolo, B., Yang, Y., Trappe, S.: TWEAK-Fn14 pathway activation after exercise in human skeletal muscle: insights from two exercise modes and a time course investigation. J. Appl. Physiol. **118**(5), 569–578 (2015)

76. Bhatnagar, S., Mittal, A., Gupta, S.K., Kumar, A.: TWEAK causes myotube atrophy through coordinated activation of ubiquitin-proteasome system, autophagy, and caspases. J. Cell. Physiol. **227**(3), 1042–1051 (2012)

77. Johnston, A.J., Murphy, K.T., Jenkinson, L., Laine, D., Emmrich, K., Faou, P., et al.: Targeting of Fn14 prevents cancer-induced cachexia and prolongs survival. Cell. **162**(6), 1365–1378 (2015)

78. Wang, G., Biswas, A., Ma, W., Kandpal, M., Cocker, C., Grandgenett, P., et al.: Metastatic cancers promote cachexia through ZIP14 upregulation in skeletal muscle. Nat. Med. **24**, 770–781 (2018)

79. Shakri, A., Zhong, T., Ma, W., Coker, C., Kim, S., Calluori, S., et al.: Upregulation of ZIP14 and altered zinc homeostasis in muscles in pancreatic cancer cachexia. Cancers. **18**(12), 3 (2019)

80. Tsai, V., Husaini, Y., Sainsbury, A., Brown, D., Breit, S.: The MIC-1/GDF15-GFRAL pathway in energy homeostasis: implications for obesity, cachexia, and other associated diseases. Cell Metab. **28**, 353–368 (2018)

81. Sadasivan, S., Chen, Y., Gupta, N., Taneja, K., Maresh, S., Gonzalez, A., et al.: The role of GDF15 (growth/differentiation factor 15) during prostate carcinogenesis. Cancer Res. **78**, 421 (2018)

82. Borner, T., Shaulson, E., Ghidewon, M., Barnett, A., Horn, C., Doyle, R., et al.: GDF15 induces anorexia through nausea and emesis. Cell Metab. **31**(2), 351–362 (2020)

83. Pirruccello-Straub, M., Jackson, J., Wawersik, S., Webster, T., Salta, L., Long, K., et al.: Blocking extracellular activation of myostatin as a strategy for treating muscle wasting. Sci. Rep. **8**, 2292 (2018)

84. Chen, J., Walton, K., Hagg, A., Colgan, T.: Specific targeting of TGF-β family ligands demonstrates distinct roles in the regulation of muscle mass in health and disease. PNAS. **114**(26), 5266–5275 (2017)

85. Ding, H., Zhang, G., Sin, K., Liu, Z.: Activin A induces skeletal muscle catabolism via p38 mitogen activated protein kinase. J. Cachexia. Sarcopenia Muscle. **8**(2), 202–212 (2017)

86. Morvan, F., Rondeau, J., Zou, C., Minetti, G.: Blockade of activin type II receptors with a dual anti-ActRIIA/IIB antibody is critical to promote maximal skeletal muscle hypertrophy. Proc. Natl. Acad. Sci. USA. **114**(47), 12448–12453 (2017)

87. Golan, T., Geva, R., Richards, D., Madhusudan, S.: LY2495655, an antimyostatin antibody, in pancreatic cancer: a randomized, phase 2 trial. J. Cachexia. Sarcopenia Muscle. **9**(5), 871–879 (2018)

88. Becker, C., Lord, S., Studenski, S., Warden, S.: Myostatin antibody (LY2495655) in older weak fallers: a proof-of-concept, randomised, phase 2 trial. Lancet Diabetes Endocrinol. **3**(12), 948–957 (2015)

89. Brink, M., Wellen, J., Delafontaine, P.: Angiotensin II causes weight loss and decreases circulating insulin-like growth factor I in rats through a pressor-independent mechanism. J. Clin. Invest. **97**(11), 2509–2516 (1996)

90. Adigun, A.Q., Ajayi, A.A.: The effects of enalapril-digoxin-diuretic combination therapy on nutritional and anthropometric indices in chronic congestive heart failure: preliminary findings in cardiac cachexia. Eur. J. Heart Fail. **3**(3), 359–363 (2001)

91. Russell, S.T., Sanders, P.M., Tisdale, M.J.: Angiotensin II directly inhibits protein synthesis in murine myotubes. Cancer Lett. **231**(2), 290–294 (2006)

92. Sanders, P.M., Russell, S.T., Tisdale, M.J.: Angiotensin II directly induces muscle protein catabolism through the ubiquitin–proteasome proteolytic pathway and may play a role in cancer cachexia. Br. J. Cancer. **93**(4), 425–434 (2005)

93. Yoshida, T., Tabony, A.M., Galvez, S., Mitch, W.E., Higashi, Y., Sukhanov, S., et al.: Molecular mechanisms and signaling pathways of angiotensin II-induced muscle wasting: potential therapeutic targets for cardiac cachexia. Int. J. Biochem. Cell Biol. **45**(10), 2322–2332 (2013)

94. Penafuerte, C.A., Gagnon, B., Sirois, J., Murphy, J., MacDonald, N., Tremblay, M.L.: Identification of neutrophil-derived proteases and angiotensin II as

biomarkers of cancer cachexia. Br. J. Cancer. **114**(6), 680–687 (2016)

95. Anker, S.D., Negassa, A., Coats, A.J., Afzal, R., Poole-Wilson, P.A., Cohn, J.N., et al.: Prognostic importance of weight loss in chronic heart failure and the effect of treatment with angiotensin-converting-enzyme inhibitors: an observational study. Lancet. **361**(9363), 1077–1083 (2003)

96. Marinho, R., Alcantara, P., Ottoch, J., Seelaender, M.: Role of exosomal Micro-RNAs and myomiRs in the development of cancer cachexia-associated muscle wasting. Front. Nutr. **4**, 69 (2017)

97. Miller, J.: Characterisation and mechanisms of altered body composition and tissue wasting in cancer cachexia. University of Edinburgh (2020)

98. Anindo, M., Yaqinuddin, A.: Insights into the potential use of microRNAs as biomarker in cancer. Int. J. Surg. **10**(9), 443–449 (2012)

99. Lan, H., Lu, H., Wang, X., Jin, H.: MicroRNAs as potential biomarkers in cancer: opportunities and challenges. Biomed. Res. Int. 125094

100. Narasimhan, A., Ghosh, S., Stretch, C., Greiner, R., Bathe, O., Baracos, V., et al.: Small RNAome profiling from human skeletal muscle: novel miRNAs and their targets associated with cancer cachexia. J. Cachexia. Sarcopenia Muscle. **8**(3), 405–416 (2017)

101. He, W., Calore, F., Londhe, P., Canella, A., Guttridge, D., Croce, C.: Microvesicles containing miRNAs promote muscle cell death in cancer cachexia via TLR7. Proc. Natl. Acad. Sci. USA. **111**(12), 4525–4529 (2014)

102. White, J.: IL-6, cancer and cachexia: metabolic dysfunction creates the perfect storm. Transl Cancer Res. **6**(2), 280–285 (2017)

103. Yoshikwa, T., Takano, M., Kouta, H., Horikoshi, M., Asakawa, T., Kudoh, K.: Can serum IL-6 be a sentinel biomarker for cancer cachexia in gynaecologic cancer patients? Gynaecol. Cancer. **36**(15_suppl), e17544–e17544 (2018)

104. Miller, A., McLeod, L., Alhayyani, S., Szczepny, A., Watkins, D., Chen, W., et al.: Blockade of the IL-6 trans-signalling/STAT3 axis suppresses cachexia in Kras-induced lung adenocarcinoma. Oncogene. **36**(21), 3059–66 (2017)

105. Ramsey, M., Talbert, E., Ahn, D., Bekaii-Saab, T., Badi, N., Bloomston, P., et al.: Circulating interleukin-6 is associated with disease progression, but not cachexia in pancreatic cancer. Pancreatology. **19**(1), 80–87 (2019)

106. Bayliss, T., Smith, J., Schuster, M., Dragnev, K., Rigas, J.: A humanized anti-IL-6 antibody (ALD518) in non-small cell lung cancer. Expert. Opin. Biol. Ther. **11**(12), 1663–1668 (2011)

107. Tournadre, A., Pereira, B., Dutheil, F., Giraud, C., Courteix, D., Sapin, V., et al.: Changes in body composition and metabolic profile during interleukin

6 inhibition in rheumatoid arthritis. J. Cachexia. Sarcopenia Muscle. **8**, 639–646 (2017)

108. Callaway, C., Delitto, A., D'Lugos, A., Patel, R., Nosacka, R., Delitto, D., et al.: IL-8 released from human pancreatic cancer and tumor-associated stromal cells signals through a CXCR2-ERK1/2 axis to induce muscle atrophy. Cancers. **11**(12), 1863 (2019)

109. Antunes, J., Ferreira, R., Moreira-Goncalves, D.: Exercise training as therapy for cancer-induced cardiac cachexia. Trends Mol. Med. **24**(8), 709–727 (2018)

110. Hall, D., Griss, T., Ma, J., Sanchez, B., Sadek, J., Tremblay, A., et al.: The AMPK agonist 5-aminoimidazole-4-carboxamide ribonucleotide (AICAR), but not metformin, prevents inflammation-associated cachectic muscle wasting. EMBO Mol. Med. **10**(7), e8307 (2018)

111. VanderVeen, B., Fix, D., Carson, J.: Disrupted skeletal muscle mitochondrial dynamics, mitophagy and biogenesis during cancer cachexia: a role for inflammation. Oxidative Med. Cell. Longev. 2017;Article ID 3292087

112. Miller, J., Skipworth, R.: Novel molecular targets of muscle wasting in cancer patients. Curr. Opin. Clin. Nutr. Metab. Care. **22**(3), 196–204 (2019)

113. Brzeszczynska, J., Johns, N., Schlib, A., Degen, S., Langen, R., Schols, A., et al.: Loss of oxidative defense and potential blockade of satellite cell maturation in the skeletal muscle of patients with cancer but not in the healthy elderly. Aging. **8**(8), 1690–1702 (2016)

114. Temel, J.S., Abernethy, A.P., Currow, D.C., Friend, J., Duus, E.M., Yan, Y., et al.: Anamorelin in patients with non-small-cell lung cancer and cachexia (ROMANA 1 and ROMANA 2): results from two randomised, double-blind, phase 3 trials. Lancet Oncol. **17**(4), 519–531 (2016)

115. Currow, D., Temel, J.S., Abernethy, A., Milanowski, J., Friend, J., Fearon, K.C.: ROMANA 3: a phase 3 safety extension study of anamorelin in advanced non-small-cell lung cancer (NSCLC) patients with cachexia. Ann. Oncol. Off. J. Eur. Soc. Med. Oncol. **28**(8), 1949–1956 (2017)

116. Crawford, J., Johnston, M.A., Taylor, R.P., Dalton, J. T., Steiner, M.S.: Enobosarm and lean body mass in patients with non-small cell lung cancer. J. Clin. Oncol. [Internet]. (2014) [cited 2017 May 15];**32**, 5s (suppl; abstr 9618). Available from: http://meetinglibrary.asco.org/content/128938-144

117. ClinicalTrials.gov. Bethesda (MD): National Library of Medicine (US). 2000 Feb 29 – Identifier NCT02330926 Multimodal Intervention for Cachexia in Advanced Cancer Patients Undergoing Chemotherapy – 05-Jan-2015, Cited 04-Sept-2016 [Internet]. ClinicalTrials.gov. 2015 [cited 2016 Sep 4]. Available from: https://clinicaltrials.gov/ct2/show/NCT02330926

118. Solheim, T.S., Laird, B.J.A., Balstad, T.R., Stene, G. B., Bye, A., Johns, N., et al.: A randomized phase II feasibility trial of a multimodal intervention for the management of cachexia in lung and pancreatic cancer. J. Cachexia. Sarcopenia Muscle. (2017) Jun 14

119. Evans, W.J., Hellerstein, M., Orwoll, E., Cummings, S., Cawthon, P.M.: D3-Creatine dilution and the importance of accuracy in the assessment of skeletal muscle mass. J. Cachexia. Sarcopenia Muscle. **10**(1), 14–21 (2019)

120. Costa, R., Caro, P., de Matos-Neto, E., Lima, J., Radloff, K., Alves, M., et al.: Cancer cachexia induces morphological and inflammatory changes in the intestinal mucosa. J. Cachexia. Sarcopenia Muscle. **10**(5), 1116–1127 (2019)

121. Morrison, D., Preston, T.: Formation of short chain fatty acids by the gut microbiota and their impact on human metabolism. Gut Microbes. **7**, 189–200 (2016)

122. Miller, J., Laird, B., Skipworth, R.: The immunological regulation of cancer cachexia and its therapeutic implications. J. Cancer Metastasis Treat. **5**, 68 (2019)

123. Roberts, E., Deonarine, A., Jones, J., Denton, A., Feig, C., Lyons, S., et al.: Depletion of stromal cells expressing fibroblast activation protein-α from skeletal muscle and bone marrow results in cachexia and anemia. J. Exp. Med. **3**(210), 1137–1157 (2013)

124. Zhang, G., Liu, Z., Ding, H., Zhou, Y., Doan, H., Sin, K., et al.: Tumor induces muscle wasting in mice through releasing extracellular Hsp70 and Hsp90. Nat. Commun. **8**(589) (2017)

125. Burfeind, K., Zhu, X., Levasseur, P., Michaelis, K., Norgard, M., Marks, D.: TRIF is a key inflammatory mediator of acute sickness behavior and cancer cachexia. Brain Behav. Immun. **73**, 364–374 (2018)

Exercise: A Critical Component of Cachexia Prevention and Therapy in Cancer

Emidio E. Pistilli, Hannah E. Wilson, and David A. Stanton

Abstract

Cachexia is a condition characterized by loss of body weight as a result of chronic disease. In the setting of cancer, cachexia is more commonly observed in late stage and metastatic disease and the loss of body weight is attributed to varying combinations of muscle wasting and loss of adipose tissue. Cancer cachexia is a clinically relevant medical issue and is estimated to account for up to 20% of cancer mortality. Additionally, cachexia in cancer patients is associated with reduced tolerance to tumor-directed therapies.

Cross-sectional studies support the observation that people with greater levels of physical activity and exercise habits have lower risk of a cancer diagnosis, as well as a lower risk for cancer recurrence. Exercise recommendations for patients with cancer do not differ from those of healthy adults, and patients should strive to be as physically active as their condition allows. The purpose of this chapter is to define cachexia and identify clinical parameters used to diagnose this condition, discuss recommendations for exercise prescription in cancer patients, and provide an overview of the expected physiological responses to both aerobic and resistance training in cancer patients, in order to have a positive impact on the incidence and prevalence of cachexia. In general, aerobic exercise training is not associated with significant increases in body mass or in reversal of body weight loss in cancer patients. In contrast, studies do suggest that resistance exercise training has a role in maintenance or prevention of the onset of cachexia in cancer patients, although differing responses may occur between different cancer types. Therefore, we suggest that structured exercise programs in patients with cancer include resistance training to take advantage of these effects on lean body mass and muscle functional capacity.

E. E. Pistilli (✉)
Division of Exercise Physiology, Department of Human Performance, West Virginia University School of Medicine, Morgantown, WV, USA

Cancer Institute, West Virginia University School of Medicine, Morgantown, WV, USA

Department of Microbiology, Immunology, and Cell Biology, West Virginia University School of Medicine, Morgantown, WV, USA

West Virginia Clinical and Translational Sciences Institute, West Virginia University School of Medicine, Morgantown, WV, USA
e-mail: epistilli2@hsc.wvu.edu

H. E. Wilson
Cancer Institute, West Virginia University School of Medicine, Morgantown, WV, USA

D. A. Stanton
Division of Exercise Physiology, Department of Human Performance, West Virginia University School of Medicine, Morgantown, WV, USA

© Springer Nature Switzerland AG 2022
S. Acharyya (ed.), *The Systemic Effects of Advanced Cancer*,
https://doi.org/10.1007/978-3-031-09518-4_11

Learning Objectives

- Define the term cachexia and understand the clinical symptoms used to diagnose cachexia in cancer patients.
- Understand the differential effects that tumor growth can have on skeletal muscle remodeling, including effects on muscle mass as well as muscle function.
- Understand the different modes of exercise that can safely be performed by cancer patients and the recommendations for exercise prescription in cancer patients.
- Understand the physiological responses that can be expected in cancer patients when performing aerobic exercise training.
- Understand the physiological responses that can be expected in cancer patients when performing resistance training.

1   Introduction

Growth and proliferation of tumor cells alter normal physiological function of organ systems, both adjacent to and distant from the site of the primary tumor (reviewed in [1, 2]). There is a growing appreciation for the ability of tumors to induce alterations in gene and protein expression patterns, cell metabolism, etc., as a way to fuel tumor cell proliferation and growth, at the expense of the hosts' tissues [1]. Skeletal muscle, as an organ system, represents approximately 40% of total body mass, and muscle breakdown can provide substrates to fuel tumor cell metabolism, in the form of amino acids, glucose and glycogen, and fats [3, 4]. It is common to observe significant decreases in overall body mass in the late stages of cancer as well as decreases in muscle functional ability, a syndrome that has been termed cachexia [5]. Different tumor types appear to induce differing responses in skeletal muscle, including varying degrees of muscle wasting and muscle fatigue [6]. The term cancer cachexia has been used to describe these tumor-induced alterations in skeletal muscle over many millennia

[7]. Cancer cachexia is a clinically relevant medical issue, is estimated to account for up to 20% of cancer mortality and is associated with reduced tolerance to tumor-directed therapies [8]. It is our belief that exercise training has the capacity to not only improve muscle mass and muscle function in cancer patients, but may also help reduce overall cancer mortality by allowing patients to better tolerate tumor-directed therapies. Therefore, the purpose of this chapter is to define cachexia and identify clinical parameters used to diagnose this condition, discuss recommendations for exercise prescription in cancer patients, and provide an overview of the expected physiological responses to both aerobic and resistance training in cancer patients, in order to have a positive impact on the incidence and prevalence of cachexia.

2   Defining Cancer Cachexia

The defining characteristic of cancer cachexia is the loss of body mass, which can be the result of both loss of skeletal muscle mass as well as adipose tissue mass [6, 8]. The prevalence of cachexia and average percentage of weight loss varies between cancer types, with breast and hematological cancers showing relatively low prevalence and pancreatic and colon cancers showing a high prevalence [6]. For a patient to be considered cachectic by published standards [9, 10], a 5% or greater loss in body mass within a 6 month period of time, or 0.027% average daily weight loss, must occur [11]. Calculation of the skeletal muscle index of specific skeletal muscles using CT scans can also be used to determine the extent of muscle loss and the appearance of cachexia [12]. Recently, it has been suggested that in addition to body weight loss, additional factors including chronic inflammation, fatigue, decreased muscle strength, and reduced food intake are important in predicting overall survival in patients with cachexia [13]. Once cachexia becomes apparent, nutritional interventions, such as increasing caloric intake, are not successful at regaining or attenuating the loss of body mass [9]. This underscores the need to be proactive in maintaining body mass in the face of a cancer diagnosis and treatment, and exercise can be a useful tool.

Connected with the loss of muscle mass is the idea of a decrease in muscle functional capacity in cancer patients. Muscle weakness is a symptom of muscle mass loss, as smaller muscles produce less force (i.e., decreased maximal strength). For example, in the late stages of cancer when patients have lost a significant percentage of muscle mass, the existing muscle mass will produce less force simply due to the fact that less muscle tissue is available to produce force. This has led to the dogma that cancer-induced muscle dysfunction is linked to muscle wasting. However, the decreased maximal strength that results from muscle wasting is different from muscle fatigue, which can be defined as a decrease in muscle force due to repeated contractions or extended periods of exertion and can occur in the absence of muscle wasting [14, 15]. For example, early-stage breast cancer patients commonly report a significant degree of overall fatigue while typically remaining weight stable [6, 11, 16–18]. Clinicians should be aware that not all patients with cancer will exhibit dramatic changes in body weight and/or muscle mass during tumor growth or tumor-directed therapies. However, this does not necessarily imply that muscle dysfunction, in the form of persistent fatigue, is not adversely affecting the quality of life of the patient. Based on recent data in early-stage breast cancer patients, skeletal muscle is responding at the early stages of tumor growth, and these changes would predict the dysregulation of pathways associated with muscle fatigue [19, 20].

In addition to noted differences in cachexia between cancer types, there are reported differences in cachexia when comparing men and women (reviewed in [21]). In terms of overall body composition, males have greater skeletal muscle mass than females, while females have a greater percentage of adipose tissue. Within these tissues themselves, there are significant differences in the proteomic, transcriptomic, and metabolomics landscapes. In fact, skeletal muscle and subcutaneous adipose tissue are two of the three most sexually dimorphic human tissues, with mammary tissue unsurprisingly ranking as the most sexually dimorphic tissue [22]. In general, men with cancer report approximately double the incidence of cachexia as women [23, 24]. In addition, multiple studies have shown that in men, loss of muscle function correlates with loss of muscle mass, but this does not appear to be the case in women. That is, women experience muscle dysfunction without a commensurate loss of muscle mass [25]. An interesting study by Rune et al. [26] showed sexual dimorphism in glucose and lipid metabolism in human muscle biopsies, but these differences disappeared when primary human myoblasts were cultured in vitro, suggesting that this metabolic dimorphism results from systemic factors rather than intrinsic differences between male and female skeletal muscle. While it may be tempting to attribute these differences entirely to sex hormones, an analysis of the murine skeletal muscle proteome by Metskas et al. [27] found no difference in abundance or activation state of proteins that are known to be hormonally regulated, such as hormone-sensitive lipase, sex steroid receptors, or GLUT4. Collectively, it appears that cachexia incidence and prevalence differ based on sex; however, the exact mechanisms that underlie these observations are not fully understood.

3   Associations of Exercise, Physical Activity and Risk of Developing Cancer

Strong experimental evidence exists to support the notion that individuals with greater levels of physical activity have a reduced risk for the development of multiple types of cancer [28–30]. In a large meta-analysis that included 1.44 million adults, those with physical activity levels at the 90th percentile had a 7% lower risk of total cancer, with some cancers having as high as a 20% lower risk. The seven cancer types with the strongest inverse association between risk and activity levels were esophageal, liver, lung, kidney, gastric, endometrium, and myeloid leukemia, while a moderately strong association was found for myeloma, colon, head and neck, rectal, bladder, and breast. Interestingly, this study suggested that these associations with physical activity and cancer risk maybe independent of obesity levels and smoking history, presenting a strong argument for physical activity and exercise in cancer prevention [31]. Studies focused on specific cancer types are largely supportive of these findings. For example, greater physical activity is associated with a 25% reduction in the risk of colon cancer, and this association may be independent of changes in body composition. The Melbourne Collaboration Cohort Study found that greater pre-diagnosis exercise was associated with enhanced disease-specific survival [32]. The Collaborative Women's Longevity Study found a 51% decrease in breast cancer mortality in the most physically active study participants [33]. Strong evidence for this relationship existed for estrogen receptor (ER) and progesterone receptor (PR) positive tumors, while recent data now suggests this relationship exists for hormone receptor negative breast cancers [31]. A similar mortality risk reduction of 25% was observed for endometrial cancer and those with greater physical activity levels [34]. Interestingly, the mode of exercise does not seem to make a difference, as performance of resistance training also seems to reduce risk of specific cancers, including colon and kidney cancers [35].

While the data are certainly compelling for lifetime physical activity and lower cancer risk, data also exist to support the association of post-cancer diagnosis physical activity and risk of cancer recurrence. Specifically, increasing physical activity levels may lead to more positive prognoses as well as a decrease in overall mortality and cancer-specific mortality for patients diagnosed with breast, colon, and pancreatic cancers [33, 36, 37]. This benefit may be observed in response to as little as three hours of physical activity accumulated over the course of one week, with one study observing no additional benefits with greater levels of activity [38, 39]. Despite these clear associations between physical activity and lower cancer mortality, one study of over 1200 breast cancer survivors report that only 32% of participants engage in the currently recommended amount of physical activity

[40]. Additionally, data from the Centers of Disease Control and Prevention show that only 22.9% of Americans engage in the daily recommended levels of aerobic and resistance-based exercise, with 27.2% of men and 18.7% of women meeting the guidelines [41]. Collectively, the data present a very strong argument for greater amounts of exercise and physical activity in reducing the risk of cancer [31, 42–45]. However, efforts to determine the barriers to being physically active would be beneficial, since inverse relationships are observed between pre-diagnosis exercise levels and overall risk of developing cancer as well as post-diagnosis exercise levels and risk of cancer recurrence.

4 Exercise and Physical Activity in Preventing and Treating Cancer Cachexia

Exercise prescription guidelines for cancer patients have been established, and interestingly, do not differ from the guidelines for healthy adults. The US Department of Health and Human Services provided the Physical Activity Guidelines for Americans (PAGA), with recommendations on aerobic, resistance, and flexibility exercise [46]. The guidelines recommend accumulating 150 min of moderate-intensity or 75 min of vigorous-intensity aerobic exercise per week, moderate-intensity resistance training for all major muscles of the body at least 2 days per week, and flexibility exercises for all major muscle groups on days that physical activity is performed. In general, this exercise prescription for healthy adults is not modified for cancer patients; in other words, cancer patients are encouraged to be as physically active as their individual condition allows. Additionally, individuals that have a history of exercise training and are diagnosed with cancer are not encouraged to reduce their exercise levels following the diagnosis. With regards to specific cancer types, the American College of Sports Medicine (ACSM) provided specific adaptations or modifications to these guidelines for the cancer patient [47]. However, the conclusions from both of these reports are that any physical activity is beneficial in patients with cancer, and patients should strive to be as physically active as possible through treatment and survivorship. While the health benefits of regular physical activity by cancer patients, in the form of aerobic, resistance and flexibility training are accepted, the ability of exercise training to prevent the onset of cachexia or reverse the appearance of cachexia once it is diagnosed is currently not as well established. Therefore, in the following sections, we will provide an overview of studies conducted in cancer patients utilizing both aerobic-based and resistance-based exercise training and the observations on overall body mass, muscle mass, and muscle function.

4.1 Aerobic Exercise Training and Cancer Cachexia

Aerobic exercise training consists of continuous or intermittent exercise that is performed at a low relative intensity for greater than 2 min (typically greater than 20 min), and relies on oxidative metabolism for energy demands. The expected adaptations to aerobic exercise training in healthy adults include increased maximal oxygen uptake (i.e., VO_{2max}), increased respiratory capacity, increased muscle levels of mitochondria and aerobic enzyme activities, increased muscle capillary density, and reduced body fat [48]. While some studies have shown an increase in the muscle fiber area of type I muscle fibers following aerobic exercise training [49], increases in muscle mass in general are not an expected adaptation to aerobic exercise training. Therefore, the question that will be addressed in this section is: Does aerobic exercise training prevent the onset of or reverse the appearance of cachexia in cancer patients?

Aerobic exercise training has been extensively studied in patients with breast cancer. Among studies that have utilized extended periods of aerobic exercise training (>12 weeks), there were minimal effects in measures of overall body weight or lean body mass. Specifically, training that included walking/running [50–53] or moderate-intensity tai chi [54] showed no

change in body weight, body composition, or lean body mass. However, one study did report significant decreases in breast-associated pain and swelling in the training group [50], and a second study showed improvements in measures of exercise performance, such that jumping power and agility running tests were improved in the training group compared to a control group [53]. These results are consistent with the fact that breast cancer has one of the lowest rates and prevalence of cachexia and women with cancer typically report a significant degree of muscle fatigue while remaining weight stable [6, 11].

In one particularly interesting study [55], physically inactive post-menopausal breast cancer survivors, ranging from 1 to 10 years post-breast cancer diagnosis, were randomized to either a standard-of-care group or aerobic training group. Aerobic training consisted of three supervised training sessions per week and two at-home sessions per week, progressing to 30-minute sessions at 60–80% maximum heart rate. The aerobic training group had a significant decrease in body fat percentage, while also showing a significant increase in lean body mass compared to the standard-of-care group. While this study included patients with a wide range of post-diagnosis durations, it does suggest that aerobic exercise training can affect skeletal muscle and lean body mass in post-menopausal survivors of breast cancer.

Aerobic exercise training in men with prostate cancer has not been associated with maintenance of lean body mass, although there are fewer studies in this specific cancer type. In men receiving androgen deprivation therapy (ADT) for at least 6 months, 30 min of brisk walking for 5 days per week resulted in significant decreases in overall body weight and body mass index (BMI), as well as decreases in fat mass, compared to the non-exercise group. This study also reported a non-significant increase in lean body mass in the exercise training group [56]. In a shorter 12-week intervention study, men with prostate cancer receiving ADT and healthy controls performed cycle ergometer training for 3 days per week in which participants performed intervals that approached 100% VO_{2max}. In support of the prior observations, body weight and body fat decreased with exercise, in both prostate cancer patients and healthy controls, with no change in lean body mass. It should be noted that this study did not include a non-exercise control group [57]. From these studies, it seems as though aerobic exercise training for up to 6 months in men diagnosed with prostate cancer and receiving ADT has minimal effects at either maintaining or improving lean body mass.

In cancers of the gastrointestinal tract, including colorectal and esophageal cancers, both cycle ergometer training and supervised walking had little impact on maintaining or improving lean body mass. Specifically, cycle ergometer training in patients with stage II/III colorectal cancer post-chemotherapy was associated with reductions in both body mass and body fat percentage [58]. Nurse-supervised walking 3 days per week in patients with esophageal cancer receiving neoadjuvant chemotherapy was associated with less of a decrease in body weight, as well as less of a decrease in walking distance and handgrip strength [59]. Collectively, aerobic exercise training in patients with cancer appears to have minimal effects on preserving or improving lean body mass, and is actually associated with decreases in overall body weight and often body fat percentage. However, it should be noted that aerobic exercise has other potential benefits for the cancer patient, related to immune system surveillance and general mood and well-being [60]. Therefore, despite minimal effects on lean body mass, aerobic exercise training should be encouraged, as tolerated, as part of the total exercise prescription in patients and survivors.

4.2   Resistance Training and Cancer Cachexia

Resistance training is a term used to collectively classify exercise training that focuses on contracting skeletal muscles against an external resistance, with the goals to increase muscle mass and/or muscle strength. Muscle mass can be increased following resistance training through a process of training-induced protein synthesis, in which skeletal muscle proteins are synthesized and added to existing muscle tissue [61]. Consistent resistance training that results in greater muscle size will also induce increases in muscle strength, since a larger muscle is a stronger muscle. However, resistance training can also be performed with the goal to increase resistance to fatigue. In general, resistance training will induce both structural and functional changes within skeletal muscle, in the form of greater muscle fiber size, alterations in muscle fiber type composition, changes in muscle fiber architecture, and alterations in enzyme and substrate concentrations and utilization [62]. Therefore, the question that will be addressed in this section is: Does anaerobic/strength training prevent the onset of or reverse the appearance of cachexia in cancer patients?

A significant number of studies utilizing resistance training alone or in combination with aerobic training have been completed in men with prostate cancer and in women with breast cancer. These studies collectively show an association between resistance exercise training and improvements in lean body mass, muscle strength, and/or cancer-related fatigue. In a study by Segal et al. [63], men with prostate cancer that received ADT performed 3 days per week of RT for 12 total weeks that included training sets of 8–12 repetitions at 60–80% one repetition maximum (1RM). Following the intervention, the participants demonstrated significant improvements in muscular fitness, with no decreases in overall body mass. Similarly, 12 weeks of combined aerobic and resistance training in men with prostate cancer receiving ADT was associated with increased lean muscle mass, increased strength, and reductions in cancer-related fatigue [64]. Twenty-four weeks of combined aerobic and circuit-style strength training in men with prostate cancer receiving ADT was also associated with improvements in body composition and measures of metabolism, including oxidation of fats and glucose utilization [65]. Collectively, these studies demonstrate that muscle adaptations to strength training can be expected in men with cancer, even in the face of androgen deprivation therapy.

Exercise interventions in patients with breast cancer, using resistance training alone or in combination with aerobic exercise, are associated with both gains in lean muscle mass and strength while also helping to improve cancer-related fatigue and measures of quality of life. For example, resistance training for as little as 8 weeks was associated with greater strength measures following the intervention [66]. Longer interventions were also associated with significant gains in lean muscle mass [67–69]. In addition, numerous studies report improvements in measures of breast cancer-related fatigue [66, 68, 70]. Improvements in cancer-related fatigue were evident in patients receiving either standard adjuvant chemotherapy [71] or adjuvant radiotherapy [72]. In a study by Adamsen et al. [73], 269 patients receiving chemotherapy for numerous solid and hematological cancers showed improvements in strength, measures of quality of life and medical outcomes in response to high-intensity resistance training performed 3 days per week for only 6 weeks. Collectively, these data highlight the positive influence resistance training can have in patients with cancer on measures of lean body mass and muscle strength.

A potential side effect of resistance training in patients with breast cancer is the appearance of lymphedema, especially following axillary lymph node dissection or sentinel lymph node biopsy. Lymphedema is characterized by the accumulation of fluid within tissues of the upper arm attributed to the disruption of lymphatic pathways following surgery [74]. Recent studies have provided compelling data that progressive strength training of the upper body does not increase the risk of lymphedema or exacerbate lymphedema-associated symptoms (reviewed in [75]). In fact, the intermittent muscular contractions performed during strength training may actually improve the flow of lymph through active limbs [76]. Additionally, the data acquired in breast cancer patients that performed resistance training, using either low-intensity or moderate-high-intensity training methods, demonstrate gains in muscle strength of the trained muscle groups [70, 77–79]. Lack of physical activity and obesity are associated with greater risk of

lymphedema [80], providing support for the inclusion of resistance exercise training for the breast cancer patient. Importantly, the data strongly suggest that strength training can be performed without risk of developing or exacerbating symptoms of lymphedema. Recommendations for prescribing resistance exercise training in patients with either risk of developing lymphedema or currently experiencing lymphedema include 2–3 days per week of resistance exercises targeting all the major muscle groups of the body, performed on non-consecutive days, and with conservative increments in training loads [75].

5 Conclusions and Perspectives

Cancer cachexia remains a clinically relevant syndrome, which is estimated to account for up 20% of cancer-related mortality, especially in late-stage disease and metastatic settings. Despite this well-known statistic, therapeutic interventions aimed at preventing the onset or attenuating the rate of cachexia have largely been unsuccessful. The goal of this chapter was to present data to support the inclusion of structured exercise training in patients with cancer, as a means to preserve both lean muscle mass as well as muscle function. Collectively, the data seem to suggest that resistance training, performed either alone or in combination with aerobic training, represents a safe and effective strategy to maintain muscle function and in some cases, increase lean muscle mass in patients with cancer. Both moderate-intensity and high-intensity resistance training performed at least 2 days per week is associated with increases in muscle strength, improvements in cancer-related fatigue and in maintenance or improvement in muscle mass. Interestingly, improvements in muscle mass have been observed in men with prostate cancer that are receiving androgen deprivation therapy and in women with breast cancer that are receiving both adjuvant chemotherapy and radiotherapy. Therefore, we propose that patients diagnosed with cancer engage in regular structured exercise training as their individual condition allows, which includes sessions dedicated to resistance training exercises targeting all the major skeletal muscles of the body. Resistance training, based on PAGA and ACSM recommendations, will potentially provide the optimal stimulus for the prevention of cachexia and associated muscle dysfunction that can occur in the setting of cancer.

References

1. DeBerardinis, R.J., Chandel, N.S.: Fundamentals of cancer metabolism. Sci. Adv. **2**, e1600200 (2016)
2. Hanahan, D., Weinberg, R.A.: The hallmarks of cancer. Cell. **100**, 57–70 (2000)
3. Heymsfield, S.B., Waki, M., Kehayias, J., Lichtman, S., Dilmanian, F.A., Kamen, Y., Wang, J., Pierson Jr., R.N.: Chemical and elemental analysis of humans in vivo using improved body composition models. Am. J. Phys. **261**, E190–E198 (1991)
4. Pietrobelli, A., Heymsfield, S.B., Wang, Z.M., Gallagher, D.: Multi-component body composition models: recent advances and future directions. Eur. J. Clin. Nutr. **55**, 69–75 (2001)
5. Lok, C.: Cachexia: The last illness. Nature. **528**, 182–183 (2015)
6. Baracos, V.E., Martin, L., Korc, M., Guttridge, D.C., Fearon, K.C.H.: Cancer-associated cachexia. Nat. Rev. Dis. Primers. **4**, 17105 (2018)
7. Katz, A.M., Katz, P.B.: Diseases of the heart in the works of Hippocrates. Br. Heart J. **24**, 257–264 (1962)
8. Argiles, J.M., Busquets, S., Stemmler, B., Lopez-Soriano, F.J.: Cancer cachexia: understanding the molecular basis. Nat. Rev. Cancer. **14**, 754–762 (2014)
9. Evans, W.J., Morley, J.E., Argiles, J., Bales, C., Baracos, V., Guttridge, D., Jatoi, A., Kalantar-Zadeh, K., Lochs, H., Mantovani, G., Marks, D., Mitch, W.E., Muscaritoli, M., Najand, A., Ponikowski, P., Rossi Fanelli, F., Schambelan, M., Schols, A., Schuster, M., Thomas, D., Wolfe, R., Anker, S.D.: Cachexia: a new definition. Clin. Nutr. **27**, 793–799 (2008)
10. Fearon, K., Strasser, F., Anker, S.D., Bosaeus, I., Bruera, E., Fainsinger, R.L., Jatoi, A., Loprinzi, C., MacDonald, N., Mantovani, G., Davis, M., Muscaritoli, M., Ottery, F., Radbruch, L., Ravasco, P., Walsh, D., Wilcock, A., Kaasa, S., Baracos, V.E.: Definition and classification of cancer cachexia: an international consensus. Lancet Oncol. **12**, 489–495 (2011)
11. Wilson, H.E., Stanton, D.A., Montgomery, C., Infante, A.M., Taylor, M., Hazard-Jenkins, H., Pugacheva, E.N., Pistilli, E.E.: Skeletal muscle reprogramming by breast cancer regardless of treatment history or tumor molecular subtype. NPJ Breast Cancer. **6**, 18 (2020)
12. Delitto, D., Judge, S.M., George Jr., T.J., Sarosi, G.A., Thomas, R.M., Behrns, K.E., Hughes, S.J., Judge, A.

R., Trevino, J.G.: A clinically applicable muscular index predicts long-term survival in resectable pancreatic cancer. Surgery. **161**, 930–938 (2017)
13. Vanhoutte, G., van de Wiel, M., Wouters, K., Sels, M., Bartolomeeussen, L., De Keersmaecker, S., Verschueren, C., De Vroey, V., De Wilde, A., Smits, E., Cheung, K.J., De Clerck, L., Aerts, P., Baert, D., Vandoninck, C., Kindt, S., Schelfhaut, S., Vankerkhoven, M., Troch, A., Ceulemans, L., Vandenbergh, H., Leys, S., Rondou, T., Dewitte, E., Maes, K., Pauwels, P., De Winter, B., Van Gaal, L., Ysebaert, D., Peeters, M.: Cachexia in cancer: what is in the definition? BMJ Open Gastroenterol. **3**, e000097 (2016)
14. Evans, W.J., Lambert, C.P.: Physiological basis of fatigue. Am. J. Phys. Med. Rehabil. **86**, S29–S46 (2007)
15. Neefjes, E.C.W., van den Hurk, R.M., Blauwhoff-Buskermolen, S., van der Vorst, M., Becker-Commissaris, A., de van der Schueren, M.A.E., Buffart, L.M., Verheul, H.M.W.: Muscle mass as a target to reduce fatigue in patients with advanced cancer. J. Cachexia. Sarcopenia Muscle. **8**, 623–629 (2017)
16. Bower, J.E., Ganz, P.A., Desmond, K.A., Rowland, J.H., Meyerowitz, B.E., Belin, T.R.: Fatigue in breast cancer survivors: occurrence, correlates, and impact on quality of life. J. Clin. Oncol. **18**, 743–753 (2000)
17. Cella, D., Lai, J.S., Chang, C.H., Peterman, A., Slavin, M.: Fatigue in cancer patients compared with fatigue in the general United States population. Cancer. **94**, 528–538 (2002)
18. Curt, G.A.: The impact of fatigue on patients with cancer: overview of FATIGUE 1 and 2. Oncologist. **5**(Suppl 2), 9–12 (2000)
19. Bohlen, J., McLaughlin, S.L., Hazard-Jenkins, H., Infante, A.M., Montgomery, C., Davis, M., Pistilli, E.E.: Dysregulation of metabolic-associated pathways in muscle of breast cancer patients: preclinical evaluation of interleukin-15 targeting fatigue. J. Cachexia. Sarcopenia Muscle. **9**, 701–714 (2018)
20. Wilson, H.E., Rhodes, K.K., Rodriguez, D., Chahal, I., Stanton, D.A., Bohlen, J., Davis, M., Infante, A.M., Hazard-Jenkins, H., Klinke, D.J., Pugacheva, E.N., Pistilli, E.E.: Human breast cancer xenograft model implicates peroxisome proliferator-activated receptor signaling as driver of cancer-induced muscle fatigue. Clin. Cancer Res. **25**, 2336–2347 (2019)
21. Montalvo, R.N., Counts, B.R., Carson, J.A.: Understanding sex differences in the regulation of cancer-induced muscle wasting. Curr. Opin. Support. Palliat. Care. **12**, 394–403 (2018)
22. Gershoni, M., Pietrokovski, S.: The landscape of sex-differential transcriptome and its consequent selection in human adults. BMC Biol. **15**, 7 (2017)
23. Anderson, L.J., Liu, H., Garcia, J.M.: Sex differences in muscle wasting. Adv. Exp. Med. Biol. **1043**, 153–197 (2017)

24. Wallengren, O., Iresjo, B.M., Lundholm, K., Bosaeus, I.: Loss of muscle mass in the end of life in patients with advanced cancer. Support Care Cancer. **23**, 79–86 (2015)

25. Stephens, N.A., Gray, C., MacDonald, A.J., Tan, B.H., Gallagher, I.J., Skipworth, R.J., Ross, J.A., Fearon, K.C., Greig, C.A.: Sexual dimorphism modulates the impact of cancer cachexia on lower limb muscle mass and function. Clin. Nutr. **31**, 499–505 (2012)

26. Rune, A., Salehzadeh, F., Szekeres, F., Kuhn, I., Osler, M.E., Al-Khalili, L.: Evidence against a sexual dimorphism in glucose and fatty acid metabolism in skeletal muscle cultures from age-matched men and post-menopausal women. Acta Physiol (Oxf.). **197**, 207–215 (2009)

27. Metskas, L.A., Kulp, M., Scordilis, S.P.: Gender dimorphism in the exercise-naive murine skeletal muscle proteome. Cell. Mol. Biol. Lett. **15**, 507–516 (2010)

28. Friedenreich, C.M., Neilson, H.K., Lynch, B.M.: State of the epidemiological evidence on physical activity and cancer prevention. Eur. J. Cancer. **46**, 2593–2604 (2010)

29. Neilson, H.K., Friedenreich, C.M.: Lifestyle factors associated with cancer incidence, recurrence and survival. In: Irwin, M.L. (ed.) ACSM's guide to exercise and cancer survivorship, pp. 29–47. Human Kinetics, Illinois (2012)

30. Thune, I., Furberg, A.S.: Physical activity and cancer risk: dose-response and cancer, all sites and site-specific. Med. Sci. Sports Exerc. **33**, S530–50 (2001); discussion S609–10

31. Moore, S.C., Lee, I.M., Weiderpass, E., Campbell, P.T., Sampson, J.N., Kitahara, C.M., Keadle, S.K., Arem, H., Berrington de Gonzalez, A., Hartge, P., Adami, H.O., Blair, C.K., Borch, K.B., Boyd, E., Check, D.P., Fournier, A., Freedman, N.D., Gunter, M., Johannson, M., Khaw, K.T., Linet, M.S., Orsini, N., Park, Y., Riboli, E., Robien, K., Schairer, C., Sesso, H., Spriggs, M., Van Dusen, R., Wolk, A., Matthews, C.E., Patel, A.V.: Association of leisure-time physical activity with risk of 26 types of cancer in 1.44 million adults. JAMA Intern. Med. **176**, 816–825 (2016)

32. Haydon, A.M., Macinnis, R.J., English, D.R., Giles, G.G.: Effect of physical activity and body size on survival after diagnosis with colorectal cancer. Gut. **55**, 62–67 (2006)

33. Holick, C.N., Newcomb, P.A., Trentham-Dietz, A., Titus-Ernstoff, L., Bersch, A.J., Stampfer, M.J., Baron, J.A., Egan, K.M., Willett, W.C.: Physical activity and survival after diagnosis of invasive breast cancer. Cancer Epidemiol. Biomark. Prev. **17**, 379–386 (2008)

34. Cust, A.E.: Physical activity and gynecologic cancer prevention. Recent Results Cancer Res. **186**, 159–185 (2011)

35. Mazzilli, K.M., Matthews, C.E., Salerno, E.A., Moore, S.C.: Weight training and risk of 10 common types of cancer. Med. Sci. Sports Exerc. **51**, 1845–1851 (2019)

36. Irwin, M.L., Smith, A.W., McTiernan, A., Ballard-Barbash, R., Cronin, K., Gilliland, F.D., Baumgartner, R.N., Baumgartner, K.B., Bernstein, L.: Influence of pre- and postdiagnosis physical activity on mortality in breast cancer survivors: the health, eating, activity, and lifestyle study. J. Clin. Oncol. **26**, 3958–3964 (2008)

37. Meyerhardt, J.A., Heseltine, D., Niedzwiecki, D., Hollis, D., Saltz, L.B., Mayer, R.J., Thomas, J., Nelson, H., Whittom, R., Hantel, A., Schilsky, R.L., Fuchs, C.S.: Impact of physical activity on cancer recurrence and survival in patients with stage III colon cancer: findings from CALGB 89803. J. Clin. Oncol. **24**, 3535–3541 (2006)

38. Holmes, M.D., Chen, W.Y., Feskanich, D., Kroenke, C.H., Colditz, G.A.: Physical activity and survival after breast cancer diagnosis. JAMA. **293**, 2479–2486 (2005)

39. Kenfield, S.A., Stampfer, M.J., Giovannucci, E., Chan, J.M.: Physical activity and survival after prostate cancer diagnosis in the health professionals follow-up study. J. Clin. Oncol. **29**, 726–732 (2011)

40. Irwin, M.L., McTiernan, A., Bernstein, L., Gilliland, F.D., Baumgartner, R., Baumgartner, K., Ballard-Barbash, R.: Physical activity levels among breast cancer survivors. Med. Sci. Sports Exerc. **36**, 1484–1491 (2004)

41. Blackwell, D.L., Clarke, T.C.: State variation in meeting the 2008 federal guidelines for both aerobic and muscle-strengthening activities through leisure-time physical activity among adults aged 18–64: United States, 2010–2015. Natl Health Stat Report, 1–22 (2018)

42. Bernstein, L., Patel, A.V., Ursin, G., Sullivan-Halley, J., Press, M.F., Deapen, D., Berlin, J.A., Daling, J.R., McDonald, J.A., Norman, S.A., Malone, K.E., Strom, B.L., Liff, J., Folger, S.G., Simon, M.S., Burkman, R.T., Marchbanks, P.A., Weiss, L.K., Spirtas, R.: Lifetime recreational exercise activity and breast cancer risk among black women and white women. J. Natl. Cancer Inst. **97**, 1671–1679 (2005)

43. Friedenreich, C.M., Gregory, J., Kopciuk, K.A., Mackey, J.R., Courneya, K.S.: Prospective cohort study of lifetime physical activity and breast cancer survival. Int. J. Cancer. **124**, 1954–1962 (2009)

44. McTiernan, A., Kooperberg, C., White, E., Wilcox, S., Coates, R., Adams-Campbell, L.L., Woods, N., Ockene, J., Women's Health Initiative Cohort, S.: Recreational physical activity and the risk of breast cancer in postmenopausal women: the Women's Health Initiative Cohort Study. JAMA. **290**, 1331–1336 (2003)

45. Moorman, P.G., Jones, L.W., Akushevich, L., Schildkraut, J.M.: Recreational physical activity and ovarian cancer risk and survival. Ann. Epidemiol. **21**, 178–187 (2011)

46. Committee, P. A. G. A.: Physical activity guidelines advisory committee report: US Department of Health and Human Services (2008)
47. Schmitz, K.H., Courneya, K.S., Matthews, C., Demark-Wahnefried, W., Galvao, D.A., Pinto, B.M., Irwin, M.L., Wolin, K.Y., Segal, R.J., Lucia, A., Schneider, C.M., von Gruenigen, V.E., Schwartz, A. L., American College of Sports, M: American College of Sports Medicine roundtable on exercise guidelines for cancer survivors. Med. Sci. Sports Exerc. **42**, 1409–1426 (2010)
48. Swank, A., Sharp, C.: Adaptations to aerobic endurance training programs. In: Haff, G.G., Triplett, N.T. (eds.) Essentials of strength training and conditioning, pp. 115–133. Human Kinetics, Illinois (2016)
49. Costill, D.L., Daniels, J., Evans, W., Fink, W., Krahenbuhl, G., Saltin, B.: Skeletal muscle enzymes and fiber composition in male and female track athletes. J. Appl. Physiol. **40**, 149–154 (1976)
50. Backman, M., Wengstrom, Y., Johansson, B., Skoldengen, I., Borjesson, S., Tarnbro, S., Berglund, A.: A randomized pilot study with daily walking during adjuvant chemotherapy for patients with breast and colorectal cancer. Acta Oncol. **53**, 510–520 (2014)
51. DeNysschen, C.A., Burton, H., Ademuyiwa, F., Levine, E., Tetewsky, S., O'Connor, T.: Exercise intervention in breast cancer patients with aromatase inhibitor-associated arthralgia: a pilot study. Eur. J. Cancer Care (Engl). **23**, 493–501 (2014)
52. Matthews, C.E., Wilcox, S., Hanby, C.L., Der Ananian, C., Heiney, S.P., Gebretsadik, T., Shintani, A.: Evaluation of a 12-week home-based walking intervention for breast cancer survivors. Support Care Cancer. **15**, 203–211 (2007)
53. Nikander, R., Sievanen, H., Ojala, K., Oivanen, T., Kellokumpu-Lehtinen, P.L., Saarto, T.: Effect of a vigorous aerobic regimen on physical performance in breast cancer patients – a randomized controlled pilot trial. Acta Oncol. **46**, 181–186 (2007)
54. Janelsins, M.C., Davis, P.G., Wideman, L., Katula, J. A., Sprod, L.K., Peppone, L.J., Palesh, O.G., Heckler, C.E., Williams, J.P., Morrow, G.R., Mustian, K.M.: Effects of Tai Chi Chuan on insulin and cytokine levels in a randomized controlled pilot study on breast cancer survivors. Clin. Breast Cancer. **11**, 161–170 (2011)
55. Irwin, M.L., Alvarez-Reeves, M., Cadmus, L., Mierzejewski, E., Mayne, S.T., Yu, H., Chung, G.G., Jones, B., Knobf, M.T., DiPietro, L.: Exercise improves body fat, lean mass, and bone mass in breast cancer survivors. Obesity (Silver Spring). **17**, 1534–1541 (2009)
56. O'Neill, R.F., Haseen, F., Murray, L.J., O'Sullivan, J. M., Cantwell, M.M.: A randomised controlled trial to evaluate the efficacy of a 6-month dietary and physical activity intervention for patients receiving androgen deprivation therapy for prostate cancer. J. Cancer Surviv. **9**, 431–440 (2015)
57. Hvid, T., Winding, K., Rinnov, A., Dejgaard, T., Thomsen, C., Iversen, P., Brasso, K., Mikines, K.J., van Hall, G., Lindegaard, B., Solomon, T.P., Pedersen, B.K.: Endurance training improves insulin sensitivity and body composition in prostate cancer patients treated with androgen deprivation therapy. Endocr. Relat. Cancer. **20**, 621–632 (2013)
58. Piringer, G., Fridrik, M., Fridrik, A., Leiherer, A., Zabernigg, A., Greil, R., Eisterer, W., Tschmelitsch, J., Lang, A., Frantal, S., Burgstaller, S., Gnant, M., Thaler, J., Austrian, B., Colorectal Cancer Study, G: A prospective, multicenter pilot study to investigate the feasibility and safety of a 1 year controlled exercise training after adjuvant chemotherapy in colorectal cancer patients. Support Care Cancer. **26**, 1345–1352 (2018)
59. Xu, Y.J., Cheng, J.C., Lee, J.M., Huang, P.M., Huang, G.H., Chen, C.C.: A walk-and-eat intervention improves outcomes for patients with esophageal cancer undergoing neoadjuvant chemoradiotherapy. Oncologist. **20**, 1216–1222 (2015)
60. Idorn, M., Hojman, P.: Exercise-dependent regulation of NK cells in cancer protection. Trends Mol. Med. **22**, 565–577 (2016)
61. Tsika, R.: The muscular system: the control of muscle mass. In: Farrell, P.A., Joyner, M.J., Caiozzo, V.J. (eds.) ACSMs advanced exercise physiology, pp. 152–170. Lippincot Williams and Wilkens (2012)
62. French, D.: Adaptations to anaerobic training programs. In: Haff, G.G., Triplett, N.T. (eds.) Essentials of strength training and conditioning, pp. 87–113. Human Kinetics, Illinois (2016)
63. Segal, R.J., Reid, R.D., Courneya, K.S., Malone, S.C., Parliament, M.B., Scott, C.G., Venner, P.M., Quinney, H.A., Jones, L.W., D'Angelo, M.E., Wells, G.A.: Resistance exercise in men receiving androgen deprivation therapy for prostate cancer. J. Clin. Oncol. **21**, 1653–1659 (2003)
64. Galvao, D.A., Taaffe, D.R., Spry, N., Joseph, D., Newton, R.U.: Combined resistance and aerobic exercise program reverses muscle loss in men undergoing androgen suppression therapy for prostate cancer without bone metastases: a randomized controlled trial. J. Clin. Oncol. **28**, 340–347 (2010)
65. Wall, B.A., Galvão, D.A., Fatehee, N., Taaffe, D.R., Spry, N., Joseph, D., Hebert, J.J., Newton, R.U.: Exercise improves V O2max and body composition in androgen deprivation therapy-treated prostate cancer patients. Med. Sci. Sports Exerc. **49**, 1503–1510 (2017)
66. Herrero, F., San Juan, A.F., Fleck, S.J., Balmer, J., Perez, M., Canete, S., Earnest, C.P., Foster, C., Lucia, A.: Combined aerobic and resistance training in breast cancer survivors: a randomized, controlled pilot trial. Int. J. Sports Med. **27**, 573–580 (2006)
67. Battaglini, C., Bottaro, M., Dennehy, C., Rae, L., Shields, E., Kirk, D., Hackney, A.C.: The effects of an individualized exercise intervention on body

composition in breast cancer patients undergoing treatment. Sao Paulo Med. J. **125**, 22–28 (2007)

68. Courneya, K.S., McKenzie, D.C., Mackey, J.R., Gelmon, K., Friedenreich, C.M., Yasui, Y., Reid, R. D., Cook, D., Jespersen, D., Proulx, C., Dolan, L.B., Forbes, C.C., Wooding, E., Trinh, L., Segal, R.J.: Effects of exercise dose and type during breast cancer chemotherapy: multicenter randomized trial. J. Natl. Cancer Inst. **105**, 1821–1832 (2013)

69. Kolden, G.G., Strauman, T.J., Ward, A., Kuta, J., Woods, T.E., Schneider, K.L., Heerey, E., Sanborn, L., Burt, C., Millbrandt, L., Kalin, N.H., Stewart, J.A., Mullen, B.: A pilot study of group exercise training (GET) for women with primary breast cancer: feasibility and health benefits. Psychooncology. **11**, 447–456 (2002)

70. Courneya, K.S., Segal, R.J., Mackey, J.R., Gelmon, K., Reid, R.D., Friedenreich, C.M., Ladha, A.B., Proulx, C., Vallance, J.K., Lane, K., Yasui, Y., McKenzie, D.C.: Effects of aerobic and resistance exercise in breast cancer patients receiving adjuvant chemotherapy: a multicenter randomized controlled trial. J. Clin. Oncol. **25**, 4396–4404 (2007)

71. Schmidt, M.E., Wiskemann, J., Armbrust, P., Schneeweiss, A., Ulrich, C.M., Steindorf, K.: Effects of resistance exercise on fatigue and quality of life in breast cancer patients undergoing adjuvant chemotherapy: a randomized controlled trial. Int. J. Cancer. **137**, 471–480 (2015)

72. Steindorf, K., Schmidt, M.E., Klassen, O., Ulrich, C. M., Oelmann, J., Habermann, N., Beckhove, P., Owen, R., Debus, J., Wiskemann, J., Potthoff, K.: Randomized, controlled trial of resistance training in breast cancer patients receiving adjuvant radiotherapy: results on cancer-related fatigue and quality of life. Ann. Oncol. **25**, 2237–2243 (2014)

73. Adamsen, L., Quist, M., Andersen, C., Moller, T., Herrstedt, J., Kronborg, D., Baadsgaard, M.T., Vistisen, K., Midtgaard, J., Christiansen, B., Stage, M., Kronborg, M.T., Rorth, M.: Effect of a multimodal high intensity exercise intervention in cancer patients undergoing chemotherapy: randomised controlled trial. BMJ. **339**, b3410 (2009)

74. Mortimer, P.S.: The pathophysiology of lymphedema. Cancer. **83**, 2798–2802 (1998)

75. Nelson, N.L.: Breast cancer-related lymphedema and resistance exercise: a systematic review. J. Strength Cond. Res. **30**, 2656–2665 (2016)

76. Gashev, A.A.: Physiologic aspects of lymphatic contractile function: current perspectives. Ann. N. Y. Acad. Sci. **979**, 178–87 (2002); discussion 188–96

77. Ahmed, R.L., Thomas, W., Yee, D., Schmitz, K.H.: Randomized controlled trial of weight training and lymphedema in breast cancer survivors. J. Clin. Oncol. **24**, 2765–2772 (2006)

78. Schmitz, K.H., Ahmed, R.L., Troxel, A., Cheville, A., Smith, R., Lewis-Grant, L., Bryan, C.J., Williams-Smith, C.T., Greene, Q.P.: Weight lifting in women with breast-cancer-related lymphedema. N. Engl. J. Med. **361**, 664–673 (2009)

79. Schmitz, K.H., Ahmed, R.L., Troxel, A.B., Cheville, A., Lewis-Grant, L., Smith, R., Bryan, C.J., Williams-Smith, C.T., Chittams, J.: Weight lifting for women at risk for breast cancer-related lymphedema: a randomized trial. JAMA. **304**, 2699–2705 (2010)

80. Helyer, L.K., Varnic, M., Le, L.W., Leong, W., McCready, D.: Obesity is a risk factor for developing postoperative lymphedema in breast cancer patients. Breast J. **16**, 48–54 (2010)

Printed in the United States
by Baker & Taylor Publisher Services